Interdisciplinary Applied Mathematics

Volume 49

Editors

Anthony Bloch, University of Michigan, Ann Arbor, MI, USA
Charles L. Epstein, University of Pennsylvania, Philadelphia, PA, USA
Alain Goriely, University of Oxford, Oxford, UK
Leslie Greengard, New York University, New York, USA

Advisors

L. Glass, McGill University, Montreal, QC, Canada
R. Kohn, New York University, New York, USA
P. S. Krishnaprasad, University of Maryland, College Park, MD, USA
Andrew Fowler, University of Oxford, Oxford, UK
C. Peskin, New York University, New York, USA
S. S. Sastry, University of California, Berkeley, CA, USA
J. Sneyd, University of Auckland, Auckland, New Zealand
Rick Durrett, Duke University, Durham, NC, USA

More information about this series at http://www.springer.com/series/1390

Frithjof Lutscher

Integrodifference Equations in Spatial Ecology

 Springer

Frithjof Lutscher
Mathematics and Statistics
University of Ottawa
Ottawa, ON, Canada

ISSN 0939-6047   ISSN 2196-9973   (electronic)
Interdisciplinary Applied Mathematics
ISBN 978-3-030-29296-6   ISBN 978-3-030-29294-2   (eBook)
https://doi.org/10.1007/978-3-030-29294-2

Mathematics Subject Classification: 92D40, 92D25, 39A60, 37L15, 37N25, 35K57

This Springer imprint is published by the registered company Springer Nature Switzerland AG.
The registered company address is: Gewerbestrasse 11, 6330 Cham, Switzerland

To my teachers.

Foreword

Integrodifference equations are relative newcomers to the world of mathematical ecology. First emerging in the 1970s and 1980s, these discrete-time, continuous-space models with generalized growth and dispersal components have increased in popularity, and the number of variations has grown over time. Much of this increase in popularity has been driven by the realization, on the biological side, that such models can be more closely aligned with real ecological systems than can the more traditional reaction–diffusion models. Growth is often seasonal, so discrete-time dynamics makes sense, and dispersal is often complex and nondiffusive, so generalized dispersal kernels give added realism. From a broader perspective, the move toward integrodifference equations reflects a desire to forge more intimate connections between theory and biological observation than has been previously found. This desire, shared by ecologists and mathematicians alike, reflects the healthy state of the field of theoretical ecology.

While originally designed to model the space-time dynamics of a single species with discrete, nonoverlapping generations, integrodifference equations have now been extended to cover a range of additional features, such as age-structure, interactions between different species, stochasticity, heterogeneous environments, and complex, multi-staged dispersal strategies, to name a few. The importance of getting the details of ecology exactly right has grown. Ecologists are no longer satisfied with crude caricatures of the systems they study. There is a move to ensure that models will now provide faithful reflections of detailed and specific quantitative hypotheses about the biological processes. Integrodifference equations have a role to play by injecting much-needed realism into space-time models in ecology.

Although typically providing more ecological realism than reaction–diffusion models, integrodifference models nonetheless retain many qualitative features found in reaction–diffusion models. The critical patch-size problem, describing the minimum area needed for population persistence, exists for both models, as do the traveling wave and spreading speeds problems, describing the wave of advance of invasive populations. Pattern formation via dispersal-driven instabilities can also be found in both models, as can many other features. In other words, integrodifference equations have much in common with parabolic partial differential equations. Both couple growth and dispersal in a similar way, and this gives rise to deeper mathematical properties that are shared and govern fundamental properties of solutions. For example, under reasonable assumptions, each equation has a so-called comparison principle, which states that the operator is order preserving: initially ordered solutions remain ordered for all time. In turn, this feature allows for the construction of super- and sub-solutions, which can be very useful in determining qualitative behaviors such as population persistence or traveling waves. Thus, the study of integrodifference equations has many mathematical features that will resonate with researchers versed in the theory of reaction–diffusion equations.

Despite these similarities, integrodifference models also exhibit differences with reaction–diffusion models. Many of the differences arise from the very general natures of the growth and the dispersal models found in the integrodifference framework. The growth term, essentially a discrete-time map, can exhibit a hump-shaped feature, ecologically describing the process of overcompensation, which may lead to chaos, even for a single species. This property of chaotic dynamics cannot be found for single-species differential equation models. In fact, it is well known that three interacting species are required before chaos can arise in a system of differential equations. The dispersal kernel, expressed as a probability density function, can take any number of exotic forms, including heavy tails. These can lead to unusual outcomes, for example, giving rise to an invasion processes that continually accelerate with time. What a wonderful mathematical playground in which to undertake quantitative analysis!

The timing of this book is perfect. While the number of papers developing and analyzing integrodifference equations is growing quickly, it is still just possible to compile a reference list that includes every significant publication written in the area. In this volume, Frithjof Lutscher has attempted the monumental task, and in doing so has created a comprehensive survey of the field. The book is, however, much more than a survey of integrodifference equations. There are significant new results in the book, particularly in later sections. These new results mesh seamlessly with existing knowledge, filling in gaps and extending analyses.

Taken in its entirety, the book provides a synthesis, a woven tapestry of ideas, based on the mathematical structures underlying integrodifference equations, and on the biological insights that these mathematical structures yield. Imprinted on the book is the perspective of an applied mathematician, embracing structure and striving for clarity of thought. As such, the book is a joy to read.

Opportunities of this type, to write a book that carves out a developing field, are rare and valuable. When I read the book, I am glad that Frithjof Lutscher undertook the writing because he has done a tremendous job of it. This will be the perfect text for both graduate students who are new to the field and for seasoned researchers who need to look up specific results. This text is timely and will define the field of integrodifference equations for many years to come.

Pender Island, BC, Canada   Mark Lewis
June 2019

Preface

Ecosystems are marvelous assemblages of individuals that grow, reproduce, interact with one another, move about in space, and eventually die. Ecosystems also provide essential services to humans, from oxygen production and recycling of organic matter to food provision and pollination of agricultural crops. At the same time, ecosystems are in peril from human activity, such as overexploitation, landscape fragmentation, and various effects of global change. Spatial ecology aims to understand the role that individual movement, population interaction, and landscape characteristics play in generating the patterns of species distributions that we observe in space and time, in particular questions of population persistence, population spread, and stability.

Dynamic mathematical models are powerful tools for understanding natural phenomena in the physical sciences and increasingly also in the life sciences. Mathematical methods, from the rigor of model formulation to the depth of analysis and the power of computation, are indispensable when dealing with the wide range of spatial and temporal scales that are inherent in many of the most fundamental questions of spatial ecology. Dynamical systems are also fascinating objects to study in their own right. The interaction between dynamical systems models and their applications in ecology and other fields is a constant source of mutual challenges and inspiration. The vision of this book is to introduce the reader to a class of models known as *integrodifference equations* and show how the fascinating, interdisciplinary circle of observation, modeling, analysis, and interpretation enhances ecological understanding and mathematical theory at the same time.

Integrodifference equations are models for the temporal evolution of the density of one or more populations from one generation to the next. These models are tailor-made to adequately represent the dynamics of organisms with a particular life cycle, where population growth and individual dispersal occur separated in time but synchronized in the population. Ecological examples contain many plant and insect species, particularly in temperate climates. Mathematically speaking, integrodifference equations define discrete dynamical systems (recursions) on some appropriate function space. Their mathematical analysis and ecological application have seen great progress in the past three decades. This book provides the first comprehensive introduction to the subject and serves as a reference guide to the, by now, sizable literature on all aspects of these equations.

The book is divided into three parts:

I Basics and Foundations.   Chapters 1–8 contain the most important aspects of modeling with and analysis of integrodifference equations, using the simplest possible scenarios. The focus is on extinction, persistence, stability, and spread of a single species in simple landscapes. Basic ideas for numerical approaches are provided.

II Applications and Approximations.   Chapters 9–12 consider various aspects of real-world applications. The simplifying assumptions from the first part are modified where necessary to more adequately describe realistic situations. Various methods of model simplification are presented.

III Extensions and Challenges.   Chapters 13–17 discuss substantial and significant extensions to the simpler models in order to tackle more challenging biological questions, such as the dynamics of stage-structured populations, interacting populations, and spatial and temporal variation in the environment.

Throughout this book, questions of population dynamics and their application to real-world systems motivate all models and mathematical analysis, and all theoretical and computational results are discussed in relation to ecological theory and implications. The greatest progress is made where ecology and mathematics come together to inspire each other toward deeper understanding in each discipline and their interplay.

Acknowledgments

This book could not have been written without the inspiration, support, and encouragement of many. First and foremost, I thank Mark Lewis, who continues to be a great mentor, collaborator, and friend. I deeply appreciate his guidance and support throughout my career. He introduced me to the fascinating world of integrodifference equations and encouraged me to write this book. I am also deeply grateful to my wonderful long-time friend and collaborator Christina Cobbold for her inspiration and enthusiasm and the many hours spent together at work and play. I am forever indebted to my PhD supervisor, Karl Peter Hadeler, for inspiring me to explore the fascinating worlds that can be found between mathematics and biology. I hope that he would have liked this book and would have seen some of the seeds that he planted come to fruition in it.

Mark Kot, together with various collaborators, wrote the foundational articles that have inspired so much of the research on integrodifference equations, including my own. He has since produced many more beautiful and insightful articles on the subject. I admire the clarity, precision, and elegance in his works. He and Mike Neubert have graciously shared with me many insights and suggestions on integrodifference equations over the years. I also thank all of the participants of the BIRS workshop "Integrodifference equations in ecology: 30 years and counting" (event number 16w5121) for their feedback on a preliminary version of the book, in particular, Bingtuan Li, Roger Lui, Xiao-Qiang Zhao, and Joy Zhou.

Several students have contributed to this book in various ways. Jeff Musgrave meticulously collected the relevant literature until 2014; Dominic Brass provided an updated comprehensive list in 2017. The students of my topics courses on integrodifference equations in 2015 and 2018 provided feedback for drafts of some chapters, in particular Adèle Bourgeois, Jason Bramburger, and Alessandro Selvitella. I am also grateful to the Technische Universität München for awarding me a John von Neumann visiting professorship in 2017. Several chapters of the book were written and revised during my time there and with the help of the students who attended my classes. Mark Kot and his students Nora Gilbertson, Benjamin Liu, and Kelsey Marcinko read the penultimate version of the book, gave many helpful suggestions, and helped me fix some inconsistencies.

I am also deeply grateful to my editor, Julia Cochrane, for suggestions and corrections regarding everything, including grammar and expression. Not only did she turn my mathematical ramblings into clear English, she also taught me a lot about editing in the process. I thank the editorial and production team at Springer for all the support and care that they provided for my book, in particular, Danielle Walker and Donna Chernyk, whose patience and flexibility made the logistics easy for me.

Last but not least, I thank Mateja and Kaja for their company and support. They joined my joy over progress and endured me talking about regress, frustration, and seemingly insurmountable hurdles while writing this book. They reminded me that there are many other and much more important things in life than completing a scientific book.

A book on any research subject, once in print, is fixed and cannot evolve with the progress of the field that it aims to describe. Christina Cobbold and I, with the help of Wenyan Li, have created an online resource that includes a searchable reference list for integrodifference equations. We aim to maintain this website as an evolving repository for scientific publications, video presentations, computational tools, and teaching resources on integrodifference equations. It is accessible at https://integrodifference.frithjof.ca/.

We thank the University of Ottawa for hosting this site. We welcome all constructive comments and will post additions and corrections to the book at this site.

My greatest hope is that this book and the accompanying online resource will inspire new research directions in the field of integrodifference equations and create a community of mathematicians, modelers, and ecologists who transcend their disciplinary boundaries to advance knowledge in all.

Ottawa, ON, Canada   Frithjof Lutscher
June 2019

Contents

Part I
Basics and Foundations

Chapter 1
Models for Spatial Population Dynamics

Abstract We introduce some of the population dynamics questions that inspire many of the models and results in this book, and we give their ecological context. We place integrodifference equations in the context of related modeling approaches. Finally, we outline the goals of this book, its prerequisites, and the organization of its content.

1.1 Spatial Aspects of Population Dynamics

One fundamental question in spatial ecology is, what conditions are necessary for a particular species to be present at a particular location? This deceptively simple question is at the heart of modern conservation biology: how do we design nature reserves to ensure the survival of a particular species? Its economic cousin, which arises, e.g., in fisheries, is, how much of a population can we harvest, and where, without jeopardizing its survival and that of others that depend on it? And a planning perspective on the same question is, how can we design agricultural operations, and where should we place infrastructure to minimize negative effects on ecosystems?

Another striking example of spatial processes in ecology is biological invasions, where alien species spread into new territory and may disrupt ecosystem function, diminish biodiversity, and require massive investments in remediation. Human activities such as travel or international trade facilitate the arrival of alien species and their spread in new environments. Spatial ecology aims to provide theory to predict the speed of spatial spread of a species from the various underlying reproductive and dispersal mechanisms. Not all invasions are detrimental—some are even intended. The release of biological control agents against a destructive pest species can have many advantages over the massive use of pesticides. Research is needed to predict and assess the spread and efficiency of the agent and to optimally place its release locations. In a world of global change, species will have to move and colonize new territories to keep up with their preferred climatic conditions. Spatial ecology aims to predict which species will be able to do so and to develop theory to implement mitigation measures for those that may not.

© Springer Nature Switzerland AG 2019
F. Lutscher, *Integrodifference Equations in Spatial Ecology*, Interdisciplinary Applied Mathematics 49, https://doi.org/10.1007/978-3-030-29294-2_1

All of these questions inherently involve considerations of space. Whether a population persists in a given environment depends on how individuals move about, use the available resources, and avoid existing dangers. The focus of this book is on the level of populations, although the considerations begin at the individual level and continue to the community level of interacting populations. The basic processes at the individual level are survival, reproduction, and relocation in space. On the population level, these processes interact to determine whether a population persists in a given geographic location. Will individuals survive and reproduce at a high enough rate? Will they move elsewhere and die or establish new local populations and spread to other geographic locations? Will others move from elsewhere and support a local population? Persistence versus extinction considers population dynamics as a binary outcome only. More nuanced questions consider other aspects of changes in spatial distribution over time. Is a population stable over many generations? Is the population concentrated in some areas and at low density in others? Does population density reflect the distribution of the resource of this population? Is the population oscillating in a relatively predictable manner over time, or potentially showing chaotic variation? Does this variation reflect external environmental variation or is it generated by interaction and the use of space? On the community level, interactions contribute to the increase or decline of the populations involved. Can two or more populations that compete for the same essential resources coexist, e.g., by employing different strategies of using the available space? Will consumer–resource interactions destabilize populations and trigger oscillations? How do predator and prey movement strategies shape their spatial distribution? The sheer scope of these problems, that is, their spatial and temporal extent, make mathematical models indispensable tools to answer some of these questions.

1.2 Mathematical Models

Mathematical models can provide fundamental insights into the mechanisms involved in population dynamics, may serve to process the growing amount of available data, and allow us to test management strategies in simulations before implementing them in the real world. The focus of this book is on the fundamental mechanisms, but notes and reference to the other two aspects are included throughout.

Mathematical models in the form of dynamical systems have served ecological theory well for over a century and have spurred the development of mathematical theory in return. *Ordinary differential equations* for the growth of a single nonspatial population in continuous time date back to Verhulst (1838) and were extended to interacting populations by Lotka (1920) and Volterra (1926). They are now found in many textbooks in ecology as well as in mathematics. The study of spatial aspects of population spread and population persistence began with the work by Fisher (1937), Skellam (1951), and Kierstead and Slobodkin (1953).

These authors combined ordinary differential equations for population genetics and population dynamics with diffusion equations for spatial spread. The resulting *reaction–diffusion equations* in continuous space and time have yielded many deep insights into spatial phenomena in ecology as well as the mathematical structure of infinite-dimensional dynamical systems (Cantrell and Cosner 2003).

Dynamical systems models for populations in discrete generations rose to fame with the discovery that simple density-dependent growth functions could generate complex and chaotic dynamic behavior (May 1975). They had been the backbone for theoretical studies of insect host–parasitoid systems for several decades since the foundational work by Nicholson and Bailey (1935). These *difference equations* are sometimes easier to formulate, typically easier to simulate, and almost always more difficult to analyze than their continuous-time counterparts. The two foundational works that combine discrete-generation population dynamics with dispersal in continuous space are by Weinberger (1982) in genetics and by Kot and Schaffer (1986) in ecology. The resulting equations were later named *integrodifference equations* (IDEs) (Neubert et al. 1995). After the discovery of the mathematical phenomenon of accelerating invasions (Kot et al. 1996), ecologists quickly embraced these equations as their framework of choice to test models against data for species invasions (Lewis et al. 2006). Meanwhile, mathematicians took up the challenge of studying the qualitative behavior of these infinite-dimensional recursions.

Several precursors to IDEs exist in the literature. Slatkin (1973) formulates an IDE model for the density of an allele in a population to study the steady-state distribution in a cline. Roff (1974) studies the discrete-time dynamics of a population distributed over discrete patches in space. The corresponding equations are now commonly known as *coupled map lattices* and studied in many different contexts; see de Camino-Beck and Lewis (2009) and references therein. Many models exist for continuous-time population dynamics in discrete spatial patches. They are sometimes known as *patch models*, but there does not seem to be a common established terminology. All of these modeling approaches (Table 1.1) are originally deterministic but have been extended to include stochastic elements.

This book provides the first comprehensive exposition and review of the mathematical and ecological literature on IDEs. These equations constitute a mathematical framework for studying and understanding how individual dispersal characteristics and interactions within and between populations interact to generate spatio-temporal patterns of population distribution and abundance. The particular focus is on species

Table 1.1 Classification of deterministic population dynamic models according to their representation of time and space.

	No space	Continuous space	Discrete space
Continuous time	Ordinary differential equation	Reaction–diffusion equation	Patch model
Discrete time	Difference equation	**Integrodifference equation (IDE)**	Coupled map lattice

with distinct growth and dispersal phases, which include many plant, insect, and bird species in temperate climates. The vision of this book is to enable new opportunities for ecology and mathematics to meet and create synergies that lead to deeper understanding of ecological phenomena and create better tools and guidelines for management of ecosystems.

1.3   Goals and Requirements

The first goal in writing this book is to introduce the fascinating theory of IDEs and their intriguing applications in spatial ecology for modelers, mathematicians, and theoretical ecologists. The literature on this subject has grown tremendously since the foundational papers mentioned above, but it is distributed over a wide range of journals in modeling, mathematical analysis, and ecological theory and applications. The second goal is therefore to collect, sort, and coherently summarize the existing literature on IDEs in spatial ecology. The third goal is to facilitate and inspire new research projects by presenting the questions and literature in a common overarching framework and pointing out open problems and knowledge gaps. The hope is that, in keeping with the spirit of this book, new projects will emerge from scientific questions, generate novel mathematics, and lead to new insights into the dynamics of ecological systems.

Most of the material in this book has been published elsewhere before, but we present a number of results, examples, and applications here for the first time. The major novelty, however, is the unified presentation of the existing models, theory, analysis, and applications. The guiding principle in the presentation is to always begin with the simplest possible meaningful example, where many results can be obtained explicitly. The second step is then to generalize and illustrate the robust results that carry over from the simple example as well as the novel aspects that were not present in the simple example. This contrast allows us to isolate mechanisms that are responsible for certain outcomes. Some of the deeper mathematical results are stated and illustrated, but their proofs are not included, only referenced. In doing so, the book aims to strike a balance between completeness and accessibility. Most chapters end with a section titled "Further Reading," where connections to other aspects are mentioned and some open problems are indicated.

Early versions of chapters of this book were used in topic courses in mathematics on the graduate level. The minimal requirements for reading the basic chapters are a strong command of differential and integral calculus, linear algebra, and some fundamental concepts and approaches from dynamical systems and partial differential equations. The most important of these are mentioned in Chap. 2, but no formal definitions or proofs are provided for them. Recurring tools and techniques from applied mathematics are linear stability analysis, Taylor series approximation, separation of variables and Fourier analysis, and perturbation expansion. Familiarity with at least one higher-level programming language is helpful. Simple MATLAB code for simulating IDEs is provided, but any standard software package that has fast Fourier transforms implemented can be used. Some aspects of the theory are not generally taught in a mathematics curriculum, e.g., monotone systems theory. Many of these can be studied for a term project or in a directed reading course.

1.4 Outline of the Book

The first part of the book begins with an overview of some of the fundamental questions of spatial ecology and outlines the mathematical background required for understanding the material presented. Chapters 2–8 develop all aspects of the theory of IDEs from model derivation to basic mathematical analysis and numerical implementation. The guiding principle is to explain every new aspect with the simplest possible example and motivate the more general study with it. Chapter 2 carefully derives the basic model, discusses its assumptions and limitations, and summarizes some of the mathematical background required to proceed. Chapter 3 deals with the so-called critical patch-size problem, the question of how much space a population needs to persist. Chapter 4 looks at the steady-state problem and the spatial profile of the population distribution. Chapters 5 and 6 deal with spatial spread and biological invasions in the absence and presence of an Allee effect, respectively. A typical IDE contains only the outcome of the dispersal process, but in many cases it is helpful and necessary to model the actual process itself, which we do in Chap. 7. For illustration and practical applications, we present recipes and a few warnings about numerical implementations of IDEs in Chap. 8.

In the second part of the book, we present many applications of the theory from the first part to more realistic ecological problems. Including more realism often requires modifications of the models and sometimes new theory to understand their behavior. In Chaps. 9 and 10, we present various techniques for how to approximate population dynamics and spatial spread characteristics when only partial information about dispersal is available. Chapter 11 examines the intricate shapes that the fronts of invading species can take. Chapter 12 reviews many applications of IDEs to date, for example, to river ecosystems, to global change scenarios, to Reid's paradox, and more.

The third part of the book contains extensions of the theory that represent the leading edge of the theory and its applications. Chapter 13 considers stage-structured populations and presents the most recent literature connecting models to data for invasive species. Chapter 14 includes the interaction of two species and studies phenomena such as spatial pattern formation. Chapters 15 and 16 deal with population dynamics in spatially and temporarily (stochastically) varying environments. The final chapter summarizes recent developments in various directions and includes a review of connections of this theory to related approaches.

Chapter 2
Modeling with Integrodifference Equations

Abstract We derive the basic integrodifference equation and discuss its two main ingredients: the growth function and the dispersal kernel. We introduce several ecological concepts that recur throughout this book and highlight how ecological assumptions are reflected in the mathematical model. This detailed understanding will allow us to formulate ecological insights from the mathematical results and understand the limitations of these insights.

2.1 Derivation of the Basic Model

Integrodifference equations (IDEs) are best suited to describing the life cycle of annual plants or insects, as depicted in Fig. 2.1. At the beginning of a season, seeds germinate and seedlings grow, developing leaves, flowers, and seeds. During this phase, plants do not move in space. Later in the season, plants release seeds that subsequently travel to a different location, e.g., by wind or birds. As the plants die, seeds get buried in the soil, ready to germinate at the beginning of the next season. Similarly, many insects emerge from eggs at the beginning of a season, developing through larval stages and pupation to the adult form. The adults then fly, deposit eggs for the next generation, and die. One common key aspect of these life histories is the separation of a growth phase, during which spatial dispersal is negligible, and a dispersal phase, during which no growth occurs. Another is the synchrony with which these processes occur in the population. And, finally, there is no overlap between different generations. IDEs are designed to model the dynamics of populations that satisfy these characteristics.

To formulate a mathematical model corresponding to such a life cycle, we denote the spatial density of the population in year (or generation) t as $N_t(x)$. The spatial variable x is in some domain of interest, denoted by Ω. We census the population at the beginning of the growth phase. This growth phase is modeled as a discrete map $N_t \mapsto F(N_t)$, which applies at every $x \in \Omega$. There is a large body of literature on the derivation of different forms of the *growth function* F and on the analysis of the resulting dynamics of the population. We summarize the most relevant aspects below in Sect. 2.2.

F. Lutscher, *Integrodifference Equations in Spatial Ecology*, Interdisciplinary Applied Mathematics 49, https://doi.org/10.1007/978-3-030-29294-2_2

Fig. 2.1 Schematic
description of the life cycle of
an organism with distinct
growth and dispersal phase.

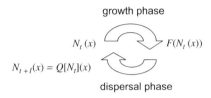

During the dispersal phase, an individual moves from one location to another
with a certain probability. We describe the outcome of this process by a *dispersal
kernel* K. The probability density of an individual's location at the end of the
dispersal phase is $K(x, y)$, where $y \in \Omega$ was its location before the dispersal phase.
Alternatively, in a one-dimensional domain Ω one can think of $K(x, y)\Delta x$ as the
probability that an individual from location y moves to the interval $[x, x + \Delta x)$.
We give several examples of dispersal kernels in Sect. 2.3 below, and we return to a
more detailed mechanistic perspective on modeling the dispersal process in Chap. 7.
We assume that individuals move independently of one another.

From one census to the next, individuals first reproduce and then move. The
population density in the following generation is obtained by adding up all the
individuals present after the growth phase that *arrive* at a location x from any y.
This process leads to the IDE

$$N_{t+1}(x) = Q[N_t](x) := \int_\Omega K(x, y) F(N_t(y), y) \mathrm{d}y . \tag{2.1}$$

Throughout, we denote by Q the *next-generation operator* that maps the population
density from one generation to the next.

The second argument in the growth function F indicates that the environment
may be heterogeneous so that growth conditions change in space. For example, an
extreme form of spatial heterogeneity is that of an isolated island: a single patch of
suitable habitat is surrounded by unsuitable habitat of very large extent so that no
individuals can reach the island from elsewhere. Determining the conditions under
which a population can persist on such an isolated island is an important question in
conservation ecology and a recurring theme in this book; see, e.g., Chap. 3.

Taking the census before the growth phase is a somewhat arbitrary choice. In
empirical studies, the optimal time of sampling a population is determined by
climatic conditions and morphological aspects of the organism. Taking a census
before, rather than after, the dispersal phase leads to an equation of the form

$$\tilde{N}_{t+1}(x) = \tilde{Q}[\tilde{N}_t](x) := F\left(\int_\Omega K(x, y) \tilde{N}_t(y) \mathrm{d}y, y \right) . \tag{2.2}$$

The two formulations are equivalent under the transformation $\tilde{N} = F(N)$, but some
care is required to pick appropriate initial conditions (Andersen 1991; Lutscher and
Petrovskii 2008).

We now take a closer look at the two main ingredients of an IDE, the growth function and the dispersal kernel, and we introduce the most common examples.

2.2   Growth Functions

Several types of growth functions with particular characteristics describe the various ecological mechanisms and processes that underlie the growth dynamics in an IDE. They can be studied as nonspatial models for discrete-time ecological processes in their own right. Recursion equations of the form

$$N_{t+1} = F(N_t) \tag{2.3}$$

have a long history in ecological modeling (Hassell 1975; May 1975). For a comprehensive review and unifying approach to modeling discrete-time growth functions, see Sandefur (2018).

Function F describes the total number or density of individuals in generation $t + 1$ that are produced from individuals at generation t. Consequently, we always require the following two properties of the growth function:

$$F(0) = 0, \quad \text{and} \quad F(N) \geq 0 \quad \text{for} \quad N \geq 0. \tag{2.4}$$

The first condition ensures that no individuals are produced where there are none to start with. It excludes the possibility of immigration from elsewhere. The latter condition ensures that the population remains nonnegative.

The most basic aspects of the long-term dynamics of the recursion in (2.3) are determined by its *steady states* or *fixed points* and their stability. Steady states are defined by

$$F(N^*) = N^*. \tag{2.5}$$

A steady state N^* is *locally stable* if all solutions starting near N^* stay near N^*. It is *locally asymptotically stable* if, in addition, all solutions starting near N^* converge to N^*. Finally, N^* is *unstable* if it is not stable. If the stability properties of a fixed point change depending on parameter value, we say that the point undergoes a *bifurcation*.

Linear stability analysis provides criteria for stability by expanding the right-hand side of (2.3) in a Taylor series near a fixed point. We write $N_t = N^* + n_t$ and assume that $|n_t|$ is small. Then

$$N^* + n_{t+1} = N_{t+1} = F(N_t) = F(N^* + n_t) = F(N^*) + F'(N^*)n_t + \text{h.o.t.}, \tag{2.6}$$

where h.o.t. indicates higher-order terms in n_t, here at least quadratic. After canceling and dropping higher-order terms, we arrive at a linear equation for the perturbation n_t as

$$n_{t+1} = F'(N^*)n_t. \tag{2.7}$$

If $|F'(N^*)| < 1$, then all solutions of this recursion converge to zero (see (2.10) below), which implies that solutions of (2.3) that start near N^* converge to N^*. Hence, we have the sufficient condition that a fixed point N^* is locally asymptotically stable for the difference equation in (2.3) if

$$|F'(N^*)| < 1 \,. \tag{2.8}$$

Beyond approaching a stable fixed point, solutions of difference equations may exhibit more complicated behavior, such as cyclic, quasi-periodic, or chaotic dynamics (May 1975). We will not discuss the qualitative dynamics of difference equations here but rather explain them and relate them to the properties of the growth functions as they arise throughout the book. We refer the reader to some of the many textbooks on dynamical systems that give good introductions to the study of discrete-dynamical systems and their applications in ecology, e.g., Murray (2001), Kot (2001), Edelstein-Keshet (2005). We will simply write "stable" for "locally asymptotically stable" and use the long terminology only if confusion could arise otherwise.

   While the growth function determines the total number or density of individuals generated, it is sometimes convenient to consider the per capita growth rate, $F(N)/N$. It represents the number of individuals produced by a single individual. If the per capita growth rate exceeds unity, then the population grows, if it is less than unity, it declines. We present four important examples of growth functions, sometimes also called *updating functions*, and discuss their basic dynamic properties.

A Linear Growth Function

The linear function

$$F(N) = RN \tag{2.9}$$

models the case that the per capita number of offspring, R, is independent of population density. The only steady state is $N^* = 0$. Recursion (2.3) with a linear growth function has the solution

$$N_t = R^t N_0 \,. \tag{2.10}$$

Hence, the assumption of a constant per capita growth rate leads to exponential growth ($R > 1$) or decay ($0 < R < 1$). Negative values of R cannot occur for the growth function but may arise as the linearization of some nonlinear updating function at a positive steady state. Then the solutions in (2.10) show oscillations that decay ($-1 < R < 0$) or explode ($R < -1$). A constant per capita growth function seems unrealistic when individuals in a population interact. We turn to nonlinear growth functions next.

The Beverton–Holt Growth Function

For a very simple nonlinear growth function, we assume that the per capita number of offspring decreases with population density as individuals compete for resources such as food or space. The Beverton–Holt function,

$$F(N) = \frac{R}{1 + \kappa N}\, N \,, \tag{2.11}$$

was originally proposed for recruitment in fish stocks (Beverton and Holt 1957). The per capita number of offspring is inversely related to density. Parameter κ measures the strength of density dependence.

We can reduce the number of parameters by nondimensionalizing the model. We write $N_t = \hat{N} n_t$, where $\hat{N}$ is a constant with the same physical dimensions as N_t and n_t is a nondimensional quantity. Then we find

$$n_{t+1} = \frac{R n_t}{1 + \kappa \hat{N} n_t} \,. \tag{2.12}$$

Different choices of $\hat{N}$ will yield different but equivalent model formulations. Choosing $\hat{N} = 1/\kappa$ results in a particularly simple form of the denominator, whereas choosing $\hat{N} = (R - 1)/\kappa$ gives a particularly simple form of the positive fixed point, namely $n^* = 1$. The latter choice requires $R > 1$. We will use various forms, depending on which is most convenient, and we will still write N_t instead of n_t for the nondimensionalized model.

We study the dynamics of the (scaled) Beverton–Holt function

$$N_{t+1} = \frac{R N_t}{1 + (R - 1) N_t} \,, \qquad R > 1 \,, \tag{2.13}$$

in more detail. The fixed points are $N^* = 0$, the extinction state or trivial state, and $N^* = 1$, the persistence state. The respective derivatives are $F'(0) = R$ and $F'(1) = 1/R$. Hence, there is a unique, locally stable positive fixed point; the zero state is unstable. We visualize the dynamics in a *cobweb diagram* in Fig. 2.2. Starting from some initial condition N_0, we find $F(N_0)$, then project onto the diagonal to find N_1, and continue by finding $F(N_1)$, and so on, see, e.g., Kot (2001).

The Beverton–Holt updating function has two important properties. First, it is monotone increasing; i.e., if $N > \tilde{N}$, then $F(N) > F(\tilde{N})$. Consequently, solutions of recursion (2.13) are monotone, either increasing (i.e., $N_{t+1} > N_t$ for all t) or

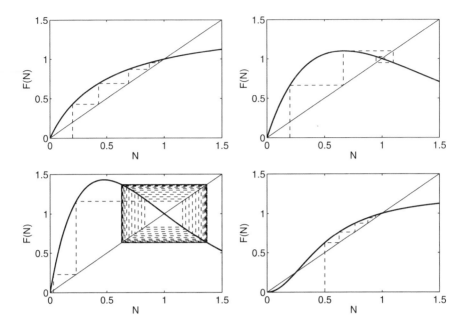

Fig. 2.2 Growth functions and the cobwebbing process. **Top left:** The Beverton–Holt function (2.13) has monotone dynamics (parameter $R = 3$, initial condition $N_0 = 0.2$). **Top right:** The Ricker function (2.19) with stable positive state and nonmonotone dynamics (parameter $r = 1.5$, initial condition $N_0 = 0.2$). **Bottom left:** The Ricker function with a stable two-cycle (parameter $r = 2.1$, initial condition $N_0 = 0.03$). **Bottom right:** The Allee function with a solution that converges to the stable positive state (parameters $R = 5$, $\gamma = 2$, initial condition $N_0 = 0.5$).

decreasing (i.e., $N_{t+1} < N_t$ for all t). In particular, solutions cannot oscillate. Second, it is concave down. Consequently, there can only be one positive steady state. When we combine the two properties, we see that the positive state is globally asymptotically stable. Monotonicity is an important concept with far reaching consequences in dynamical systems. We shall see in the next section that a lack of monotonicity can lead to highly complex dynamics. In subsequent chapters, we shall also define what it means for the next-generation operator Q to be monotone (or order preserving) and see how important this property is in many contexts.

It turns out that we can explicitly solve the iteration in (2.13). The new variable $\tilde{N}_t = 1/N_t$ satisfies the linear difference equation

$$\tilde{N}_{t+1} = \frac{1}{R}\tilde{N}_t + \frac{R-1}{R},\qquad (2.14)$$

which can be solved to obtain

$$\tilde{N}_t = \frac{1}{R^t}(\tilde{N}_0 - 1) + 1.\qquad (2.15)$$

Inverting the solution gives

$$N_t = \frac{R^t N_0}{(R^t - 1)N_0 + 1}.$$ (2.16)

The Beverton–Holt function can also be derived from a mechanistic model in continuous time, namely the Verhulst equation

$$\frac{d}{dt}N = rN(1 - N).$$ (2.17)

Separation of variables gives the solution

$$N(t) = \frac{e^{rt} N_0}{1 + (e^{rt} - 1)N_0}.$$ (2.18)

Evaluating the solution at integer times t, we obtain precisely the solution of the Beverton–Holt map with $R = e^r$.

The Ricker Growth Function

Also in the context of fish stocks and around the same time, Ricker (1954) proposed that the per capita growth rate decays exponentially with population density. In nondimensional form, the updating function can be written as

$$F(N) = N \exp(r(1 - N)), \qquad r > 0.$$ (2.19)

The fixed points are $N^* = 0$ and $N^* = 1$, and the derivatives are $F'(0) = e^r$ and $F'(1) = 1 - r$. The zero state is unstable, but the dynamics around the positive state can be more complicated. The positive state is globally asymptotically stable if $0 < r < 2$. Solutions are monotone when $0 < r < 1$ but show decaying oscillations when $1 < r < 2$; see Fig. 2.2. At $r = 2$, a flip bifurcation generates two-cycles (periodic orbits of length 2), and for increasing values of r also of length 4, 8, and so forth, eventually leading to chaotic dynamics (Kot 2001). In Chap. 4, we shall see how these complicated dynamics can appear in spatial models.

The poster child of chaotic dynamics is the discrete logistic equation (Kot 2001). It is typically written as $F(N) = rN(1 - N)$. By scaling the positive steady state to $N^* = 1$, it can be written in the form

$$F(N) = (r + 1)N - rN^2.$$ (2.20)

Since the logistic function can take on negative values, it is not an ideal choice for our modeling purposes. Function F is positive for $0 < N < (r + 1)/r$ and has its maximum value of $(r + 1)^2/(4r)$ at $(r + 1)/(2r)$. If this maximum is no larger than the positive zero, then the interval $[0, (r + 1)/r]$ is invariant under the logistic map.

This is the case when $r \leq 3$. Hence, if we limit parameter values to $0 < r < 3$ and initial conditions to $0 \leq N_0 \leq (r + 1)/r$, we avoid biologically unreasonable negative densities. The advantage of the logistic map is that a number of calculations can be carried out explicitly. The dynamics it generates are qualitatively the same as for the Ricker map (Kot 2001).

The difference in the qualitative behavior of solutions between the Beverton–Holt and the Ricker dynamics can be related to their ecological interpretation in terms of intra-specific competition. Beverton–Holt dynamics model *contest competition,* where all individuals are ranked and resources are distributed according to ranking. Ricker dynamics result from *scramble competition,* where resources are distributed equally among all individuals. The latter scenario can lead to the "boom-and-bust" behavior, where the population explodes when resources are abundant and crashes when resources are rare. This effect is known as *overcompensation.* When resources are distributed according to a ranking, overcompensation is avoided and the population stabilizes, which is known as compensation.

An Allee Effect

An *Allee effect* occurs when the per capita growth rate may increase with population density, at least in some range (Allee 1949). Some mechanisms that can lead to an Allee effect include mate finding, group hunting, or group defense (Courchamp et al. 2008). An Allee effect is said to be *strong* if a population will grow only above a certain positive threshold and decline when below this threshold. It is said to be *weak* if the population can grow even at low density, just not as fast as at intermediate densities.

Mathematically, we can express these conditions by saying (Wang et al. 2002)

$$F(N) > F'(0)N \quad \text{for some} \quad N , \tag{2.21}$$

with $F'(0) < 1$ for a strong Allee effect and $F'(0) \geq 1$ for a weak Allee effect. We can nondimensionalize and assume that the largest fixed point is $F(1) = 1$. We shall also assume that F is monotone increasing on $[0, 1]$, so that there is no overcompensation. A model function for the strong Allee effect is (Wang et al. 2002; Musgrave et al. 2015)

$$F(N) = \frac{RN^\gamma}{1 + (R - 1)N^\gamma} , \tag{2.22}$$

with $R > \gamma > 1$. In addition to the two steady states $N^* = 0$, $N^* = 1$, there is a second positive steady state $0 < N_a^* < 1$. This state represents the *Allee threshold.* Populations below N_a^* will go extinct, whereas initial populations above N_a^* will grow to $N^* = 1$. The states $N^* = 0$ and $N^* = 1$ are locally stable; N_a^* is unstable. The dynamics of this map are depicted in Fig. 2.2.

2.3 Dispersal Kernels

A dispersal kernel does not describe the *process* of dispersal but rather summarizes the *outcome* of this process. It is the probability density function of the location of an individual after the dispersal process. In plant ecology, the dispersal kernel is also known as the seed shadow (Neubert et al. 1995).

We denote the probability density of the location of an individual after dispersal by $K(x, y)$ if the location prior to dispersal was y. Naturally, K is nonnegative. If there is no dispersal-induced mortality, a dispersal kernel in one spatial dimension must satisfy the condition

$$\int_{\mathbb{R}} K(x, y)\mathrm{d}x = 1 \quad \text{for all} \quad y \in \mathbb{R}, \tag{2.23}$$

and similarly in higher dimensions. If dispersal carries a risk of mortality or if the domain of interest, Ω, does not represent all locations to which individuals can disperse, then we only have the inequality

$$\int_{\Omega} K(x, y)\mathrm{d}x \leq 1 \quad \text{for all} \quad y \in \Omega. \tag{2.24}$$

For example, wind-dispersed seeds from plants on an isolated island may be carried off the island into the water.

By far the majority of studies, analytical and empirical, consider dispersal kernels to depend only on the distance moved. In that case, $\widetilde{K}(|x - y|) = K(x, y)$ gives the distribution of dispersal distances. One implicitly assumes that dispersal is isotropic, i.e., identical in all directions. Much of the empirical literature is concerned with measuring and appropriately describing these distributions.

We mention the Gaussian and the Laplace dispersal kernels here. We will introduce other dispersal kernels throughout (see, e.g., Table 3.1 in Chap. 3) and devote Chap. 7 to deriving dispersal kernels from models that describe the movement process.

Gaussian Kernel

The most widely used kernel is the *Gaussian kernel* or normal distribution, which, in one spatial dimension, is given by

$$K(x, y) = K_G(x, y) = \frac{1}{\sqrt{2\pi\sigma^2}} \exp\left(-\frac{(x - y)^2}{2\sigma^2}\right), \tag{2.25}$$

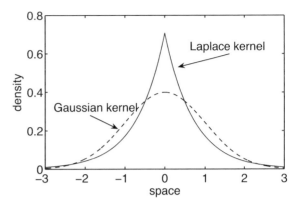

Fig. 2.3 The Gaussian and the Laplace dispersal kernel for equal variance $\sigma^2 = 1$.

with variance σ^2 and $x, y \in \mathbb{R}$; see Fig. 2.3. The two-dimensional analogue is

$$K(\mathbf{x}, \mathbf{y}) = \frac{1}{2\pi\sigma^2} \exp\left(-\frac{\|\mathbf{x} - \mathbf{y}\|^2}{2\sigma^2}\right), \tag{2.26}$$

again with variance σ^2 but $\mathbf{x}, \mathbf{y} \in \mathbb{R}^2$.

Laplace Kernel

According to a Gaussian kernel, the distance traveled decays exponentially with the square of the distance. Many datasets on dispersal report a slower decay, e.g., in insects (Neubert et al. 1995). An alternative dispersal kernel that respects this slower decay and is frequently used in calculations is the *Laplace kernel*, also known as the back-to-back exponential kernel. It is given by

$$K(x, y) = K_L(x, y) = \frac{1}{\sqrt{2\sigma^2}} \exp\left(-\sqrt{\frac{2}{\sigma^2}}|x - y|\right), \tag{2.27}$$

again with variance σ^2. To simplify notation, we will also write the Laplace kernel with dispersal parameter $a = \sqrt{\frac{2}{\sigma^2}}$. The quantity $1/a$ has units of length and denotes the mean absolute deviation or mean dispersal distance.

For a two-dimensional version of this kernel, we could simply replace the absolute value in the exponent with the norm and adjust the scaling factor accordingly, as we did for the Gaussian kernel (Etienne et al. 2002). However, the resulting kernel is *not* the two-dimensional analogue of the Laplace kernel. It is sometimes referred to as the negative exponential kernel (Nathan et al. 2012). We derive the correct two-dimensional Laplace kernel in Sect. 7.2. We discuss some of the issues in comparing dispersal in one and two dimensions in Sect. 12.7.

Compared to the Gaussian kernel for equal variance, the Laplace kernel has a higher density near the release point and far away: it is leptokurtic; see Fig. 2.3 and also Table 10.1.

One of the many nice features of the Laplace kernel is that, similarly to the Beverton–Holt growth function, it can be derived from a continuous-time process (Neubert et al. 1995). We will delve into mechanistic movement models in detail in Chap. 7 and derive dispersal kernels from first principles. Here we briefly explain the ideas. We consider a single individual and make the most simple assumption that this individual performs a random walk, uncorrelated and unbiased, in a homogeneous landscape. This individual could be an adult insect flying around and laying eggs, or it could be the seed of a plant sailing in the wind. The next, also very simplistic assumption could be that there is a constant probability per unit time that the individual stops the random walk. The adult insect lays the egg; the seed drops to the ground. Then we write an equation for the spatial distribution of locations where the individual stops. This distribution is the dispersal kernel that emerges from the process of random walk and constant stopping probability. For this simple process, it turns out to be the Laplace kernel (Neubert et al. 1995).

Deriving dispersal kernels from mechanistic movement models is one option; another is incorporating empirical results directly. The modeling framework of IDEs is ideally suited to utilizing certain measurements of dispersal patterns in dynamic models. Fujiwara et al. (2006) develop estimators for the parameters of dispersal kernels in the presence of mortality and incomplete census. Lewis et al. (2006) provide an estimator for the moment-generating function of a dispersal kernel in an application to predicting spread spatial rates from data. Incorporating empirical data for the outcome of dispersal into reaction–diffusion equations for the dispersal process is arguably more complicated.

The flexibility to accommodate a wide range of dispersal patterns in IDEs is at the same time an appealing feature to empiricists and a challenge for analysts. It is not easy to give general, empirically motivated conditions that a function $K(x, y)$ of two variables should satisfy in order to be a dispersal kernel; it seems a bit easier to do this for the distribution of dispersal distances $\tilde{K}(|x - y|)$. In subsequent chapters, we will address the question of how certain characteristics of the dispersal kernel influence the patterns observed at the population level.

2.4   Motivating Example: Dispersal on and from an Island

We close this chapter with a simplified example that illustrates several concepts that we explore in detail in subsequent chapters. We imagine a scenario where the suitable habitat for a population is an isolated island. It could be an actual island surrounded by water or, more generally, an area of suitable habitat surrounded by an unsuitable environment, such as a meadow in a forest or a park in a city. We assume that no individuals of the population arrive from elsewhere to the island.

In the simplest possible case, when the island is small compared to the dispersal ability of individuals and if the island is relatively homogeneous, one could imagine that individuals move everywhere on the island with equal probability. The kernel describing this situation is the uniform kernel

$$K(x, y) = \frac{1}{|\Omega|}, \qquad \text{where} \qquad |\Omega| = \int_{\Omega} dx \qquad (2.28)$$

denotes the total area of the island. The resulting IDE (2.1) collapses to a nonspatial difference equation. Even if the initial population is not spatially uniform, the first generation is. Subsequently, population dynamics simply follow the one-dimensional discrete map given by the averaged growth function

$$N_{t+1} = \overline{F}(N_t) := \frac{1}{|\Omega|} \int_{\Omega} F(N_t(y)) dy. \qquad (2.29)$$

Somewhat more interesting is the case where a certain proportion of individuals leaves the island during dispersal. For example, wind-borne seeds are carried away into the surrounding water. The remaining individuals distribute evenly on the island. Then the model becomes

$$N_{t+1} = s\overline{F}(N_t), \qquad (2.30)$$

where s denotes the probability that an individual initially on the island successfully stays on the island during one dispersal period. The probability that an individual leaves the island is $1 - s$. When $F(N) = RN$ is the linear function, then the island model with departure from the island becomes $N_{t+1} = sRN_t$. The trivial steady state $N^* = 0$ is stable if $sR < 1$ and unstable if $sR > 1$.

When $F(N)$ is the (scaled) Beverton–Holt function, then the spatial model with departure from the island also collapses to a Beverton–Holt model, multiplied by s. The linearization of this model at the trivial state is the previous linear model. The nontrivial steady state is given by

$$N^* = \frac{sR - 1}{R - 1}. \qquad (2.31)$$

This point is positive, and therefore biologically relevant, only if $sR > 1$, i.e., if the trivial state is unstable. When the point is positive, it is stable. Hence, if we choose s, the probability to stay on the island, as a bifurcation parameter, we observe a transcritical bifurcation at $s = 1/R$; see Fig. 2.4.

In more ecological terms, if the trivial state is unstable, then a small initial population can grow and persist on the island. A sufficient proportion of offspring has to remain on the island for the population to be viable. This observation is, of course, neither new nor surprising. It is also clear that the probability of staying on the island depends on the size and shape of the island and on the movement behavior.

Fig. 2.4 Transcritical bifurcation in the simple island dispersal model. When $s < 1/R$, the trivial state is stable (thick solid line); when $s > 1/R$, it is unstable (dashed line), while the positive state is stable (thick solid line).

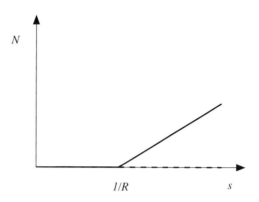

The question is how. What are the effects of the shape of the island? What are the effects of nonuniform dispersal? What are the effects of a nonuniform population distribution? These questions will be the focus of attention in subsequent chapters.

2.5 Further Reading

The literature on (nonspatial) difference equations in ecological applications is vast, as is the literature on their mathematical properties. Many other discrete maps can be derived from mechanistic growth models in continuous time. For example, Bellows (1981) uses seven different mortality functions to derive various maps, among them the Ricker map. Veit and Lewis (1996) use a pair-formation model together with the Beverton–Holt model to derive an updating function with Allee effect; see Sect. 12.6. Geritz and Kisdi (2004) use several continuous-time consumer–resource models to derive many different updating functions and to determine ecological conditions and scenarios for the emergence of overcompensation. Brännström and Sumpter (2005) derive a variety of intra-specific competition models and discuss their dynamic properties. Schreiber (2003) combines an Allee effect with overcompensation and finds chaotic transients and extinction dynamics.

Estimating parameter values for these functions from observations can be challenging. For example, the Beverton–Holt and the Ricker function both have decreasing per capita reproduction rates. Even when empirical curves for these per capita rates are available, it may be impossible to statistically distinguish between the two alternatives. Yet, the difference in dynamical behavior between the two is striking.

While there are many different growth functions, certain properties (e.g., monotonicity) are crucial to determining the dynamic behavior. Similarly, depending on the question that we want to study, certain properties of dispersal kernels are crucial, whereas others have little effect. For example, we will see in Chap. 5 that the rate at which the tails of a dispersal kernel approach zero is crucial to determining the speed at which a population will spread in space. Accordingly, theoretical models typically use few kernels, each representing a different set of properties. Empirically, however, many different forms have been suggested and fit to observations. Bullock and Clarke (2000) use an exponential and a power law distribution to fit measurements of heather seeds (*Calluna vulgaris*) but conclude that neither of them individually gives a good fit. Rather, a linear combination between the two, a so-called mixed model, gives the best fit. We consider such mixed kernels in Sect. 12.5. Kot et al. (1996) fit, among others, a Weibull distribution, a gamma distribution, or a log-normal distribution to the same dataset of fruit fly dispersal data and compare the resulting predictions for population spread rates. We turn to the question of spatial spread in Chap. 5. Tufto et al. (2005) present the gamma-binomial distribution that can interpolate between the Gaussian kernel and the Laplace kernel. We use this interpolation in Sect. 10.1 to illustrate the value of approximations.

Chapter 3
Critical Patch-Size

Abstract One of the basic questions in spatial ecology is: how much space does a population need to persist? The *critical patch-size* is the size of the suitable habitat where population gain through reproduction balances population loss through dispersal. The question of how large a certain habitat has to be to support a given population has important applications in conservation biology, e.g., when designing a protected area to ensure the survival of an endangered population. The analysis in this chapter is based on linearization, thereby implicitly assuming that the population growth function has no Allee effect. We explicitly compute the critical size when dispersal is described by a Laplace kernel. We then compare how different dispersal patterns affect this critical size. At the end of the chapter, we consider the class of separable kernels and introduce an approximation method.

3.1 Linearizing the Equation

If the habitat of a population consists of a single island or patch, denoted by Ω, we can expect that dispersing individuals are likely to leave this patch when it is small and likely to remain in the patch when it is large. Since the area of a patch grows quadratically with length or diameter and the boundary grows only linearly, we can expect that there is a *critical patch-size*, below which a population will go extinct and above which a population will persist. This *critical patch-size problem* dates back to the work by Skellam (1951) and Kierstead and Slobodkin (1953) for reaction–diffusion equations and is reviewed comprehensively by Cantrell and Cosner (2003). For IDEs, the critical patch-size problem was first addressed by Kot and Schaffer (1986). We begin this chapter with their ideas.

The density $N_t(x)$ of the population satisfies IDE (2.1) on some bounded domain of interest, Ω :

$$N_{t+1}(x) = Q[N_t](x) = \int_\Omega K(x, y) F(N_t(y), y) \mathrm{d}y. \tag{3.1}$$

© Springer Nature Switzerland AG 2019

F. Lutscher, *Integrodifference Equations in Spatial Ecology*, Interdisciplinary Applied Mathematics 49, https://doi.org/10.1007/978-3-030-29294-2_3

We assume that there is no Allee effect. We first define what it means for the population to persist and then determine the conditions, on Ω, on K, and on F, such that the population persists.

In the absence of immigration, new individuals can only be born from existing individuals. This fact is reflected in the property $F(0) = 0$; see Sect. 2.2. It implies that $N^*(x) = 0$ is a fixed point of IDE (3.1). This trivial solution corresponds to the absence of the population. If this fixed point is stable, then solutions with initial conditions close to zero will converge to zero; if it is unstable, at least one solution will grow. In the former case, we say that the population will go extinct, and in the latter, we say that it can persist. Hence, we study how the stability properties of the trivial solution depend on patch-size, dispersal behavior, and growth function. This definition of persistence does not indicate any long-term behavior; i.e., whether the population converges to a positive steady state or shows more complicated dynamics. We will address these questions in subsequent chapters.

If N^* is any steady state, we write $N_t(x) = N^*(x) + n_t(x)$, expand the growth function in (3.1) in a Taylor series, and drop the higher-order terms; i.e., we approximate it by its linearization. We write

$$N_{t+1}(x) = N^*(x) + n_{t+1}(x)$$

$$= \int_\Omega K(x, y) F(N_t(y), y) dy$$

$$= \int_\Omega K(x, y) \left[F(N^*(y), y) + \frac{\partial F}{\partial N}(N^*(y), y) n_t(y) + \text{h.o.t.} \right] dy$$

$$= N^*(x) + \int_\Omega K(x, y) R(y) n_t(y) dy + \text{h.o.t.}, \tag{3.2}$$

where $R(y) = \frac{\partial F}{\partial N}(N^*(y), y)$. After canceling $N^*(x)$ and the higher order terms, we obtain a linear equation for n_t in the form of an integral operator. In analogy with the finite-dimensional case, we call the resulting operator the "derivative" of Q and write Q'. (It turns out to be the Fréchet derivative of Q; see, e.g., Keener 2000.) Just as in the finite-dimensional case, the derivative depends on where we linearize, here at $N^*(x)$. We denote this dependency as $Q'[N^*]$. Also as in the finite-dimensional case, it is a linear operator, here applied to n_t. We write the linearization of (3.1) at $N^* = 0$ as

$$n_{t+1}(x) = Q'[0] n_t(x) = \int_\Omega K(x, y) R(y) n_t(y) dy, \qquad R(y) = \frac{\partial F}{\partial N}(0, y).$$

$$\tag{3.3}$$

We expect the linear equation to have exponential solutions of the form $n_t(x) = \lambda^t \phi(x)$, where λ is an eigenvalue and ϕ an eigenfunction of the linear integral operator in (3.3). Solutions of this form will grow, i.e., the population will persist, if $\lambda > 1$, and they will decay, i.e., the population will decline, if $0 \le \lambda < 1$. For the solution to be biologically meaningful, we must ensure that ϕ is nonnegative. We are hence led to study the eigenvalue problem

$$\lambda\phi(x) = Q'[0]\phi(x) = \int_\Omega K(x, y)R(y)\phi(y)\mathrm{d}y \tag{3.4}$$

in some appropriately chosen function space.

Before we turn to an example where exact persistence conditions can be obtained, we put our calculations on a solid mathematical footing. The following theorems summarize some of the most important properties of the linear integral operator in (3.4) and its relation to the nonlinear IDE (3.1) for our purposes. We begin with some abstract stability theory.

Theorem 3.1 (Compactness and Linearized Stability) *Assume that Ω is bounded and connected and that K is continuous and positive. Assume that F is smooth and satisfies $F(0, y) = 0$. Then the operators in (3.1) and (3.4) are compact in the space of continuous functions. In particular, the spectrum of (3.4) consists of eigenvalues that can only accumulate at zero. If the spectrum is contained in the interior of the unit circle, then $N^* = 0$ is asymptotically stable for (3.1). If at least one eigenvalue of (3.4) is outside the unit circle, then $N^* = 0$ is unstable.*

The proof of compactness is standard (see also Sect. 3.6), as is the statement about the spectrum (Keener 2000). The stability and instability results can be found as Theorems 1 and 2 in Iooss (1979). Next, we present a generalization of the Perron–Frobenius theorem for positive matrices (Caswell 2001) and a special case of the Krein–Rutman theorem for positive operators (Krasnosel'skii 1964). We discuss several cases and further details at the end of this chapter. An outline of the proof and certain extensions can be found in Chap. 13.

Theorem 3.2 (Dominant Eigenvalue) *Assume that Ω is bounded and connected and that K and R are continuous and positive. Then the integral operator in (3.4) in the space of continuous functions has a dominant eigenvalue, i.e., a real, positive eigenvalue that is larger in modulus than all other eigenvalues. The corresponding dominant eigenfunction is positive.*

3.2 Critical Patch-Size for the Laplace Kernel

We consider the domain of interest to be the interval $\Omega = [-L/2, L/2]$ of length L, and we assume that the growth function is independent of space, so that $R(y) = R = F'(0) > 0$. The Laplace kernel in (2.27) is positive and continuous so that all of the conditions in Theorem 3.2 are satisfied. Hence, we study the eigenvalue problem

$$\lambda\phi(x) = R\int_{-L/2}^{L/2} \frac{a}{2}\exp(-a|x - y|)\phi(y)\mathrm{d}y, \quad R = F'(0), \tag{3.5}$$

where $1/a$ is the mean dispersal distance. We begin with a simple observation that holds for general dispersal kernels.

Lemma 3.1 *If $R(y) = R$ is constant, and if $\int_\Omega K(x, y)dx < 1$ for some y, then the dominant eigenvalue satisfies $\lambda < R$.*

Proof From (2.24), we always have the inequality $\int_\Omega K(x, y)dx \leq 1$. We integrate (3.5) with respect to x and use the inequality for the kernel to get

$$\lambda \int_\Omega \phi(x)dx = R \int_\Omega \int_\Omega K(x, y)dx\phi(y)dy < R \int_\Omega \phi(y)dy. \tag{3.6}$$

The lemma can be extended to the case when $R = R(x)$ is not constant. We then obtain the upper bound $\lambda < \max_{y \in \Omega} R(y)$ since K is nonnegative.   □

The integral equation in (3.5) can be turned into an equivalent second-order differential equation. We split the integral into two parts to eliminate the absolute value as

$$\lambda \phi(x) = \frac{Ra}{2} \left[\int_{-L/2}^{x} e^{-a(x-y)}\phi(y)dy - \int_{L/2}^{x} e^{a(x-y)}\phi(y)dy \right]. \tag{3.7}$$

Differentiating this equation once, we obtain

$$\lambda \phi'(x) = -\frac{Ra^2}{2} \left[\int_{-L/2}^{x} e^{-a(x-y)}\phi(y)dy + \int_{L/2}^{x} e^{a(x-y)}\phi(y)dy \right], \tag{3.8}$$

and after a second time, we get

$$\lambda \phi''(x) = -Ra^2\phi(x) + \frac{Ra^3}{2} \int_{-L/2}^{L/2} e^{-a|x-y|}\phi(y)dy. \tag{3.9}$$

Substituting (3.5) for the integral and rearranging terms, we find the linear differential equation

$$\phi''(x) + a^2 \left(\frac{R}{\lambda} - 1 \right) \phi(x) = 0, \qquad x \in \left(-\frac{L}{2}, \frac{L}{2} \right). \tag{3.10}$$

To solve this equation, we need to supply boundary conditions. Evaluating (3.8) at the boundary points, we find

$$\phi'(-L/2) = a\phi(-L/2) \qquad \text{and} \qquad \phi'(L/2) = -a\phi(L/2). \tag{3.11}$$

Hence, the eigenvalue problem in (3.5) is equivalent to a regular Sturm–Liouville problem (see, e.g., Keener (2000) for details on such problems).

Fig. 3.1 Graphs of the left- (solid) and right- (dashed) hand sides of the transcendental equation (3.13) as a function of $1/\lambda > 1/R$. Each intersection corresponds to an eigenvalue of the problem. Parameters are $R = 5$ and $aL = 2$. The smallest intersection point corresponds to the largest eigenvalue.

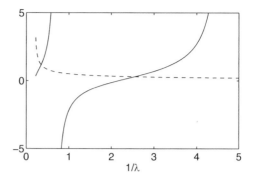

Using the symmetry of the problem, we find solutions of the form

$$\phi(x) = \cos(Ax) \quad \text{with} \quad A^2 = a^2 \left(\frac{R}{\lambda} - 1 \right). \tag{3.12}$$

Here, we require the relation $\lambda < R$ that we proved above. Substituting this ansatz into the boundary conditions, we find that eigenvalues of (3.5) are given by those $\lambda < R$ that satisfy the transcendental equation

$$\tan \left(\frac{aL\sqrt{R/\lambda - 1}}{2} \right) = \frac{1}{\sqrt{R/\lambda - 1}}. \tag{3.13}$$

This equation has infinitely many solutions. We illustrate these in Fig. 3.1 by plotting both sides of (3.13) as a function of $1/\lambda$.

The trivial solution of the IDE is stable if the dominant eigenvalue is less than unity and unstable if it is larger. The critical patch-size corresponds to the bifurcation point where the dominant eigenvalue equals unity. Setting $\lambda = 1$, we solve (3.13) for L and obtain the threshold value for the critical patch-size.

Proposition 3.1 *The dominant eigenvalue of (3.5) is given by the largest value of λ that solves the transcendental equation in (3.13). The critical patch-size, L^*, is the value of L for which the dominant eigenvalue equals one. It is given by*

$$L^* = \frac{2}{a\sqrt{R - 1}} \arctan \left(\frac{1}{\sqrt{R - 1}} \right). \tag{3.14}$$

If the patch is shorter than L^*, then dispersal loss exceeds reproductive gain in the patch, and the population will die out. If the patch is longer than L^*, then the situation is reversed and the population can persist. The critical patch-size decreases as the growth rate, R, increases or the mean dispersal distance, $1/a$, decreases; see Fig. 3.2.

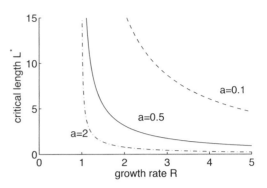

Fig. 3.2 Critical patch-size L^* as a function of growth rate R according to formula (3.14) for different values of $a = 0.1$ (dashed), $a = 0.5$ (solid), and $a = 2$ (dash-dot).

As for linear matrix population models, the dominant eigenvalue can be interpreted as an overall or global population growth rate (Caswell 2001). Then the condition $\lambda < R$ is the mathematical equivalent to the ecological insight that the global growth rate on a bounded patch with loss of individuals by dispersal is always smaller than the local growth rate at every point.

On a more technical note, we have tacitly assumed that the eigenfunction is twice continuously differentiable. We shall not be concerned with questions of regularity of solutions of IDEs here but instead assume the degree of differentiability necessary for calculations. Regularity is, of course, an important question to study in (partial) differential equations such as reaction–diffusion equations (Cantrell and Cosner 2003). Integral equations are much more "forgiving" in that the integral operator will "smooth" out solutions (as long as the kernel is smooth enough), whereas in differential equations, derivatives of solutions have to exist (at least in some weak sense) for the equations to make sense.

3.3 Scaling

According to (3.13), the eigenvalues of the integral operator in (3.4) depend only on R and the nondimensional parameter combination $\hat{a} = aL$. Recall that $1/a$ has units of length. We formalize this observation by applying the same nondimensionalization method as in (2.12) to the IDE

$$N_{t+1}(x) = \int_{-L/2}^{L/2} K(x, y) F(N_t(y), y) \mathrm{d}y . \tag{3.15}$$

We set $x = Lw$ and $N_t(x) = \hat{N} n_t(w)$, where w and n_t are nondimensional quantities. Substituting, we obtain

$$\hat{N} n_{t+1}(w) = \int_{-1/2}^{1/2} K(Lw, Lz) F(\hat{N} n_t(z), Lz) \mathrm{d}(Lz) , \tag{3.16}$$

where we wrote $y = Lz$. Hence, using the nondimensional growth function $\hat{F}(n, z) = \frac{1}{N}F(\hat{N}n, Lz)$ and the scaled kernel $\hat{K}(w, z) = LK(Lw, Lz)$, we can write IDE (3.15) in the equivalent form

$$n_{t+1}(w) = \int_{-1/2}^{1/2} \hat{K}(w, z)\hat{F}(n_t(z), z)dz . \tag{3.17}$$

The substitution rules for integration ensure that the properties of the kernel remain unchanged. For the Laplace kernel, we find

$$\hat{K}(w, z) = aLe^{-aL|w-z|} = \hat{a}e^{-\hat{a}|w-z|} , \tag{3.18}$$

as expected from (3.13).

Even when $R(x) > 0$ is not constant, one can still reduce integral equation (3.5) to differential equation (3.10) with boundary conditions (3.11). In that case, an explicit solution for L^* is not available, but the resulting eigenvalue problem can be studied analogously to eigenvalue problems that result from reaction–diffusion equations on bounded domains, e.g., by variational methods (Cantrell and Cosner 2003).

The deeper reason for why we can reduce the integral equation to a differential equation is that the Laplace kernel is the Green's function of a certain differential operator; see Chap. 7. The same idea applies to more complicated mechanistic movement models formulated in heterogeneous domains and including boundary conditions (Van Kirk and Lewis 1997; Musgrave and Lutscher 2014a). We return to these questions in Chaps. 7 and 15.

The symmetry assumption on K means that movement from x to y is equally likely as from y to x. Such an assumption seems reasonable in a homogeneous landscape but can be violated when the environment is heterogeneous or when dispersal is biased due to external forces. We will address some of these points in Sects. 7.5 and 12.2 and in Chap. 15.

3.4 Effects of Dispersal Kernels on Critical Patch-Size

The Laplace kernel is one of the very few dispersal kernels for which the critical patch-size can be calculated explicitly. How would the habitat requirement for population persistence change if dispersal followed a different kernel? We illustrate this question by numerically calculating and comparing eigenvalues of a number of commonly used kernels, listed in Table 3.1.

If the growth rate $R(x) = R$ is constant within a patch, the dominant eigenvalue in (3.4) is given by $\lambda = R\tilde{\lambda}$, where $\tilde{\lambda}$ is the dominant eigenvalue of the operator

$$\phi \mapsto \int_{-L/2}^{L/2} K(x, y)\phi(y)dy . \tag{3.19}$$

Table 3.1 Dispersal kernels for comparison of critical patch-size; see Fig. 3.3 for illustration.

Name	Formula	Constraint
Gaussian	$\frac{1}{\sqrt{2\pi\sigma^2}} \exp\left(-\frac{x^2}{2\sigma^2}\right)$	$\sigma^2 = 1$
Laplace	$\frac{a}{2} \exp\left(-a\lvert x\rvert\right)$	$a = \sqrt{2}$
Top-hat	$\frac{1}{2\beta}, \quad x \in [-\beta, \beta]$	$\beta = \sqrt{3}$
Tent	$\frac{1}{\eta} - \frac{\lvert x\rvert}{\eta^2}, \quad x \in [-\eta, \eta]$	$\eta = \sqrt{6}$
Double gamma	$\frac{1}{2\Gamma(k)\theta} \left\lvert\frac{x}{\theta}\right\rvert^{k-1} \exp\left(-\left\lvert\frac{x}{\theta}\right\rvert\right)$	$k(k+1)\theta^2 = 1$
Double Weibull	$\frac{k}{2\theta} \left\lvert\frac{x}{\theta}\right\rvert^{k-1} \exp\left(-\left\lvert\frac{x}{\theta}\right\rvert^k\right)$	$\theta^2\Gamma(1+2/k) = 1$

The gamma function is denoted by $\Gamma(k)$.

We evaluate the dominant eigenvalue of this operator numerically for several kernels and compare. We choose symmetric kernels of the form $K(x, y) = \widetilde{K}(x - y)$. When a kernel has only one parameter, we scale it such that the variance is equal to one. When there are two parameters, we reduce them to a single parameter by the same constraint.

The kernels are plotted in the left panel in Fig. 3.3; the corresponding dominant eigenvalues $\tilde{\lambda}$ are plotted in the right panel. The differences between the dominant eigenvalues for the Gaussian, Laplace, top-hat, and tent kernels are quite small (top row). The double gamma kernel with $k = 1$ is identical to the Laplace kernel. The Laplace kernel has the largest dominant eigenvalue for small patch-sizes. Its mass is the most concentrated near zero. Individuals typically stay close to their previous location and therefore many can successfully stay in a small domain.

Overall, we see that the critical patch-size is relatively insensitive to the general shape of the kernel. Lockwood et al. (2002) study the dependence of the critical patch-size on the kernel, and specifically on the shape of the tails of the kernels, via the average dispersal success approximation, that we will see in Chap. 9. They arrive at the same conclusion that the decay rate of the tail of a dispersal kernel as $\lvert x\rvert \to \infty$ has insignificant effects on the critical patch-size. This conclusion, however, holds only for symmetric dispersal. When dispersal is asymmetric, the critical patch-size can depend quite sensitively on the shape of the tails of the dispersal kernel, as we shall see in Sect. 12.2.

3.5   Separable Kernels

When the dispersal kernel is *separable*, the infinite-dimensional eigenvalue problem collapses to a finite-dimensional problem, i.e., a matrix eigenvalue problem. Such a simplification can offer some explicit calculations and some special insights

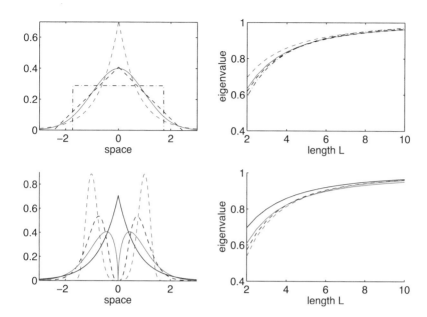

Fig. 3.3 Dispersal kernels and corresponding dominant eigenvalues as function of patch-size. **Top left:** Comparison of the Laplace (blue, dashed), Gaussian (blue, solid), top-hat (black, dashdot), and tent (black, dashed) kernels with variance $\sigma^2 = 1$. **Top right:** Corresponding dominant eigenvalue $\tilde{\lambda}$. **Bottom left:** Double gamma kernel (black) with parameters $k = 1$ (solid) and $k = 5$ (dashed); and double Weibull kernel (blue) with parameters $k = 1.5$ (solid) and $k = 5$ (dashed). The variance is $\sigma^2 = 1$ in all cases. **Bottom right:** Corresponding eigenvalues. The double gamma kernel with $k = 1$ is identical to the Laplace kernel.

(Kot and Schaffer 1986; Bramburger and Lutscher 2019). We assume for a moment that the dispersal kernel can be written as a product, $K(x, y) = K_1(x)K_2(y)$. For example, if individuals can explore the entire patch and decide where to settle based on local habitat conditions, then K_2 could be a constant and $K_1(x)$ would indicate the probability that an individual moves to location x (Robertson and Cushing 2011). The resulting eigenvalue equation for the critical patch-size reads

$$\lambda\phi(x) = \int_\Omega K_1(x)K_2(y)R(y)\phi(y)\mathrm{d}y . \tag{3.20}$$

In this case, the problem is really only one-dimensional. Since the left-hand side of the equation is a multiple of $\phi(x)$ and the right-hand side is a multiple of $K_1(x)$, we may assume $\phi = K_1$ and calculate the eigenvalue as

$$\lambda = \int_\Omega K_1(y)K_2(y)R(y)\mathrm{d}y . \tag{3.21}$$

More generally, a kernel is called *separable* if it can be written as a finite sum of products in the form

$$K(x, y) = \sum_{j=1}^{m} K_1^{(j)}(x) \, K_2^{(j)}(y) \, . \tag{3.22}$$

In this case, any eigenfunction has to be a linear combination of the functions $K_1^{(j)}$. Consequently, the eigenvalue problem can be reduced to an m-dimensional linear system and solved by matrix methods. Some kernels are separable; many others can be approximated arbitrarily closely by a separable kernel. We give examples for either case.

Finite Radius of Dispersal

In the simplest case, we can link a kernel of uniform dispersal to the fraction of individuals staying on an island in the simple model in Sect. 2.4. Let us assume that individuals on an island $\Omega = [-L/2, L/2]$ disperse uniformly with a maximum distance $L_m > L$. The latter condition ensures that individuals from one end of the island can reach the other end of the island in one dispersal period. The corresponding kernel is the top-hat kernel as in Table 3.1 with parameter $\beta = L_m$. Since $L_m > L$, we have $K(x, y) = 1/(2L_m)$ for all $x, y \in \Omega$. We can choose K_1 to be this constant and $K_2 = 1$. Hence, the positive eigenfunction is constant and the dominant eigenvalue is

$$\lambda = \frac{1}{2L_m} \int_{-L/2}^{L/2} R(y) dy = \frac{L}{2L_m} \frac{1}{L} \int_{-L/2}^{L/2} R(y) dy \, . \tag{3.23}$$

This expression has a simple interpretation. The fraction $L/(2L_m)$ is the relative size of the domain to the dispersal area. This quantity denotes the probability of an individual to stay on the island. The remaining expression is the spatially averaged growth rate. The population can persist if $\lambda > 1$, i.e., if the per capita number of offspring that stay on the island exceeds unity.

A more interesting example of a separable kernel is the cosine kernel

$$K(x, y) = \begin{cases} \frac{\pi}{4L_m} \cos\left(\frac{\pi}{2L_m}(x - y)\right), & y - L_m \leq x \leq y + L_m, \\ 0, & \text{otherwise,} \end{cases} \tag{3.24}$$

with maximum dispersal distance L_m (Kot and Schaffer 1986). As before, we consider the spatial domain to be the island $\Omega = [-L/2, L/2]$, and we assume

$L_m > L$. We can simplify the notation by scaling space as $x = 2L_m w$. The domain becomes $\Omega = [-l, l]$ with $l = L/(4L_m)$. The resulting kernel

$$K(w, z) = \begin{cases} \frac{\pi}{2} \cos(\pi(w - z)), & z - 0.5 \le w \le z + 0.5, \\ 0, & \text{otherwise}, \end{cases} \quad (3.25)$$

can be separated using the trigonometric identity

$$\cos(\pi(w - z)) = \cos(\pi w)\cos(\pi z) + \sin(\pi w)\sin(\pi z). \quad (3.26)$$

In terms of the above notation, we have $K_1^{(j)} = K_2^{(j)}$ with

$$K_1^{(1)}(w) = \sqrt{\pi/2}\cos(\pi w) \quad \text{and} \quad K_1^{(2)}(w) = \sqrt{\pi/2}\sin(\pi w). \quad (3.27)$$

We write an eigenfunction of (3.4) as the linear combination $\phi(w) = c_1 K_1^{(1)}(w) + c_2 K_1^{(2)}(w)$. Substituting this ansatz into (3.4) leads to the matrix eigenvalue problem

$$\lambda \begin{bmatrix} c_1 \\ c_2 \end{bmatrix} = \frac{\pi}{2} \begin{bmatrix} \int_{-l}^{l} \cos^2(\pi z) R(z) dz & \int_{-l}^{l} \cos(\pi z) R(z) \sin(\pi z) dz \\ \int_{-l}^{l} \cos(\pi z) R(z) \sin(\pi z) dz & \int_{-l}^{l} \sin^2(\pi z) R(z) dz \end{bmatrix} \begin{bmatrix} c_1 \\ c_2 \end{bmatrix}. \quad (3.28)$$

When $R(z) = R$ is a constant, the above integrals can be evaluated explicitly. The off-diagonal elements are zero by symmetry. The resulting matrix is

$$\frac{R}{4} \begin{bmatrix} 2\pi l + \sin(2\pi l) & 0 \\ 0 & 2\pi l - \sin(2\pi l) \end{bmatrix}. \quad (3.29)$$

Since $l = L/(4L_m) < 1/4$, we have $0 < \sin(2\pi l) < 1$, so that the dominant eigenvalue of this matrix is

$$\lambda = \frac{R}{4}[2\pi l + \sin(2\pi l)]. \quad (3.30)$$

Hence, we can write the persistence condition $\lambda > 1$ as

$$R > R^* = \frac{4}{2\pi l + \sin(2\pi l)}, \quad 0 < l < 1/4. \quad (3.31)$$

Since the denominator on the right-hand side is an increasing function of l, the required growth rate R decreases as the patch-size L increases or the dispersal radius L_m decreases; see Fig. 3.4.

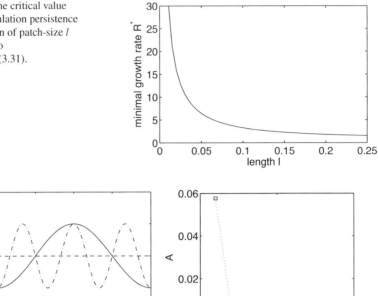

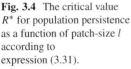

Fig. 3.4 The critical value R^* for population persistence as a function of patch-size l according to expression (3.31).

Fig. 3.5 **Left:** Heterogeneous growth rate profiles $R(w)$ according to (3.32) for $k = 1$ (solid) and $k = 2$ (dash-dot). The constant function is plotted for comparison (dashed). **Right:** Value of A in formula (3.35) as a function of k with $l = 0.2$.

Habitat Heterogeneity

We can also explicitly evaluate the matrix entries in (3.28) for some carefully chosen, nonconstant growth functions and explore how habitat heterogeneity affects persistence conditions. For example, we might choose

$$R(z) = \bar{R}\left[1 + \epsilon \cos\left((2k - 1)\pi z/l\right)\right] \tag{3.32}$$

for parameters $0 \leq \epsilon \leq 1$ and $k \in \mathbb{N}^+$. The condition on ϵ ensures that $R(z)$ is nonnegative. Parameter k determines how resources are distributed in the domain. Small values of k correspond to fewer but wider peaks, whereas large values of k indicate more but narrower peaks; see Fig. 3.5. The integral of $R(z)$ over the domain $\Omega = [-l, l]$ is equal to $2\bar{R}l$, independent of parameters ϵ and k. Hence, the total growth rate over the domain is a constant. Consequently, if dispersal were uniform, the dominant eigenvalue would be independent of parameters, according to the expression in (3.23).

The off-diagonal entries of the matrix in (3.28) with $R(z)$ as in (3.32) are zero by symmetry. The eigenvalues are the diagonal entries

$$\lambda^{\pm} = \frac{\bar{R}}{4}[2\pi l + \sin(2\pi l) \pm 2\pi \epsilon A], \qquad (3.33)$$

where, after some applications of trigonometric identities, we find

$$A = \int_{-l}^{l} \cos(2\pi z) \cos((2k-1)\pi z/l) \, dz. \qquad (3.34)$$

This integral can be evaluated explicitly with $\alpha = (2k-1)\pi/l$ as

$$A = \frac{2}{4\pi^2 - \alpha^2} [2\pi \cos(\alpha l)\sin(2\pi l) + \alpha \sin(\alpha l)\cos(2\pi l)], \qquad (3.35)$$

as long as $\alpha \neq \pm 2\pi$. From Fig. 3.5, we see that $A > 0$, so that λ^+ in (3.33) is the dominant eigenvalue, and it is larger than the dominant eigenvalue for the homogeneous landscape ($\epsilon = 0$). Hence, this form of heterogeneity, where the growth rate has a global maximum at the center of the patch, increases the population growth rate. We also see that as k increases, the gain over the homogeneous landscape gets smaller and smaller. Hence, few and wide peaks in R are more beneficial for the population than many and narrow peaks with the same total growth rate. Qualitatively similar results exist even more abstractly for reaction–diffusion models; see Sect. 2.3.1 in Cantrell and Cosner (2003).

Approximation by Separable Kernels

Many kernels are not separable but can be approximated arbitrarily closely by separable kernels (Keener 2000). In some cases, such approximations can be calculated explicitly. Latore et al. (1998) write the Gaussian kernel as an infinite sum of the form (3.22) and then truncate this sum at some finite value to obtain an approximation of the critical patch-size for the Gaussian kernel on the symmetric domain $\Omega = [-L/2, L/2]$. More precisely, the Gaussian kernel (2.25) can be written as

$$K(x, y) = \frac{1}{\sqrt{2\pi\sigma^2}} e^{-\frac{(x-y)^2}{2\sigma^2}} = \frac{1}{\sqrt{2\pi\sigma^2}} e^{-\frac{x^2}{2\sigma^2}} e^{-\frac{y^2}{2\sigma^2}} e^{\frac{xy}{\sigma^2}}. \qquad (3.36)$$

Using the series representation of the exponential function for the last term,

$$\exp\left(\frac{xy}{\sigma^2}\right) = \sum_{j=0}^{\infty} \frac{(x/\sigma)^j (y/\sigma)^j}{j!}\,, \tag{3.37}$$

we can formally write the infinite sum $K(x, y) = \sum_{j=0}^{\infty} K_1^{(j)}(x) K_2^{(j)}(y)$, with

$$K_1^{(j)}(x) = K_2^{(j)}(x) = \frac{1}{\sqrt{\sqrt{2\pi\sigma^2}\, j!}} \left(\frac{x}{\sigma}\right)^j \exp\left(-\frac{x^2}{2\sigma^2}\right). \tag{3.38}$$

We truncate this infinite sum at some number m. Then the eigenfunction can be written as the linear combination $\phi(x) = \sum_{j=0}^{m} c_j K_1^{(j)}(x)$. Functions $K_1^{(j)}$ are linearly independent. Inserting the expression into the eigenvalue equation (3.20) with constant $R(y) = R$ and rearranging gives the linear system for c_j as

$$\lambda c_j = R \sum_{k=0}^{m} K_{jk} c_k\,, \tag{3.39}$$

where

$$K_{jk} = \int_{-L/2}^{L/2} K_1^{(j)}(y) K_2^{(k)}(y) \mathrm{d}y = \frac{1}{\sqrt{2\pi\sigma^2}\, j! k!} \int_{-L/2}^{L/2} \left(\frac{y}{\sigma}\right)^{(j+k)} \exp\left(-\frac{y^2}{\sigma^2}\right) \mathrm{d}y\,. \tag{3.40}$$

Matrix (K_{jk}) is symmetric and has entries $K_{jk} = 0$ if $j + k$ is odd.

The persistence condition $\lambda = 1$ determines the minimal growth rate required for population persistence as $R^* = 1/\tilde{\lambda}$, where $\tilde{\lambda}$ is the dominant eigenvalue of matrix (K_{jk}). We plot R^* as a function of patch-size in Fig. 3.6 for different values of m. We observe that even small values of m give a very good approximation on small domains, whereas on large domains, we need to include more and more terms to obtain a valid approximation.

Fig. 3.6 Approximation of the critical growth rate R^* for the Gaussian kernel as a function of domain length. The truncation values are $m = 1$ (upper solid curve), $m = 3$ (dashed), $m = 7$ (dash-dot), and $m = 21$ (lower solid curve).

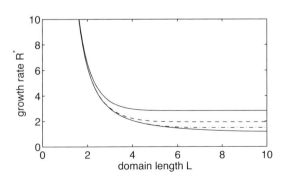

3.6 Further Reading

For a finite-dimensional matrix eigenvalue problem, the Perron–Frobenius theorem states that a positive matrix has a simple, real eigenvalue with positive eigenfunction, and this eigenvalue is larger than the modulus of any other eigenvalue (Caswell 2001). It is called the dominant eigenvalue. Theorem 3.2 is the infinite-dimensional analogue of this theorem. It is a special case of the Krein–Rutman theorem (see, e.g., Du 2006). The most important ingredient in the proof is to show that the integral operator is compact and positive (Krasnosel'skii 1964; Krasnosel'skii and Zabreiko 1984). Positivity results directly from our model assumptions. We can achieve compactness in appropriate function spaces under relatively mild conditions on the dispersal kernel. Detailed proofs are given by Hardin et al. (1990) in the space of continuous functions and by Van Kirk and Lewis (1997) and Lutscher and Lewis (2004) in the space of square-integrable functions. We present most of this proof in Chap. 13.

Many standard books provide overviews and detailed results on integral operators, e.g., Keener (2000). Several relevant cases in our context are the following. If Ω is a bounded interval and if K is continuous on $\Omega \times \Omega$, then the operator in (3.4) is compact in the space of continuous functions on Ω by the Arzelá–Ascoli theorem. If $K(x, y)$ is square integrable on $\Omega \times \Omega$, then the operator defined in (3.4) is a Hilbert–Schmidt operator and is compact in the space of square-integrable functions on Ω. The spectrum of a compact operator consists of countably many eigenvalues that may only accumulate at zero. All nonzero eigenvalues have finite multiplicity. Specializing even more, if we choose $R(y) = R$ to be constant and the dispersal kernel as positive and symmetric, i.e., $K(x, y) = K(y, x) > 0$, then the operator is positive and self-adjoint. Such an operator has at least one nonzero eigenvalue, all eigenvalues are real, and there is a dominant eigenvalue. When $R(x)$ is not constant but positive, then eigenvalue problem (3.4) is not symmetric but can be transformed via

$$\tilde{\phi}(x) = \phi(x)\sqrt{R(x)} \tag{3.41}$$

into the symmetric eigenvalue problem

$$\lambda\tilde{\phi}(x) = \int_{\Omega} K(x, y)\sqrt{R(x)R(y)}\tilde{\phi}(y)\mathrm{d}y . \tag{3.42}$$

The existence of (real) eigenvalues is again guaranteed. The condition $R(x) > 0$ on Ω is natural for the linearization at zero, but not for the linearization at a positive steady state, as we shall see in Chap. 4.

Symmetric (or symmetrizable) integral operators have many more useful properties. One of them is the existence of a variational characterization of the dominant eigenvalue. Specifically, if we assume that $R(y) = R$ is a constant and that K is symmetric, continuous, and positive, then the dominant eigenvalue λ in (3.4) can be obtained as

$$\lambda = \max_{u \in L^2(\Omega), \|u\|=1} \left\{ R \int_\Omega \int_\Omega u(x) K(x, y) u(y) \mathrm{d}y \mathrm{d}x \right\}, \tag{3.43}$$

where the maximum is taken over all square-integrable functions u on Ω with norm equal to unity. For a proof, we refer to Chap. 11 in Keener (2000) or to Theorem 3.1 by Hutson et al. (2003). The variational formula is not usually applicable in explicit calculations but is often helpful in proofs. We will use it in Sect. 9.2 for an estimate.

Zhou and Kot (2013) present several approximations for the dominant eigenvalue based on approximation by orthogonal polynomials. We assume that $\{X_i(x)\}_{i=1}^\infty$ is a family of orthogonal polynomials. Then the kernel can be expanded as the double sum

$$K(x, y) = \sum_{i,j=1}^\infty A_{ij} X_i(x) X_j(y). \tag{3.44}$$

Eigenvalue equation (3.4) with constant $R(x) \equiv R$ becomes

$$\lambda \phi(x) = R \sum_{i=1}^\infty \left(\sum_{j=1}^\infty A_{ij} \int X_j(y) \phi(y) \mathrm{d}y \right) X_i(x). \tag{3.45}$$

If we also expand the eigenfunction in terms of $\{X_i\}$, say $\phi(x) = \sum_{i=1}^\infty a_i X_i(x)$, we obtain the infinite matrix eigenvalue problem

$$\lambda a_i = R \sum_{k=1}^\infty \left(\sum_{j=1}^\infty A_{ij} \int X_j(y) X_k(y) \mathrm{d}y \right) a_k. \tag{3.46}$$

We can then truncate this system at any finite number of equations and solve for the eigenvalue. Zhou and Kot (2013) use Legendre polynomials and realize—just as with the approximation by separable kernels—that an expansion using the first one or two terms is often quite good. In Chap. 9, we present an ecologically motivated approximation, which, in some cases, is equivalent to using only the first term of the infinite matrix problem.

Symmetric or not, there are a number of numerical methods to calculate the dominant eigenvalue of the linear operator in (3.4). They have been collected, reviewed, and compared recently by Kot and Phillips (2015). We touch on this subject in Chap. 8.

Chapter 4
Positive Steady States

Abstract When a population can persist in a given region, we want to know what the long-term dynamics of the population are. Is there a positive steady state? Is it asymptotically stable? And what is the spatial distribution? From a management point of view, a unique stable steady state is the simplest: Even if the population is perturbed somewhat in one generation, it will return to its steady state over time. But many other scenarios can arise. For example, a positive state can be unstable and the population could cycle between different states. How does spatial dispersal affect these dynamics? Can it stabilize or destabilize a steady state? What are the spatial distributions throughout a cycle? In the case of an Allee effect, there could be a positive steady state even if the trivial state is locally stable. How does spatial dispersal affect the ability of a population to persist in that case? In this chapter, we present some analytical methods and numerical results that explore these questions. The effects are inherently nonlinear and therefore much harder to study completely.

4.1 The Steady-State Equation

A (nontrivial) steady state $N^*(x) \geq 0$ of (2.1) satisfies the so-called Hammerstein equation (Hammerstein 1930)

$$N^*(x) = Q[N^*](x) := \int_\Omega K(x, y) F(N^*(y), y) \mathrm{d}y . \tag{4.1}$$

While the trivial steady state always exists in the models that we consider (since $F(0) = 0$), proving the existence of a positive steady state can be a challenge. Similar to the procedure in the previous chapter, the steady-state problem (4.1) can be reduced to and studied as a two-dimensional differential equation when the dispersal kernel is the Laplace kernel (Sects. 4.2 and 4.5). For some separable kernels, explicit calculations are possible in the corresponding finite-dimensional discrete dynamical systems (Sect. 4.4). More abstract approaches are based on monotonicity and bifurcation theory (Sect. 4.3).

© Springer Nature Switzerland AG 2019

F. Lutscher, *Integrodifference Equations in Spatial Ecology*, Interdisciplinary Applied Mathematics 49, https://doi.org/10.1007/978-3-030-29294-2_4

Once the existence of a nontrivial steady state is clear, the question of its stability leads to the linear eigenvalue problem

$$\lambda \phi(x) = \int_\Omega K(x, y) R(y) \phi(y) \mathrm{d}y, \qquad R(y) = \frac{\partial F}{\partial N}(N^*(y), y). \qquad (4.2)$$

Since this equation requires the knowledge of the positive state, it can almost never be studied explicitly but is the basis for numerical methods. In contrast to the situation in the previous chapter, the term $R(y)$ can be negative here, e.g., when F is the Ricker function. If the function $R(y)$ is negative for all y, then the eigenvalue problem can be made symmetric by the substitution

$$\tilde{\phi}(x) = \phi(x) \sqrt{-R(x)}, \qquad (4.3)$$

similar to the case of positive R from the previous chapter.

4.2   A Positive Steady State with the Laplace Kernel

As in the previous chapter, the Laplace kernel offers a particularly nice example where steady states can be studied explicitly in some detail via a reduction to a system of differential equations. We follow again Kot and Schaffer (1986). We scale the space variable so that the domain of interest is $\Omega = [-1/2, 1/2]$, and we consider $1/a$ as the scaled dispersal distance (Sect. 3.3). To simplify notation, we write $N(x)$ instead of $N^*(x)$ for the steady-state density. We differentiate the steady-state equation

$$N(x) = \int_{-1/2}^{1/2} \frac{a}{2} \exp(-a|x - y|) F(N(y)) \mathrm{d}y \qquad (4.4)$$

twice and substitute to turn it into the differential equation

$$N'' + a^2[F(N) - N] = 0, \qquad x \in (-1/2, 1/2), \qquad (4.5)$$

with mixed boundary conditions

$$N'(-1/2) = aN(-1/2), \qquad N'(1/2) = -aN(1/2). \qquad (4.6)$$

Hence, the steady-state problem for the IDE is equivalent to the steady-state problem of a reaction–diffusion equation with mixed, or Robin's, boundary conditions. The only difference is in the functional form of the updating function F. We use phase-plane methods to study the existence and shape of solutions.

We write the second-order equation in (4.5) as a first-order system in N and $\hat{N} = N'$, namely

$$N' = \hat{N}, \qquad \hat{N}' = a^2[N - F(N)]. \qquad (4.7)$$

To account for the boundary conditions, we look for solutions that start on the line $\hat{N} = aN$ at $x = -1/2$ and end on the line $\hat{N} = -aN$ at $x = 1/2$. An equilibrium of system (4.7) necessarily satisfies $\hat{N} = 0$ and $N = F(N)$, i.e., we retrieve the fixed points of the corresponding nonspatial dynamics. For a linear stability analysis of the equilibrium $(N, 0)$, we calculate the Jacobian matrix of (4.7) as

$$\begin{bmatrix} 0 & 1 \\ a^2[1 - F'(N)] & 0 \end{bmatrix}. \qquad (4.8)$$

The trace of this matrix is zero; the sign of the determinant is the sign of $F'(N) - 1$. Hence, the equilibrium is a saddle point when $F'(N) < 1$ and a (linear) center if $F'(N) > 1$. We multiply the equation in (4.5) by N' and integrate to get

$$\frac{1}{2}(N')^2 + a^2 H(N) - \frac{a^2}{2}N^2 = \text{constant}, \qquad (4.9)$$

where H is an antiderivative of F, i.e., $H' = F$. This expression is a first integral for the system. Consequently, a linear center is also a nonlinear center. Using the first integral, we can numerically visualize the solution curves of (4.7) in the phase plane as the level sets of (4.9).

When $F(N)$ is the Beverton–Holt function in (2.13) with $R > 1$, we know that $F'(0) > 1$ and $F'(1) < 1$, so that the equilibria of (4.7) are $(0, 0)$, a center, and $(1, 0)$, a saddle. We can also find an antiderivative of F as

$$H(N) = \frac{R}{R - 1}\left[N - \frac{1}{R - 1}\ln(1 + (R - 1)N)\right]. \qquad (4.10)$$

The phase-plane plot in Fig. 4.1 illustrates the center at $(0, 0)$ and the saddle point at $(1, 0)$. The two straight lines from the origin correspond to the boundary conditions. The phase-plane curves are the level sets of the first integral. Solutions of the steady-state equation for IDE (4.4) correspond to the segments of the phase-plane curves of ODE (4.7) that exactly connect the boundary lines while x ranges from $-1/2$ to $1/2$. The right plot in the same figure illustrates the spatial shape of the steady state for different values of a.

When $F(N)$ is the Ricker function in (2.19), the analysis is very similar. The steady states are $(0, 0)$, a center, and $(1, 0)$, a saddle. An antiderivative is given by

$$H(N) = -\frac{e^{r(1-N)}}{r^2}[rN + 1]. \qquad (4.11)$$

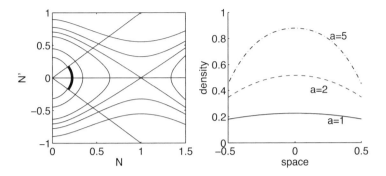

Fig. 4.1 Phase plane (left plot) and solution profile (right plot) of the steady state for IDE (4.1) with Beverton–Holt growth function. The straight lines in the left plot indicate the boundary conditions. The bold curve segment between the two lines corresponds to the solution profile with $a = 1$ in the right plot. The growth parameter is $R = 5$.

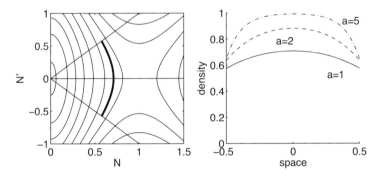

Fig. 4.2 Phase plane (left plot) and solution profile (right plot) of the steady state for IDE (4.1) with Ricker growth function. The straight lines in the left plot indicate the boundary conditions. The bold curve segment between the two lines corresponds to the solution profile with $a = 1$ in the right plot. The growth parameter is $r = 3$.

The plots in Fig. 4.2 illustrate the exact same behavior as with the Beverton–Holt function in Fig. 4.1.

We already know from the analysis of the nonspatial model in Chap. 2 that the positive state in the Beverton–Holt model is always stable, whereas the positive state in the Ricker model can lose stability through a flip bifurcation, and cyclic or even chaotic dynamics can result. The phase-plane equations for the steady state of the IDE with Beverton–Holt and Ricker function are qualitatively identical, yet we expect the stability behavior of the steady states to differ. Hence, the method we used so far to show the existence of a positive steady state in the IDE does not address the question of stability. Next, we outline a method to show the stability of the positive steady state of the IDE with Beverton–Holt growth function.

4.3 Monotone Growth Function

The explicit calculations of the positive steady state from the previous section work for a specific kernel and for (arbitrary) growth functions. They give us insight into the existence but say nothing about stability or uniqueness. Results in that direction depend on properties of the growth function, in particular monotonicity, and can be proved abstractly, independent of the particular shape of the dispersal kernel. We illustrate some of these geometric ideas here in the space of continuous functions. A more general, abstract setting with additional concavity conditions and details in the case of (systems of) square-integrable functions is discussed in Chap. 13. For a more general reference on monotone dynamical systems, see Chap. 2 in Zhao (2003), and for positive discrete dynamical systems, see Krause (2015).

We consider the usual IDE as a dynamical system

$$N_{t+1}(x) = Q[N_t](x) = \int_0^1 K(x, y) F(N_t(y)) dy \tag{4.12}$$

on the space $\mathscr{C}([0, 1])$ of (bounded) continuous functions on the unit interval. (In fact, by scaling space, the interval could be of any length.) We assume that the growth function is nonnegative and continuously differentiable and the dispersal kernel is positive and continuous. We assume that the growth function is independent of spatial location and write $R_0 = F'(0)$. Theorem 3.2 guarantees the existence of a dominant eigenvalue, λ, of the linearization of Q at zero. We denote a (positive) dominant eigenfunction by ϕ, i.e.,

$$\lambda \phi(x) = Q'[0]\phi = R_0 \int_0^1 K(x, y) \phi(y) dy. \tag{4.13}$$

We consider three properties of the growth function $F : \mathbb{R}_0^+ \to \mathbb{R}_0^+$:

(S1) F is (globally) bounded if for all $u \in \mathbb{R}_0^+$ we have $F(u) \leq C$ for some $C > 0$.
(S2) F is (linearly) bounded if for all $u \in \mathbb{R}_0^+$ we have $F(u) \leq F'(0)u$.
(S3) F is monotone if $u \geq v$ implies $F(u) \geq F(v)$.

The Beverton–Holt, Ricker, and Allee functions from Sect. 2.2 all satisfy the first condition, but only the Beverton–Holt function satisfies all three. The Ricker function is not monotone and the Allee function is not linearly bounded.

Next, we consider functions $N, \hat{N} \in \mathscr{C}([0, 1])$ and write $N \geq \hat{N}$ if $N(x) \geq \hat{N}(x)$ for every $x \in [0, 1]$. We say that operator Q is monotone (or order preserving) if $N \geq \hat{N}$ implies $Q[N] \geq Q[\hat{N}]$ in this sense. Since kernel K is positive, we find that Q is monotone if F is. Our first result shows in particular that if F is the Beverton–Holt function and zero is locally stable, then zero is globally stable. The second result indicates that if the zero state is unstable, then there is at least one positive state.

Lemma 4.1 *Assume that F satisfies properties (S2) and (S3) above, and assume that $\lambda < 1$. Then all nonnegative solutions of (4.12) converge to zero.*

Proof Since the kernel is positive, the linear boundedness of F implies the linear boundedness of Q in the sense that $Q[N] \leq Q'[0]N$. In particular, for a dominant eigenfunction, we have $Q[\phi] \leq Q'[0]\phi = \lambda\phi$. By iteration, we find $Q^t[\phi] \leq \lambda^t\phi$. Since, by assumption, we have $0 \leq \lambda < 1$, we find $Q^t[\phi] \to 0$ as $t \to \infty$.

For any given $N_0 \in \mathscr{C}([0,1])$ we can find a (multiple of a) dominant eigenfunction with $N_0 \leq \phi$. Then $Q[N_0] \leq Q[\phi]$ by monotonicity, and, by iteration, this relation holds for all powers Q^t. Hence, $Q^t[N] \to 0$. $\qquad\square$

Lemma 4.2 *Assume that F satisfies properties (S1)–(S3) above and assume that $\lambda > 1$. Then there exists at least one positive steady-state solution of (4.12).*

Proof Since, by assumption, $\lambda > 1$, we also have $R_0 > 1$ by Lemma 3.1. For any $1 < R < R_0$, we can find some $\epsilon > 0$ such that $F(z) > Rz$ for $0 \leq z < \epsilon$. Now we can pick a dominant eigenfunction (and scale if necessary) with $0 < \phi(x) < \epsilon$. Then

$$Q[\phi](x) \geq R \int_0^1 K(x,y)\phi(y)\mathrm{d}y = \frac{R}{R_0}\lambda\phi(x). \tag{4.14}$$

Since $\lambda > 1$, we can choose $R < R_0$ such that the factor $\lambda R/R_0 > 1$. Then $Q[\phi] \geq \phi$. We construct a sequence $\underline{N}_t$ via (4.12) with $\underline{N}_0 = \phi$. By the choice of R, we have $\underline{N}_1 = Q[\underline{N}_0] \geq \underline{N}_0$. By induction, we see that $\underline{N}_{t+1} \geq \underline{N}_t$ for all $t \in \mathbb{N}$, i.e., the sequence is monotone increasing. Since $F(N) \leq C$, the sequence is also bounded. Hence, the pointwise limit $N_*(x) = \lim_{t\to\infty} \underline{N}_t(x)$ exists.

Since K is continuous and the unit interval is compact, operator Q is compact on $\mathscr{C}([0,1])$. Compactness together with monotonicity implies that $\{Q[\underline{N}_t]\}_t$, and hence $\{\underline{N}_t\}_t$, converges in the space of continuous functions. By continuity of Q, we have

$$Q[N_*] = Q[\lim_{t\to\infty}\underline{N}_t] = \lim_{t\to\infty} Q[\underline{N}_t] = \lim_{t\to\infty}\underline{N}_{t+1} = N_*. \tag{4.15}$$

Hence, N_* is a fixed point of Q. $\qquad\square$

Altogether, we have shown not only the existence of a positive fixed point but also the convergence of small initial conditions to this point. In the same way, we can construct a monotone decreasing sequence, $\{\overline{N}_t\}_t$, with $\overline{N}_0 = C$ and $\overline{N}_{t+1} \leq \overline{N}_t$ for all t. Since $\underline{N}_0 < \overline{N}_0$, we apply monotonicity again and find $\underline{N}_t \leq \overline{N}_t$ for all t. In particular, the sequence $\{\overline{N}_t\}_t$ is bounded below away from zero and, hence, converges to some positive fixed point N^*. In general, $N_* \leq N^*$. Under additional assumptions on F, one can show that $N_* = N^*$ by applying the ideas of Theorem 6.3 in Krasnosel'skii (1964) as follows.

Lemma 4.3 *Assume in addition that F is concave in the sense that for all $0 < s < 1$ and $z > 0$ there exists some $\varepsilon > 0$ such that $F(sz) \geq (1 + \varepsilon)sF(z)$. Then $N^* = N_*$.*

Proof Since K is positive and $[0, 1]$ is compact, concavity of F translates into concavity of Q, i.e., for all $0 < s < 1$ and $N \in \mathscr{C}([0, 1])$ with $N(x) > 0$ for all x, there is some $\varepsilon > 0$ such that $Q[sN] \geq (1 + \varepsilon)s\,Q[N]$. We already know that $N_* \leq N^*$. Assume now that $N_* \not\geq N^*$. By positivity, there exists some $0 < s_0 < 1$ with $N_* \geq s_0 N^*$ and $N_* \not\geq sN^*$ for any $s > s_0$. Then, by monotonicity and concavity, we have

$$N_* = Q[N_*] \geq Q[s_0 N^*] \geq (1 + \varepsilon)s_0 Q[N^*] = (1 + \varepsilon)s_0 N^*, \qquad (4.16)$$

which contradicts the maximality of s_0. $\qquad\square$

4.4 Nonmonotone Growth Function

When the growth dynamics are not monotone, a positive steady state can be destabilized and give way to spatial patterns and more complicated temporal dynamics. These observations go back to the work of Kot and Schaffer (1986) and Andersen (1991), who use a discrete logistic growth function and the Ricker function, respectively. We explain and expand some of their examples here.

We begin with the (scaled) separable cosine kernel (3.25) and write $K(x, y) = K_1(x)K_1(y) + K_2(x)K_2(y)$. We can assume that the population density at any given generation is a linear combination of the two kernel functions, and we write $N_t(x) = a_t K_1(x) + b_t K_2(x)$. Substituting this ansatz into the IDE, we find

$$a_{t+1}K_1(x) + b_{t+1}K_2(x) = \int_{-l}^{l} [K_1(x)K_1(y) + K_2(x)K_2(y)]F(a_t K_1(y) + b_t K_2(y))dy.$$
$$(4.17)$$

Splitting the integral into a sum and comparing coefficients for K_1 and K_2, we derive the two-dimensional difference equation for a_t, b_t as

$$a_{t+1} = \int_{-l}^{l} K_1(y)F(a_t K_1(y) + b_t K_2(y))dy,$$

$$(4.18)$$

$$b_{t+1} = \int_{-l}^{l} K_2(y)F(a_t K_1(y) + b_t K_2(y))dy.$$

The kernel functions are the trigonometric functions from (3.27). If F is a polynomial, then we can use trigonometric identities and carry out the resulting integrations directly. For example, with the logistic growth function $F(N) = (r + 1)N - rN^2$, the integrand in the equation for a_t becomes the trigonometric polynomial

$$a_t\left((r+1)K_1^2 - a_t r K_1^3\right) - rb_t^2 K_2^2 K_1 + b_t(r+1)K_2 K_1 - a_t b_t r K_1^2 K_2 . \quad (4.19)$$

The last two terms in the sum are odd functions, so that their integral over $[-l, l]$ is zero. Carrying out the integration for this and the corresponding equation for b_t, we find (Kot and Schaffer 1986)

$$a_{t+1} = (c_1 - c_2 a_t)a_t - c_3 b_t^2 ,$$
$$b_{t+1} = (c_4 - c_5 a_t)b_t , \quad (4.20)$$

where

$$c_1 = \frac{(1+r)}{4}[2\pi l + \sin(2\pi l)] , \qquad c_2 = r\sin(\pi l)\sqrt{\frac{\pi}{2}}\left[1 - \frac{1}{3}\sin^2(\pi l)\right] ,$$

$$c_3 = \frac{r}{3}\sqrt{\frac{\pi}{2}}\sin^3(\pi l) , \qquad c_4 = \frac{(1+r)}{4}[2\pi l - \sin(2\pi l)] ,$$

$$(4.21)$$

and $c_5 = 2c_3$.

The second equation in (4.20) ensures that $b_0 = 0$ implies $b_t = 0$ for all t. In that case, the first equation becomes a logistic equation for a_t. Hence the dynamics of the IDE with the cosine kernel and the logistic growth function can exhibit all the dynamics that the nonspatial logistic equation can. At each time step, the spatial profile is that of a cosine function. But even when $b_0 \neq 0$, the solutions may still converge to the invariant set where $b_t = 0$, e.g., if we can guarantee a priori that $-1 < c_4 - c_5 a_t < 1$ for all $t > 0$ (Kot and Schaffer 1986). Bramburger and Lutscher (2019) study (4.20) in more detail and give sufficient conditions for this convergence. Whether this equation can generate other spatial patterns that are not concave, and what they might look like is currently unknown. It would require $b_t \nrightarrow 0$, yet the solution $N_t(x)$ has to remain nonnegative. We turn to another example and numerically illustrate a variety of spatial patterns emerging.

Numerical simulations of the IDE with Ricker dynamics reveal period-doubling bifurcations and chaos (Kot and Schaffer 1986). In addition, we can observe spatial population distributions that are not concave but exhibit several maxima and minima (Andersen 1991). We explore some of these phenomena numerically. We recall some dynamical properties of the nonspatial Ricker map (2.19) from Sect. 2.2. The positive state is stable for $0 < r < 2$ but becomes unstable in a flip bifurcation at $r = 2$ when a stable two-cycle appears. Increasing r beyond 2.526 destabilizes the two-cycle and, via a period-doubling bifurcation, leads to a stable four-cycle. Increasing r even further results in a cascade of period-doubling bifurcations leading to chaos; see Fig. 4.3. The *orbit diagram* in this figure displays the dynamics of the equation after transients have disappeared. In practice, we solve the equations for large times (e.g., $t = 10{,}000$) and then plot only the last few of them (here the last 50).

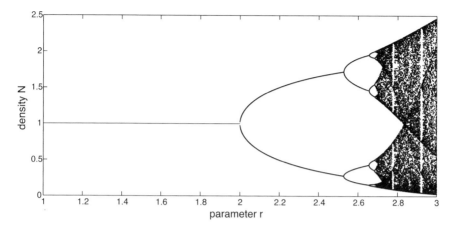

Fig. 4.3 Orbit diagram for the nonspatial Ricker model (2.19) from Sect. 2.2.

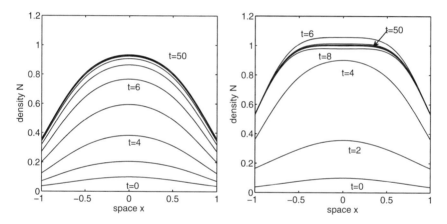

Fig. 4.4 Left plot: Solutions for the IDE with Ricker dynamics approach the steady state monotonically ($r = 0.5$). **Right plot:** Nonmonotone approach to steady state for the same model ($r = 1.5$). The dispersal kernel is a Gaussian kernel with variance $\sigma^2 = 0.1$.

For comparison and exploration, we simulated the IDE with Gaussian dispersal kernel and scaled Ricker growth function on the (scaled) domain $\Omega = [-1, 1]$. Of the two remaining parameters, we always chose r large enough and σ^2 small enough so that the population persisted according to the theory from the previous chapter. When $0 < r < 1$, the nonspatial model predicts monotone convergence to steady state, and the spatial model shows the same behavior. The steady-state profile is concave (left panel, Fig. 4.4) as we have seen before (Fig. 4.2).

When $1 < r < 2$, the nonspatial model predicts an oscillatory approach to the stable steady state. The IDE can show monotone or oscillating approaches to the steady state (right panel, Fig. 4.4). The steady-state profile may display local

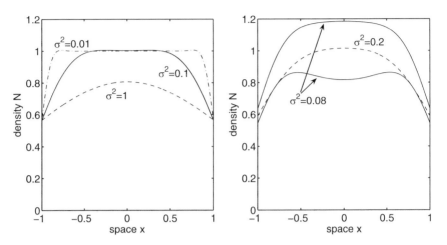

Fig. 4.5 Left plot: Steady-state profiles for the IDE with Ricker dynamics and Gaussian kernel for different variances σ^2 as indicated. The growth rate is $r = 1.8$. **Right plot:** Steady-state profile (dashed) and two-cycle profile (solid) for the same model with $r = 2.1$.

maxima near the boundary of the domain if the variance is small (Fig. 4.5). This phenomenon can be explained by the interaction of Ricker dynamics and dispersal loss at the boundary. As the variance decreases, fewer individuals leave the domain at the boundary, and the overall density increases. Near the boundary, the density is low after dispersal, and the overcompensatory dynamics produce a high density near the boundary. Most of these individuals disperse out of the domain, but some remain and create the little peak near the boundary.

When $2 < r < 2.5$ so that the nonspatial model predicts a stable two-cycle, the IDE can have a stable steady state or—for small enough variance—it can have a two-cycle (Fig. 4.5). For even larger r, when the nonspatial model predicts chaotic behavior, we observe a stable steady state in the IDE for large variance, a two-cycle for intermediate variance (plots not shown), a four-cycle and an eight-cycle (Fig. 4.6), and chaotic-looking patterns for even smaller variance (Andersen 1991). The orbit diagram for the total density (i.e., the integral of N over the domain) shows period-doubling bifurcations as the variance of the kernel decreases (Fig. 4.7).

These simulations suggest that increasing the variance, or decreasing the patch-size, has a similar effect on the stability of steady states as decreasing the growth parameter r. In Chap. 9 we will use approximations and explain why and to what extent this reasoning holds. The question of how different dispersal kernels affect the stability of steady states and the shape of their spatial profile is relatively unexplored. Replacing the Gaussian kernel with the Laplace kernel in the simulations presented reproduces the same dynamical behavior. Some differences do occur when the modal dispersal distance is away from the origin (Andersen 1991).

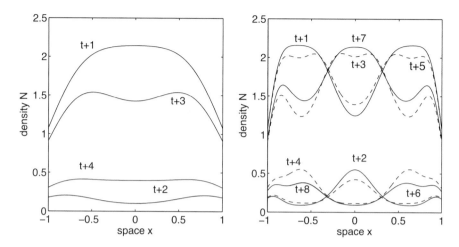

Fig. 4.6 Four-cycle (left) and eight-cycle (right) for the IDE with Ricker dynamics and Gaussian kernel. The growth rate is $r = 2.8$. The variance is $\sigma^2 = 0.1$ (left) and $\sigma^2 = 0.01$ (right).

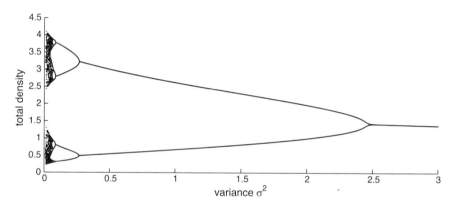

Fig. 4.7 Orbit diagram for the spatial Ricker model as a function of variance for $r = 2.8$.

4.5 Allee Growth Function

When the growth function exhibits a strong Allee effect, i.e., $F'(0) < 1$, then the zero state is locally stable in the nonspatial model and also in the corresponding IDE. There may or may not be a positive steady state; the linearization at zero gives no indication. In the nonspatial model, two positive steady states generically emerge in a saddle-node bifurcation. For the Allee growth function in (2.22), this bifurcation happens at $R = \gamma$ (Musgrave et al. 2015).

We use the phase-plane approach from above to find positive steady states for the IDE with Laplace dispersal kernel and a growth function with strong Allee effect. Equations (4.4) and (4.5) remain unchanged, and the expression in (4.9) is still a first integral. We pick the growth function $F(N) = RN^2(1 + (R - 1)N^2)^{-1}$ from (2.22) with $R > \gamma = 2$. The equation $F(N) = N$ has the (stable) trivial solution, the stable positive state $N = 1$, and Allee threshold $N_a = (R - 1)^{-1} < 1$. In the phase plane, we obtain the corresponding steady states $(0, 0)$, $(1, 0)$, and $(N_a, 0)$. The two former states are saddles; the latter is a center. We find an antiderivative of F as

$$H(N) = \frac{R}{R - 1}\left[N - \frac{1}{\sqrt{R - 1}}\arctan(N\sqrt{R - 1})\right]. \tag{4.22}$$

Since the growth function has a positive state for $R > 2$, we could expect the IDE to have a positive steady state in that case as well, provided that the variance of the dispersal kernel is small enough so that most individuals remain in the domain. Figure 4.8 reveals that this conjecture is not true. For $2 < R < R^* \approx 3.29$, there cannot be a connection between the two straight lines that correspond to the boundary conditions. The stable and unstable manifolds of the saddle at $(0, 0)$ do not cross the N-axis in the interval $(0, 1)$. They force all solutions that start at the upper boundary condition to remain above the N-axis. For $R > R^*$, the stable and unstable manifolds of the saddle at $(0, 0)$ form a homoclinic loop that crosses the N-axis. At the same time, the stable and unstable manifold of the saddle at $(1, 0)$ cross the two boundary conditions. All solutions that start between these two crossing points are forced to travel from one boundary condition to the other and thereby correspond to a steady state of the IDE (Fig. 4.9).

The critical value R^* corresponds to the situation that the stable and unstable manifolds form a heteroclinic connection between the saddle points $(0, 0)$ and $(1, 0)$. Hence, we can calculate the critical value by insisting that the expression in (4.9), which is constant along solutions, have the same value at those two points. This condition can be expressed in terms of H as $H(0) = H(1) - 0.5$ or, after some algebraic manipulation,

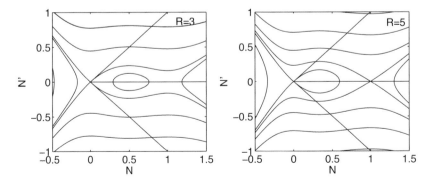

Fig. 4.8 Phase plane of the steady state for IDE (4.1) with Allee updating function (2.22) with $\gamma = 2$. The straight lines in the left plot indicate the boundary conditions. There is no connection between the boundary conditions for $R = 3$ (left plot), but there is a connection for $R = 5$ (right plot).

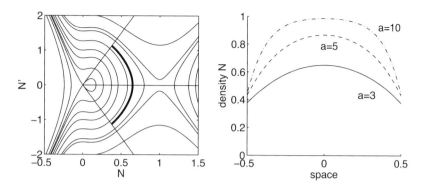

Fig. 4.9 Phase plane (left plot) and solution profile (right plot) of the steady state for IDE (4.1) with Allee updating function with $\gamma = 2$. The straight lines indicate the boundary conditions. The bold segment between the two lines corresponds to the solution profile with $a = 3$ in the right plot. The growth parameter is $R = 10$.

$$\frac{R+1}{2R} = \frac{\arctan\left(\sqrt{R-1}\right)}{\sqrt{R-1}} . \tag{4.23}$$

This transcendental equation cannot be solved explicitly. Numerically, we find the value $R^* \approx 3.29$ mentioned above.

It is somewhat surprising that, even when there is almost no dispersal-related loss from the domain, the spatial model does not have a positive steady state for $R < R^* \approx 3.29$, despite the fact that the nonspatial model has such a state for $R > 2$. To understand this observation more fully, it is helpful to study spreading phenomena, as we will do in Chap. 6.

4.6 Further Reading

Andersen (1991) studies a model for the processes of dispersal, germination, and maturation in an annual plant species. He formulates two models, depending on whether density-dependence acts on germination or maturation, and explores the resulting difference in dynamics. He uses a double Weibull kernel that may have the modal dispersal distance away from the origin, depending on parameter values. He shows that the two models are equivalent and reflect only a difference in census time. He then conducts two carefully designed numerical experiments that reveal that, despite the equivalence of the models, the observed behavior (at census time) may differ; see also Lutscher and Petrovskii (2008). In particular, he finds "pronounced spatial nonuniformity" in four- and eight-cycles but not in steady states and two-cycles.

A two-cycle of the IDE $N_{t+1} = Q[N_t]$ corresponds to a fixed point of the composite operator $Q \circ Q$. Accordingly, some of the techniques presented here can be applied to study the properties of such solutions (Kot and Schaffer 1986). The very useful reduction to a differential equation in the case of the Laplace kernel, however, cannot be extended, since the integral cannot be eliminated. Instead, we obtain a forward–backward integrodifferential equation, which is difficult to analyze. There is currently no detailed analytical study of cycles of length two or more.

A more general and abstract theory of positive discrete dynamical systems is developed in Krause (2015). Among many other results, the existence, uniqueness, and global stability of a positive fixed point in the case of a monotone and concave growth function (Lemma 4.2) follows from Theorem 5.4.3 in Krause (2015).

A rigorous computational approach to the complex dynamics that can arise under the Ricker map is based on bounds for the maximal invariant set and uses Conley index theory to calculate basins of attraction, invariant sets, and a Morse decomposition (Day et al. 2004; Day and Kalies 2013).

We present some ways to approximate the spatial profile of a steady state in Chap. 9. We extend the theory presented here to stage-structured populations in Chap. 13. Questions of existence, uniqueness, and stability of steady states recur, explicitly or implicitly, throughout this book in many applications and extensions.

Chapter 5
The Speed of Spatial Spread

Abstract When a population can persist in a certain environment, we expect that it will spread through that environment if it is initially spatially confined to some small region. How fast will this spatial spread occur? How does the speed depend on movement behavior? These questions are particularly relevant for understanding and managing biological invasions. Spreading nonnative species can cause great damage to existing local ecosystems, e.g., by replacing native species or introducing pathogens. We need insights that help us decide between different management options to slow or contain the spread of harmful species. IDEs are particularly well suited to addressing the question of how different dispersal patterns influence the speed of spread. We begin this chapter with two different scenarios for spatial spread and explicit calculations for the linear growth function. We denote these as the *point-release scenario* and the *traveling-front scenario*. Then we define the *asymptotic spreading speed* and present the results for the nonlinear theory. Throughout the chapter, we assume that there is no Allee effect; we will devote Chap. 6 to it.

5.1 Measuring Spread

Before we can investigate how fast a locally introduced population spreads, we need to find a way to measure the rate of spread. It would be natural to try to find the speed of the "edge" of the population distribution, where the density transitions from positive to zero. However, we will see in our model that the population density is typically positive everywhere from the first generation on, even if the initial density has compact support. This phenomenon of "infinite spread" also occurs in reaction–diffusion equations, thus has long been studied and criticized (Einstein 1906). It is also not clear whether we can expect such a sharp edge to exist in real biological invasions. Individuals can be extremely difficult to detect at low population densities, so that a sharp edge, even if it existed, may be impossible to find. Instead, we consider two measures: in the *point-release scenario*, we track a threshold density; in the *traveling-front scenario*, we track an entire solution profile. A third measure, the *asymptotic spreading speed*, will be mathematically

© Springer Nature Switzerland AG 2019

F. Lutscher, *Integrodifference Equations in Spatial Ecology*, Interdisciplinary Applied Mathematics 49, https://doi.org/10.1007/978-3-030-29294-2_5

most satisfying, connecting the two former measures, but also analytically most challenging to handle.

Much of the content of this chapter can be traced back to two landmark papers: Weinberger (1982) proves the existence of an asymptotic spreading speed and traveling waves for a large class of operators that includes the integral operator in (5.1) below. Kot et al. (1996) highlight the value of IDEs to invasion biology by describing the effect of different dispersal kernels on invasion speed and finding accelerating invasions. Kot (2003) gives a first review of spread phenomena in IDEs and relates the findings to various biological invasions. The recent book by Lewis et al. (2016) provides an excellent introduction to many mathematical theories (including IDEs) of biological invasions.

To consider questions of spatial spread, we envision a scenario where the potential habitat of a population is so large compared to its actual spatial extent that an increase in the extent is not limited by the availability of habitat. Therefore, we consider the potential habitat to be the entire real line. For simplicity, we also assume that the habitat is homogeneous and that dispersal is unbiased. Then the dispersal kernel is a function of distance only, i.e., $K(x, y) = \widetilde{K}(x - y)$, where $\widetilde{K}$ is an even function. For notational simplicity, we drop the tilde. Hence, the IDE reads

$$N_{t+1}(x) = Q[N_t](x) = \int_{-\infty}^{\infty} K(x - y)F(N_t(y))\mathrm{d}y. \qquad (5.1)$$

To ease notation, we frequently denote the convolution integral as $K * F(N_t)$. In this chapter, we shall assume that there is no Allee effect so that $F(N) \leq F'(0)N$. As we did for the question of population persistence, we begin our investigation of spreading phenomena with the linear model

$$N_{t+1}(x) = R \int_{-\infty}^{\infty} K(x - y)N_t(y)\mathrm{d}y = R(K * N_t)(x), \qquad (5.2)$$

where many of the relevant quantities and interesting phenomena can be calculated explicitly in special cases.

5.2   Spread from Point Release

Some biological invasions are initiated by a small number of individuals released at a single location outside their native habitat. For example, house finches (*Carpodacus mexicanus*) were illegally released in New York in the 1950s and spread westward (Veit and Lewis 1996). Shipping points for international trade are particularly prominent spots for invasions to emerge. The emerald ash borer (*Agrilus planipennis Fairmaire*) was transported in shipping and packaging material to the Detroit area from Asia and started spreading to many states and Canadian provinces from there in the late 1990s (Cappaert et al. 2005).

Mathematically, we idealize this point release of individuals by choosing a Dirac delta distribution $N_0(x) = \delta(x)$ as the initial condition. Recall that $\delta(x)$ can formally be thought of as a limit of top-hat functions (Table 3.1) when $\beta \to 0$ and the integral is constant. The delta distribution behaves particularly nicely under convolutions in that $f(x) = (f * \delta)(x)$ for all continuous functions f. More details on distributions and their properties can be found in many textbooks, e.g., Keener (2000).

Using the convolution property, we can calculate the solution of (5.2) with this initial condition explicitly as

$$N_t(x) = R^t K^{*t}(x),\tag{5.3}$$

where K^{*t} denotes the t-fold convolution integral.

While formula (5.3) is elegant in its simplicity, its usefulness is somewhat limited because we can calculate this t-fold convolution only for a few dispersal kernels. We give two examples that show markedly different behavior.

Point Release and the Gaussian Kernel

The t-fold convolution of a Gaussian kernel with mean zero and variance σ^2 is a Gaussian kernel with mean zero and variance $t\sigma^2$. Therefore, if K is a Gaussian kernel with variance σ^2, then (5.3) becomes

$$N_t(x) = \frac{R^t}{\sqrt{2t\sigma^2\pi}} e^{-\frac{x^2}{2t\sigma^2}},\tag{5.4}$$

as illustrated in Fig. 5.1. We observe that even if the initial population is concentrated at a single point, the population density in the first (and every subsequent) generation is positive everywhere. This example illustrates the phenomenon of infinitely fast propagation. As discussed above, we cannot simply define a speed of propagation by tracking the spatial location where the population density becomes positive.

A *detection threshold* is a minimal density above which one has a reasonable and consistent chance to detect the presence of a particular species in empirical work. For example, the emerald ash borer lays eggs under the bark high up in trees. The exit holes of larvae are very small. The only way to detect a small density is to fell and debark every tree in question. While this can and has been done under special circumstances on a small scale (Mercader et al. 2009), it is not feasible and is obviously harmful on large scales. At higher density, the eggs will be placed lower in the tree, and the many exit holes will be easier to spot.

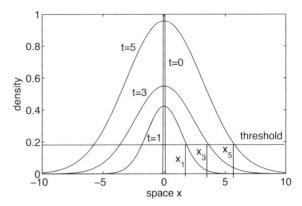

Fig. 5.1 Population spread from a point release at $x = 0$ with Gaussian dispersal kernel according to (5.3). The intersection with the detection threshold (horizontal line) defines the points x_t (see text). Parameters are $\sigma^2 = 2$ and $R = 1.5$.

In mathematical terms, we define $\widetilde{N}$ as the detection threshold, and we set x_t as the rightmost location at which the population exceeds this threshold in generation t; see Fig. 5.1. The defining equation for x_t,

$$\widetilde{N} = \frac{R^t}{\sqrt{2t\sigma^2\pi}} e^{-\frac{x_t^2}{2t\sigma^2}} , \tag{5.5}$$

can be solved (provided $\widetilde{N}$ is small enough) to get

$$x_t = \sqrt{2t^2\sigma^2 \ln R - 2t\sigma^2 \ln\left(\sqrt{2t\sigma^2\pi}\,\widetilde{N}\right)} . \tag{5.6}$$

To calculate a speed, we recall that the population was initially located at $x = 0$, so that the distance covered per unit time is given by x_t/t. This number depends on the detection threshold and changes over time, but in the limit, we obtain a speed that depends only on the two model parameters:

$$\lim_{t \to \infty} \frac{x_t}{t} = \sqrt{2\sigma^2 \ln(R)} =: c_G . \tag{5.7}$$

Hence, asymptotically, the population spreads in space with constant speed (displacement per generation). In general, this speed is denoted by c; the subscript G indicates that the expression in (5.7) is obtained specifically with the Gaussian kernel.

Point Release and the Cauchy Kernel

The *Cauchy kernel* is defined as

$$K(x) = \frac{1}{\pi} \frac{\beta}{\beta^2 + x^2} . \tag{5.8}$$

Fig. 5.2 The Gaussian
(dashed) and the Cauchy
(solid) dispersal kernel. We
set $\beta = 0.1$ and choose
$\sigma^2 = \pi\beta/2$ such that the two
kernels agree at $x = 0$.

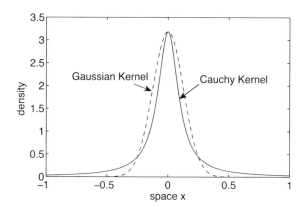

It describes a fat-tailed distribution, where dispersal distances follow a power law
decay. Hence, they decay much more slowly than for the Gaussian or the Laplace
kernel (Fig. 5.2). This kernel has no well-defined variance. In fact, all the moments
of this distribution are infinite. Nonetheless, we can calculate the t-fold convolution
for this kernel via Fourier transforms.

Following Kot et al. (1996), we use the (scaled) Fourier transform

$$\widehat{N}(\omega) = \int_{-\infty}^{\infty} N(x)e^{i\omega x}\,dx \,. \tag{5.9}$$

The Fourier transform is particularly well suited for this situation since it turns
convolutions into multiplications via a change of variables:

$$\widehat{K * K}(\omega) = \int_{-\infty}^{\infty}\int_{-\infty}^{\infty} K(x - y)K(y)e^{i\omega x}\,dy\,dx$$

$$= \int_{-\infty}^{\infty} K(y)e^{i\omega y}\,dy \int_{-\infty}^{\infty} K(z)e^{i\omega z}\,dz = \widehat{K}(\omega) \cdot \widehat{K}(\omega)\,. \tag{5.10}$$

Hence, the t-fold convolution in (5.3) becomes the product $\widehat{N}_t(\omega) = R^t\widehat{K}^t(\omega)$. So
far, the calculations are general. The challenge now is to find the inverse Fourier
transform of this expression. Numerically, this can be done using the fast Fourier
transform (FFT) and its inverse. This procedure will be the basis for one numerical
method of solving IDEs (Chap. 8). For the Cauchy kernel, the inverse can be found
analytically.

The Fourier transform of the Cauchy kernel in (5.8) is

$$\widehat{K}(\omega) = e^{-\beta|\omega|} \,. \tag{5.11}$$

Its t-fold product, $\widehat{K}^t(\omega) = \exp(-\beta t|\omega|)$, is of the same form, so that the inverse
transform of $\widehat{N}_t(\omega) = R^t\widehat{K}^t(\omega)$ must be

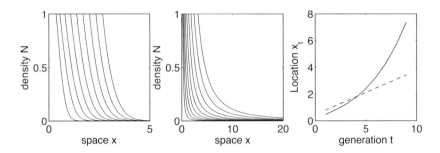

Fig. 5.3 Population spread from a point release at $x = 0$. **Left plot:** Spread with Gaussian kernel (2.25). **Middle plot:** Spread with Cauchy kernel (5.8). **Right plot:** Location x_t of the level set $N_t(x_t) = 0.2$. Parameters are $\beta = 0.1$, $\sigma^2 = \pi\beta/2$, and $R = 1.5$. Population densities are plotted for 10 generations. We note the difference in scale on the x-axis.

$$N_t(x) = \frac{R^t}{\pi} \frac{\beta t}{\beta^2 t^2 + x^2} \,. \tag{5.12}$$

The rightmost location where the population exceeds threshold $\widetilde{N}$ is given by

$$x_t = \sqrt{\frac{R^t \beta t}{\widetilde{N}\pi} - \beta^2 t^2} \,. \tag{5.13}$$

This expression grows faster than geometrically in time. In particular, there is no asymptotically constant speed of spread. The distance that the population moves per generation increases faster than linearly with each generation, so that $\lim_{t\to\infty} x_t/t = \infty$. A comparison between this behavior and the asymptotically constant speed is illustrated in Fig. 5.3. We call this phenomenon an *accelerating invasion*. As an ecological consequence, a biological invasion could speed up over time, and its location would be much harder, if not impossible, to predict.

The difference between propagation according to the Gaussian kernel with asymptotically constant speed and the Cauchy kernel with continuously accelerating speed is striking. Which properties of a dispersal kernel ensure that propagation occurs at an asymptotically constant speed? In the next section, we present a second approach to defining and calculating a measure for spatial spread. This approach will lead us to the criterion we seek.

5.3   Spread as Traveling Fronts

The preceding calculations for x_t require an explicit solution of the IDE, but that is generally not available. The right plot in Fig. 5.3 suggests the existence of solutions in the form of a *traveling front*,[1] i.e., a fixed spatial profile that is shifted by a fixed distance per generation. If we *assume* that such fronts exist, we can derive conditions on their speed from the model parameters. A traveling front describes the density of a species that is well established over a large area and continues to expand its range. This situation might occur long after an initial point release (see above) or could arise as conditions at the range boundary change, e.g., when global change opens climatic opportunities for species to expand their ranges poleward.

We denote by $N^*(x)$ the profile of the front in some generation and by c the displacement per generation. Then the profile in the next generation is $N^*(x - c)$. We substitute this expression into the linear equation (5.2) and obtain the relation

$$N^*(x - c) = R \int_{-\infty}^{\infty} K(x - y)N^*(y)\mathrm{d}y. \tag{5.14}$$

Since the equation is linear, we make the exponential ansatz $N^*(x) = \exp(-sx)$. To mimic the simulations in Fig. 5.3, we consider a rightward-moving ($c > 0$), monotone-decreasing ($s > 0$) profile. Substituting the exponential ansatz into (5.14), we obtain the relationship

$$e^{sc} = R \int_{-\infty}^{\infty} K(y)e^{sy}\mathrm{d}y =: RM(s). \tag{5.15}$$

Function M, defined in (5.15), is called the *moment-generating function* of the kernel. By definition, it satisfies $M(0) = 1$. For $s \neq 0$, the value $M(s)$ is finite only if the tails of the kernel decay to zero faster than $\exp(-sx)$. If $M(s)$ is finite for at least one nonzero value of s, we say that the kernel is *exponentially bounded*. The Gaussian kernel is exponentially bounded, but the Cauchy kernel is not. This difference will explain the different spreading behavior that we found in the previous section.

There is an alternative way to derive relation (5.15) by a so-called exponential transform: One multiplies both sides of (5.14) by e^{sx} and integrates with respect to x. This latter approach has the advantage that it does not assume a particular shape of the profile $N^*(x)$, only that it be exponentially bounded. The former approach, on the other hand, allows us to interpret parameter s as the steepness of the invasion front.

---

[1] A note on terminology is in order. The terms *traveling front*, *traveling wave*, and *traveling profile* may mean the same or slightly different things to different authors. Some authors use *constant-speed traveling front* to be more specific. Some authors require that a traveling front be bounded, while others require monotonicity in addition. In the linear equation that we study in this section, one cannot expect nontrivial bounded traveling fronts. We will use the terms interchangeably and qualify additional properties if necessary.

We can solve (5.15) for c and obtain the *dispersion relation*

$$c = c(s) = \frac{1}{s} \ln(RM(s)), \qquad s > 0, \qquad (5.16)$$

that determines the speed at which an exponential profile with parameter s travels. Since $M(0) = 1$, we have $c(s) \to \infty$ as $s \to 0$. Consequently, a flat front will travel fast. Typically, the moment-generating function increases at least exponentially in s while it exists, so that $c(s)$ is large for large values of s as well. (We give a few examples later when we discuss approximations to the spreading speed for several kernels; see Table 10.1.) Consequently, a steep front will travel fast. One can show that $c(s)$ cannot have a local maximum, so that a minimum, if it exists, must be unique.

Lemma 5.1 *Assume that K is symmetric and that the moment-generating function M of K is defined for some positive value of s. Then the function $c = c(s)$ from (5.16) is defined on some (possibly unbounded) interval. Furthermore, c has no local maximum and at most one local minimum, which—if it exists—is the global minimum.*

Proof The proof follows Weinberger (1978); see also Bourgeois (2016). Since the kernel is nonnegative and symmetric, the moment-generating function is increasing in $s > 0$. Hence, if $M(s)$ is finite for some $s > 0$, then it is finite for the entire interval $[0, s]$.

We define the function

$$\Psi(s) = \frac{\int x e^{sx} K(x) dx}{\int e^{sx} K(x) dx}.$$

By differentiation, we find

$$c'(s) = -\frac{1}{s}[c(s) - \Psi(s)] \quad \text{and} \quad c''(s) = \frac{1}{s^2}[c(s) - \Psi(s)] - \frac{1}{s}[c'(s) - \Psi'(s)].$$

Substituting and multiplying by s^2, we find $(s^2 c'(s))' = s\Psi'(s)$. We calculate

$$\Psi'(s) = \frac{\int [x - \Psi(s)]^2 e^{sx} K(x) dx}{\int e^{sx} K(x) dx} > 0.$$

In particular, the function $s^2 c'(s)$ is increasing so that $2sc' + s^2 c'' > 0$. Hence, at every critical point of c (i.e., $c' = 0$), we have $c'' > 0$. Therefore, there cannot be a local maximum. If there were two local minima, then there would have to be a local maximum in between, which is impossible. □

The slowest speed at which an exponential profile in the linear equation can travel is

$$\hat{c} = \min_{s>0} \frac{1}{s} \ln(RM(s)).$$ (5.17)

We call this speed the *minimal speed of a traveling front in the linear equation*.

The value of $\hat{c}$ can be computed explicitly for the Gaussian kernel. The moment-generating function of the Gaussian kernel (2.25) is

$$M(s) = \exp\left(\frac{\sigma^2 s^2}{2}\right).$$ (5.18)

The dispersion relation and its derivative are

$$c(s) = \frac{\ln(R)}{s} + \frac{\sigma^2 s}{2} \quad \text{and} \quad c'(s) = -\frac{\ln(R)}{s^2} + \frac{\sigma^2}{2}.$$ (5.19)

The curve $c(s)$ is strictly convex for $s > 0$. Its unique minimum occurs at the critical value $s^* = \pm\sqrt{\ln(R)/\sigma^2}$ and is given by

$$\hat{c} = c(s^*) = \sqrt{2\sigma^2 \ln(R)}.$$ (5.20)

Hence, the minimal speed for the Gaussian kernel is the same as the asymptotic speed c_G that we obtained from the point-release approach in (5.7). Since the Cauchy kernel from the previous section does not have a finite moment-generating function, it does not produce traveling profiles with constant speed.

Many other dispersal kernels have a finite moment-generating function, at least for s in some interval around zero. While the formula in (5.17) gives the minimal traveling-wave speed for those kernels, the minimization usually has to be carried out numerically. Rather than straightforward minimization, Kot et al. (1996) propose the following parametric approach to (5.15). We consider the left and right sides of equation (5.15) as curves in the (c, R)-plane, parameterized by s, and we ask when they intersect. Both curves are nondecreasing and concave up. When $c = 0$, there is no intersection since the left-hand side is the constant one, whereas on the right-hand side, we have $R > 1$ and $M(s) \geq 1$ for a symmetric kernel. As we increase c, the expression on the left-hand side increases, whereas the right-hand side remains the same. For large enough values of c, the curves will intersect. If the moment-generating function grows faster than exponentially for large s, we expect the intersection point to appear at some finite value of s. The moment-generating function of the Gaussian kernel certainly satisfies this condition. At the smallest value of c for which the curves intersect, the two curves will then be tangent, i.e., in addition to (5.15), the derivatives of both sides with respect to s are also equal. Hence, at the minimal traveling-wave speed, $\hat{c}$, the following two equations have to be satisfied simultaneously:

$$e^{s\hat{c}} = RM(s), \qquad \hat{c}e^{s\hat{c}} = RM'(s). \tag{5.21}$$

Dividing the second equation by the first and back-substituting gives us the parametric representation of the slowest speed as

$$\hat{c} = \frac{M'(s)}{M(s)}, \qquad R = \frac{e^{sM'(s)/M(s)}}{M(s)}. \tag{5.22}$$

With this representation, it is straightforward to plot $\hat{c}$ as a function of R for a given (exponentially bounded) dispersal kernel; see Fig. 5.4 below.

For the Gaussian kernel, (5.22) provides an easier way to calculate $\hat{c}$. Substituting M from (5.18), we find

$$\hat{c} = \sigma^2 s, \qquad R = e^{\sigma^2 s^2/2}. \tag{5.23}$$

Solving the second equation for s and inserting into the first results in the expression in (5.20).

For the Laplace kernel, the parametric approach leads to an explicit formula for the minimal traveling-wave speed. The moment-generating function of the Laplace kernel (2.27) with dispersal parameter a is given by

$$M(s) = \frac{a^2}{a^2 - s^2}, \qquad |s| < a. \tag{5.24}$$

The parametric representation in (5.22) becomes

$$\hat{c} = \frac{2s}{a^2 - s^2}, \qquad R = \left(1 - \frac{s^2}{a^2}\right)\exp\left(\frac{2s^2}{a^2 - s^2}\right). \tag{5.25}$$

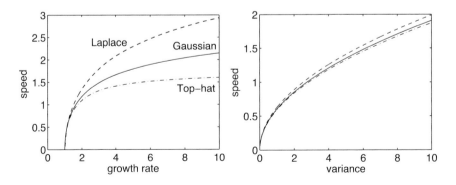

Fig. 5.4 Relationship between minimal traveling-wave speed $\hat{c}$ and growth rate R (left plot), respectively, variance σ^2 (right plot), for three different dispersal kernels: Gaussian (solid), Laplace (dashed), and top-hat (dash-dot). In the right plot, we fixed $\sigma^2 = 1$; in the left plot $R = 1.2$.

We can use this representation in two ways. One is to simplify by setting $\tilde{s} = \frac{2s^2}{a^2 - s^2}$ or $s^2 = \frac{a^2 \tilde{s}}{2 + \tilde{s}}$ to obtain

$$R = \frac{2e^{\tilde{s}}}{2 + \tilde{s}} \quad \text{and} \quad a\hat{c} = \frac{2sa}{a^2 - s^2} = \frac{\tilde{s}a}{s} = \sqrt{\tilde{s}^2 + 2\tilde{s}} \,. \tag{5.26}$$

The expression for R is now independent of a and is an increasing function of $\tilde{s}$. Hence, for each $\tilde{s}$ there is a unique R (numerically easy to find) that we can substitute into the equation for $\hat{c}$ to find the speed.

Another approach is to rewrite the equation for R in (5.25) as

$$\frac{Re^2}{2} \frac{2a^2}{a^2 - s^2} = \exp\left(\frac{2a^2}{a^2 - s^2}\right) \,. \tag{5.27}$$

This equation can be solved using the -1-branch of the Lambert W function to obtain s and subsequently $\hat{c}$ as

$$\hat{c} = -\frac{1}{a} W_{-1}\left(\frac{-2}{Re^2}\right) \sqrt{\frac{2}{W_{-1}\left(\frac{-2}{Re^2}\right)} + 1} \,. \tag{5.28}$$

Details can be found in Bourgeois (2016).

In Fig. 5.4, we illustrate how the minimal traveling-wave speeds for the Gaussian, the Laplace, and the top-hat kernel (see Table 3.1) depend on the two model parameters: the growth rate and the variance of the kernel. The speed is increasing in both parameters. The speed for the Gaussian kernel is slower than for the Laplace kernel but faster than for the top-hat kernel. When the variance is fixed, the speed for the top-hat kernel is bounded by its finite support: since no individual can move farther than $\beta = \sqrt{3\sigma^2}$, the distance moved per generation, $\hat{c}$, cannot exceed that number. In the left plot, we have $\sigma^2 = 1$, so that the minimal traveling-wave speed is bounded by $\sqrt{3} \approx 1.732$. When the growth rate is fixed, the curves for all three kernels are very close together (right plot). The curves for $\hat{c}$ as a function of R may intersect for different kernels; see Fig. 4 in Kot et al. (1996). This phenomenon can arise if one of the kernels is highly concentrated near zero but has bounded support, e.g., $K(x) = a - b \ln|x|$.

Figure 1 in Lutscher (2007) shows that the curves for $\hat{c}$ as a function of σ^2 are similar for several other kernels. In other words, the speed for the Gaussian kernel seems to approximate the speed for other kernels with the same growth rate and variance reasonably well. We will discuss this question further in Chap. 10 in the context of approximations.

5.4   Nonlinear Growth Functions

When we use a more realistic, nonlinear growth function, an explicit solution such as (5.3) for the linear model with point release is no longer available. Which of the results and insights from the linear theory still hold in the nonlinear case? It turns out that all the results that can carry over do so as long as there is no Allee effect (which we assume throughout this chapter). To begin, we consider a special class of growth functions. We shall assume that the extinction state is unstable and that there is a positive, globally stable steady state (which we can and shall assume to be unity). We shall also assume that the growth function is monotone. These assumptions are expressed as follows:

(F1) $F(0) = 0$ and $F(1) = 1$ are the only two fixed points of F.
(F2)   Fixed point $N = 0$ is unstable and $N = 1$ is stable.
(F3) F is continuous and nondecreasing.
(F4) F is differentiable at zero with $F'(0) > 1$ and linearly bounded, i.e., $F(N) \le F'(0)N$.

We shall relax the monotonicity assumption in (F3) later in the chapter. We recall that throughout the chapter, we assume that dispersal is unbiased, i.e., $K(x) = K(-x)$.

For the nonlinear IDE, a traveling front satisfies not only a shift condition similar to that in (5.14) but also asymptotic conditions. Specifically, the limits at $\pm\infty$ need to correspond to fixed points of growth function F. Hence, a traveling wave to the right with speed $c > 0$ of IDE (5.1) is a function $N^*(x)$ that satisfies the equation

$$N^*(x - c) = Q[N^*](x) = \int_{-\infty}^{\infty} K(x - y)F(N^*(y))\mathrm{d}y , \qquad (5.29)$$

with asymptotic conditions $\lim_{x \to \infty} N^*(x) = 0$ and $\lim_{x \to -\infty} N^*(x) = 1$. By symmetry, if there is a rightward traveling-front solution with speed c and asymptotic conditions as stated, then there is also a leftward traveling-front solution with speed $-c$ and asymptotic conditions interchanged. Under conditions (F1)–(F4), we can get a bound for the speed of a traveling front in the nonlinear equation from the minimal speed in the linear equation.

Lemma 5.2 *Assume that the moment-generating function of K exists in some neighborhood of zero and that the growth function satisfies $F(N) \le F'(0)N$. Then the minimum traveling-front speed of the nonlinear equation (5.29) is bounded above by $\hat{c}$ in (5.17), where $R = F'(0)$.*

Proof We assume that a traveling front exists and denote it by N^*. Since F has no Allee effect, we have the inequality

$$N^*(x - c) = \int_{-\infty}^{\infty} K(x - y)F(N^*(y))\mathrm{d}y \le F'(0) \int_{-\infty}^{\infty} K(x - y)N^*(y)\mathrm{d}y .$$
$$(5.30)$$

Taking an exponential transform on both sides leads to the bound $e^{sc} \leq F'(0)M(s)$. Therefore, the minimum c is bounded by $\hat{c}$. □

It is much harder to show, but true, that the minimal speeds in the linear and nonlinear case are equal and that traveling fronts for the nonlinear case exist under certain assumptions on F and K. To proceed in this direction, we introduce a new measure of the speed of spread, the *asymptotic spreading speed*. This concept is independent of the assumption that a traveling front exists and requires no special initial condition as in the point-release scenario. Yet, it will turn out to be closely related to both of these.

The Asymptotic Spreading Speed

The notion of an *asymptotic spreading speed* (Aronson and Weinberger 1975) can be motivated as follows. Suppose a population is newly introduced into a confined region. This population is said to spread with asymptotic speed c^* if an observer who travels at some speed $c > c^*$ will eventually be ahead of the population whereas an observer who travels at speed $c < c^*$ will eventually be surrounded by the population. Mathematically, these considerations can be expressed as follows.

Definition 5.1 The number c^* is called the *asymptotic spreading speed* if it satisfies the conditions

$$\limsup_{t\to\infty} \max_{|x|>(c^*+\epsilon)t} N_t(x) = 0, \quad \liminf_{t\to\infty} \min_{|x|<(c^*-\epsilon)t} N_t(x) \geq \beta > 0 \qquad (5.31)$$

for any small $\epsilon > 0$ and some $\beta > 0$, where $N_0 \not\equiv 0$ is compactly supported and N_t is defined by the iteration in (5.29).

A priori, it is not clear that such an asymptotic spreading speed exists. Naturally, some conditions on the dispersal kernel and growth function are necessary. For example, for the linear growth function and the Cauchy kernel, no asymptotic spreading speed exists; see Sect. 5.2. In the following, we discuss the most relevant results about the existence of an asymptotic spreading speed according to the properties of the growth function F, always assuming that the dispersal kernel is exponentially bounded.

Spread with Compensatory Growth

Weinberger (1982) proves the existence of a spreading speed and several of its properties for the general recursion operator

$$N_{t+1} = Q[N_t], \qquad (5.32)$$

defined on a space of continuous functions on $\mathbb{R}$. In particular, the theorem below applies not only to IDEs but also to other dynamic equations, e.g., to the time-1-map of an appropriate reaction–diffusion equation.

To state the assumptions and results, we will frequently identify a number with a constant function. For two continuous functions, $N, \tilde{N}$, we write $N \geq \tilde{N}$ if $N(x) \geq \tilde{N}(x)$ for all x. We denote by $\mathscr{C}_{[0,1]}$ the space of continuous functions on $\mathbb{R}$ with values in the interval $[0, 1]$. The following theorem summarizes results from Theorems 6.5 and 6.6 by Weinberger (1982).

Theorem 5.1 *Consider the following properties of operator Q on the space $\mathscr{C}_{[0,1]}$ of continuous functions.*

   (i) *Translation invariance:* $Q[N(\cdot - a)](x) = Q[N](x - a)$ *for all $a \in \mathbb{R}$.*
  (ii) *Invariance of $\mathscr{C}_{[0,1]}$:* $N \in \mathscr{C}_{[0,1]} \Rightarrow Q[N] \in \mathscr{C}_{[0,1]}$.
 (iii) *Fixed points:* $Q[0] = 0$, $Q[1] = 1$, $Q[a] > a$ *for $a \in (0, 1)$.*
  (iv) *Monotonicity:* $0 \leq N \leq \tilde{N} \leq 1 \Rightarrow Q[N] \leq Q[\tilde{N}]$.
   (v) *Continuity: If $\{f_j\} \subset \mathscr{C}_{[0,1]}$ and $f_j \to f$ uniformly on compact subsets of $\mathbb{R}$, then $Q[f_j] \to Q[f]$ pointwise as $j \to \infty$.*
  (vi) *Compactness: Every sequence $\{f_j\}$ in $\mathscr{C}_{[0,1]}$ has a subsequence $\{f_{j_i}\}$ such that $\{Q[f_{j_i}]\}$ converges uniformly on every bounded subset of $\mathbb{R}$.*

Assume that Q in (5.32) satisfies (i)–(v). Then there exists an asymptotic spreading speed $c^* > 0$ for Q. If, in addition, (vi) holds, then for every $c \geq c^*$ there exists a continuous traveling-wave solution $N_t(x) = W(x - ct)$ of Q with $W(\infty) = 0$ and $W(-\infty) = 1$. No such traveling wave exists for $c < c^*$.

For the purpose of this theorem, the definition of the asymptotic spreading speed can be strengthened in that the second condition in (5.31) can be replaced by

$$\lim_{t \to \infty} \min_{|x| < (c^* - \epsilon)t} N_t(x) = \lim_{t \to \infty} \max_{x \in \mathbb{R}} N_t(x) = 1 , \qquad (5.33)$$

provided the initial condition is bounded between zero and one. Hence, the density converges to the positive steady state behind the front.

As stated, the preceding theorem does not give a way to calculate the spreading speed. Weinberger (1982) gives a general construction to define c^* and uses it in the proof of the theorem. We present this construction in Chap. 13.8. Weinberger (1982) uses super- and sub-solutions to prove that the asymptotic spreading speed is given by the linearized formula from (5.17) if certain additional conditions are satisfied. We include this statement in the next theorem, where we apply the previous theorem to IDEs and relate the properties of the growth function F to those of operator Q.

Theorem 5.2 *Consider operator Q from (5.29). Assume that K is continuous and that its moment-generating function exists for all $s \in \mathbb{R}$. Assume that the growth function satisfies conditions (F1)–(F4). Then there exists an asymptotic spreading speed, c^*, for Q. Furthermore, c^* equals the minimal traveling-wave speed $\hat{c}$ from (5.17), where $R = F'(0)$. Finally, for all $c \geq c^*$ there exists a monotone traveling-*

front solution of (5.29), connecting zero and one with speed c, but for c < c* no such traveling-front solution exists.

Proof We will show that the assumptions on F and K guarantee that conditions (i)–(vi) in Theorem 5.1 are satisfied for Q from (5.29). We discuss the claim $\hat{c} = c^*$ separately below.

Translation invariance follows from the properties of the convolution operator by a change of variables. Since F is monotone increasing and has zero and one as fixed points, the interval $[0, 1]$ is invariant for the map F. Since K is nonnegative and integrates to unity, Q maps functions bounded between zero and one into functions with the same bounds.

Since K integrates to unity, constant functions are mapped to constant functions under Q. Since F has zero and one as fixed points, the constant functions zero and one are fixed points for Q. Since $F'(0) > 1$ and since zero and one are the only fixed points of F in $[0, 1]$, we must have $F(a) > a$ for all $0 < a < 1$. Hence, the same is true for constant functions under Q. Monotonicity of Q follows from monotonicity of F since K is nonnegative.

We prove continuity of Q under the slightly stronger assumption that F is Lipschitz. We denote by L the Lipschitz constant of F. Consider a sequence of functions $\{f_j\}$ in $\mathscr{C}_{[0,1]}$ that converges uniformly on compact subsets to f. Then for every $m > 0$ we have the estimate

$$
\begin{aligned}
|Q[f_j](x) - Q[f](x)| = {} & \left| \int K(x - y)(F(f_j(y)) - F(f(y)))\mathrm{d}y \right| \\
\leq {} & \left| \int_{|y|>m} K(x - y)(F(f_j(y)) - F(f(y)))\mathrm{d}y \right| \\
& + \left| \int_{|y|\leq m} K(x - y)(F(f_j(y)) - F(f(y)))\mathrm{d}y \right| \\
\leq {} & 2 \int_{|y|>m} K(x - y)\mathrm{d}y + L \int_{|y|\leq m} K(x - y)|f_j(y) - f(y)|\mathrm{d}y .
\end{aligned}
$$

Since the kernel integrates to unity, we can choose m large to make the first term arbitrarily small. Since f_j converges uniformly on compact subsets, we can choose j large (for fixed m) to make the second term arbitrarily small. Hence, $Q[f_j] \to Q[f]$ pointwise.

To prove that Q is compact is a little more involved but uses some of the same ideas (see, e.g., Bourgeois 2016). We choose a sequence $\{f_j\}$ in $\mathscr{C}_{[0,1]}$, a compact interval I, and some $m > 0$. Then we define the sequence

$$
\tilde{f}_j(x) = \int_{-m}^{m} K(x - y)F(f_j(y))\mathrm{d}y , \qquad x \in I .
$$

Boundedness of F gives us uniform boundedness of $\tilde{f}_j$. Since K is continuous, it is uniformly continuous on compact sets. Therefore, $\tilde{f}_j$ are equicontinuous. By the Arzelà–Ascoli theorem, there is a convergent subsequence $\tilde{f}_{j_i} \to f$. The same splitting of the integral as in the proof of continuity above then gives us uniform convergence of $Q[f_t]$ on bounded subsets of $\mathbb{R}$.   □

We have not yet proved the claim $c^* = \hat{c}$. To show that the spreading speed must be bounded above by the speed of the linear equation with $R = F'(0)$ is fairly straightforward, but the reverse inequality is much harder; we refer to Weinberger (1982). Both proofs rely on monotonicity, but the latter requires the careful construction of a sub-solution (compare the proof of Lemma 4.2).

Lemma 5.3 *Under the assumptions of the previous theorem, we have $c^* \le \hat{c}$.*

Proof If N_0 is compactly supported with $0 \le N_0(x) \le 1$ for all $x \in \mathbb{R}$, then for each $s > 0$ one can find a constant A such that $N_0(x) < A \exp(-sx)$ for all $x \in \mathbb{R}$. The solution of the nonlinear equation is bounded by the solution of the linear equation because of assumption (F4):

$$\int K(x, y) F(N_t(y)) \mathrm{d}y \le F'(0) \int K(x, y) N_t(y) \mathrm{d}y . \tag{5.34}$$

Hence, the solution N_t of the nonlinear equation with initial condition N_0 cannot spread faster than the solution of the linear equation with initial condition $A \exp(-sx)$. Choosing the value of s that corresponds to $\hat{c}$ gives the stated inequality.   □

The result that the spreading speed of the nonlinear equation is equal to the spreading speed of the linearized (at zero) equation (which, in turn, is equal to the minimal traveling wave speed of the linearized equation) is known as *linear determinacy* (Lewis et al. 2002). The *linear conjecture* is the belief that a system is linearly determinate under certain conditions (van den Bosch et al. 1990). One of these conditions is the linear boundedness condition on the growth function, which excludes an Allee effect. We consider an Allee effect in Chap. 6.

Spread with Overcompensatory Growth

The Ricker function in (2.19) with parameter $0 < r < 1$ is monotone and satisfies conditions (F1)–(F4). However, with $1 < r < 2$, it is not monotone on the interval $[0, 1]$; see Fig. 5.5. Conditions (F1), (F2), and (F4) above are satisfied, but the second part of (F3) is not. In Chap. 2, we saw that solutions of the nonspatial model approach the stable state $N = 1$ with decaying oscillations. A similar pattern can occur for the logistic growth function in (2.20).

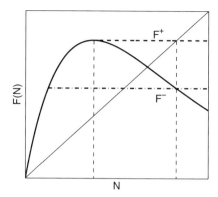

Fig. 5.5 Illustration of the definition of $F^\pm$. The solid curve is the Ricker growth function. The increasing portion of the Ricker function together with the dashed horizontal line is the function F^+. The dash-dot horizontal line together with the increasing beginning of the Ricker function is F^-. The thin dashed vertical lines indicate the location of the maximum $(\hat{N})$ and its value $(\hat{F})$.

We cannot expect the existence of a monotone traveling-front profile. Nonetheless, we can use Theorem 5.2 to show the existence of a spreading speed even with nonmonotone dynamics. The fundamental idea goes back to work by Thieme (1979) on integral equations in continuous time. Several authors apply this idea to IDEs with nonmonotone growth functions (Hsu and Zhao 2008; Li et al. 2009; Yu and Yuan 2012; Yi and Zou 2015). We present some of their ideas and results here.

We shall assume that the growth function satisfies (F1)–(F4) except the monotonicity assumption in (F3). Instead, we assume that there exists some $\hat{N} \in (0, 1)$ such that F is increasing for $0 \leq N < \hat{N}$ and decreasing for $N > \hat{N}$. Furthermore, we assume that $|F'(1)| < 1$ so that the unique fixed point is stable. We define functions $F^\pm$ as follows (see Fig. 5.5): We set $\hat{F} = \max\{F(N)|0 \leq N \leq 1\} = F(\hat{N})$ and define

$$F^+(N) = \max_{\tilde{N} \leq N}\{F(\tilde{N})\}, \qquad F^-(N) = \min\{F(N), F(\hat{F})\}. \tag{5.35}$$

Finally, we define $Q^\pm$ as in (5.29) with F replaced by $F^\pm$, respectively.

With this construction, we can bound solutions of $N_{t+1} = Q[N_t]$ from above and below by solutions of $N_{t+1} = Q^\pm[N_t]$, respectively.

Theorem 5.3 *Under the assumptions and definitions in the previous paragraph, the following hold.*

1. *$F^\pm$ satisfy conditions (F1)–(F4).*
2. *$Q^\pm$ satisfy the assumptions of Theorem 5.2; in particular, their respective spreading speeds $c_\pm^*$ exist.*
3. *Q has a spreading speed c^*, and $c_-^* = c^* = c_+^*$.*

Proof We sketch the main geometric idea of the proof but refer to Hsu and Zhao (2008), Li et al. (2009), and Yi and Zou (2015) for technical details. The construction in Fig. 5.5 guarantees that $F^{\pm}$ satisfy the conditions of Theorem 5.2, which implies that the respective spreading speeds, $c^*_{\pm}$, of operators $Q^{\pm}$ exist and are determined by the respective linearizations at zero. The construction also shows that $F^+(N) = F^-(N)$ for small enough N. The latter property implies that $(F^+)'(0) = (F^-)'(0)$ so that $c^*_- = c^*_+$.

Next, we note that $F^-(N) \leq F(N) \leq F^+(N)$, which implies that the same inequalities hold for the corresponding operators, i.e., $Q^-[N] \leq Q[N] \leq Q^+[N]$. Hence, for a given initial condition, the solution with operator Q is sandwiched between the solutions with Q^- and Q^+. Since the latter two spread at the same speed, the former has to spread at that speed as well.   □

Theorem 5.2 also guarantees the existence of monotone traveling waves that connect the zero state with the positive state for $Q^{\pm}$. The situation for Q is a bit more delicate. As previously discussed, one cannot expect a traveling wave, if it exists, to be monotone. The plots in Fig. 5.6 illustrate monotone (left) and nonmonotone (right) traveling fronts with the Ricker function for different parameter values with a characteristic function[2] as initial condition. In both cases, the profile converges to zero ahead of the front and to the positive fixed point behind the front. A general proof of this property and whether it holds for all $c \geq c^*$ turns out to be surprisingly difficult. The respective works by Hsu and Zhao (2008), Li et al. (2009), and Yu and Yuan (2012), Yi and Zou (2015) each use slightly different conditions; see also

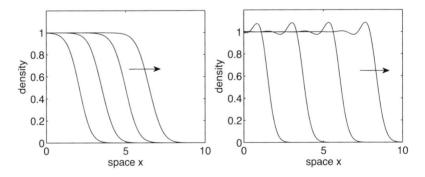

Fig. 5.6 Illustration of monotone and nonmonotone traveling fronts in the IDE with Ricker growth function and Gaussian dispersal kernel. Parameters are $\sigma^2 = 0.1$, $r = 0.8$ (left), and $r = 1.8$ (right). The initial condition is the characteristic function on $(-\infty, 0)$; the profile is plotted every four time steps.

---

[2]The *characteristic function* or indicator function of a set takes the value one on the set and zero everywhere else. It is often denoted by χ. We will frequently use the characteristic function of a half line as initial conditions in numerical simulations of traveling fronts. For example, $\chi_{(-\infty,0]}(x) = 1$ if $x \leq 0$ and $\chi_{(-\infty,0]}(x) = 0$ if $x > 0$.

Lin (2015). The conditions in Yi and Zou (2015) are particularly simple to verify geometrically. Their Theorem 4.1 implies the following result.

Theorem 5.4 *Assume that (F1), (F2), and (F4) hold. Assume further that $N = 1$ is the only fixed point of F^2. Then for all $c \geq c^*$, there exists a traveling wave, $N^*(x - c) = Q[N](x)$, with $N^*(\infty) = 0$ and $N^*(-\infty) = 1$.*

When the positive fixed point of the growth function is unstable, one cannot expect a traveling wave to converge to that point. For example, the positive fixed point of the Ricker function becomes unstable through a flip bifurcation, and a stable two-cycle emerges. Accordingly, solutions in the spatial model may show oscillations in the wake of an invasion front (Bourgeois 2016). We discuss and illustrate these phenomena in Chap. 11.

5.5 Further Reading

Spreading phenomena and traveling waves were studied in reaction–diffusion equations, starting with Fisher (1937), long before they were studied in IDEs, e.g., Weinberger (1978) and references therein. Spreading phenomena in continuous-time integral models are reviewed in Metz et al. (1999). A survey that includes continuous and discrete-time equations can be found in Zhao (2009).

The notion of the asymptotic spreading speed was also originally introduced for reaction–diffusion equations (Aronson and Weinberger 1975). It was, however, studied via the time-1-map, so that the continuous-time equation was transformed to a discrete-time recursion, similar to an IDE. Subsequent work focused on discrete-time equations and was motivated mostly by population genetics (Weinberger 1978, 1982). The traveling-front profile is unique and solutions with monotone initial data converge to this traveling front (Lui 1982a). Solutions with compactly supported initial data converge to a double traveling-front profile when (F1)–(F4) hold (Lui 1982b).

Weinberger (1982) originally restricts the dispersal range of any individual to some finite limit, which would require a compactly supported dispersal kernel. Weinberger's proof, however, when applied to IDEs, only requires that the moment-generating function of the dispersal kernel exist on the whole real line. Extensions by Hsu and Zhao (2008) and Weinberger and Zhao (2010) show that the formulas hold if the moment-generating function is finite at a single nonzero value. They also show that the asymptotic spreading speed is infinite if the moment-generating function is infinite for all nonzero values, thereby clarifying the behavior of spread with heavy-tailed kernels. Continuity of the dispersal kernel is also not necessary for these results to hold. It is usually sufficient to require that K be Lebesgue integrable to prove the existence of traveling fronts (Hsu and Zhao 2008; Yu and Yuan 2012).

The theory of spreading speeds and traveling fronts can be formulated in any finite space dimension. One then defines a spreading speed for any given direction (unit vector) and planar traveling waves moving in that direction. The results by

Weinberger (1982) and subsequent works hold in that generality. The stability of monotone traveling wave fronts with a Gaussian dispersal kernel was investigated in two dimensions by Lin et al. (2010) and in any finite dimension by Miller and Zeng (2013).

Throughout this chapter, we assumed that dispersal is unbiased so that the dispersal kernel is symmetric. Many results in this chapter carry over to asymmetric dispersal if the direction of spread is properly accounted for (Yi and Zou 2015), but some caution is necessary. We discuss biased dispersal in detail in Sect. 12.2.

The spreading speed is only an asymptotic quantity that we expect to observe for large times. It is nonetheless a useful quantity for applications since it turns out that "large" is not very large at all. In simulations, we see that a solution develops into a traveling wave with speed close to the asymptotic spreading speed after very few generations (e.g., less than 10 generations in Fig. 5.3). Accordingly, Watkinson et al. (2000) use the speed formula for the linear model to estimate the expected spatial extent of a locally introduced annual grass (*Vulpia ciliata*) within a time frame of 20 years. More examples can be found in Kot (2003) and Lewis et al. (2016).

The Gaussian and the Cauchy kernels represent two extremes: the tails of the former are exponentially bounded (thin tailed), whereas the tails of the latter decay like a power law (fat-tailed). Accordingly, the moment-generating function of the former is well defined, whereas the latter has no finite moments of any order. More generally, kernels whose tails decay slower than exponentially are called heavy tailed. While their moment-generating function does not exist, their moments

$$\int_{-\infty}^{\infty} x^n K(x) \mathrm{d}x , \qquad n = 1, 2, 3, \dots , \tag{5.36}$$

may still be finite. One example of such a kernel is the exponential square root kernel

$$K(x) = \frac{a^2}{4} \exp\left(-a\sqrt{|x|}\right) . \tag{5.37}$$

For these heavy-tailed kernels, the population density in generation t is asymptotically distributed as (Kot et al. 1996)

$$N_t(x) \sim R^t K(x) , \qquad |x|, t \to \infty . \tag{5.38}$$

From this explicit formula, we calculate the spatial extent of the population with the exponential square root kernel at time t to be

$$x_t = \frac{1}{a^2} \left[t \ln R + \ln\left(\frac{a^2}{4\tilde{N}}\right) \right]^2 . \tag{5.39}$$

Hence, the spatial extent grows quadratically in time, and therefore the speed of spatial spread grows linearly in time. Liu and Kot (2019) present a much more detailed analysis of how invasions accelerate as a function of heavy-tailed dispersal kernels, using the theory of regular variation.

The versatility with which IDEs incorporate non-Gaussian dispersal patterns, and the phenomenon of accelerating invasions for heavy-tailed kernels has attracted a lot of attention among ecologists. The sensitive dependence of spread rates on the tails of a dispersal kernel, however, poses significant challenges for empiricists. Since it is difficult, if not impossible, to track individuals that disperse very far, it is extremely challenging to decide the decay rate of an empirical dispersal kernel. Pielaat et al. (2006) propose a sampling design for seed traps, optimized to estimate invasion speeds for IDEs, i.e., for catching the density in the tails of the kernel. Nathan et al. (2003) review methods for long-distance dispersal estimates. Bianchi et al. (2009) point to the importance of selecting an appropriate kernel in the context of colonization times, i.e., when calculating the time that it takes for a certain number of individuals to arrive at one site from another.

Formula (5.17) for the minimal speed of traveling fronts in the linear equation does not actually use the dispersal kernel but rather its moment-generating function. Consequently, Clark et al. (2001b) derive an estimator for spread rates from data using the empirical moment-generating function, thereby circumventing the difficulty of having to estimate the decay rate of the tails. This idea was later expanded by Lewis et al. (2006), who also discuss differences between one- and two-dimensional estimates.

In reality on a bounded planet, of course, no dispersal kernel has truly heavy or even infinite tails. Similarly, no invasion can continue forever, so that the asymptotic speed can never be observed. Clark et al. (2001a) consider discrete individuals and track the location of the farthest-forward individual. They obtain a finite speed of spread even when dispersal distances are drawn from heavy-tailed kernels. Following up, Clark et al. (2003) explore the influence of uncertainty and estimate speeds for potentially heavy-tailed kernels. Demographic stochasticity may also lead to bounded spread rates even for kernels that produce accelerating waves for the deterministic mean-field model (Jacobs and Sluckin 2015).

Chapter 6
Spatial Spread with Allee Effect

Abstract Many species exhibit an Allee effect, where population growth rates are highest at intermediate rather than low density, and small populations may even decline. Determining the spread rates of these species turns out to be much more difficult than the theory in the preceding chapter, where there was no Allee effect. Mathematically, this difficulty arises since—just as in the case of steady states—we cannot expect the linearization at zero to give useful information about the behavior of solutions for larger density, and hence we cannot expect the linearization-based spread formulas from the previous chapter to hold. One of the most interesting biological results here is that with the Allee effect, a population may spread or retreat. Hence, eradication of an invading pest species seems possible if management measures could turn an invasion into a retreat. We begin this chapter with a caricature model for which all relevant quantities can be explicitly calculated. Then we present a general condition for whether a population will spread or retreat. Finally, we present a theorem about the existence of traveling waves and the uniqueness of their speed.

6.1 Allee Effects and Biological Invasions

Allee effects (see Sect. 2.2) are ubiquitous in nature, and some are particularly relevant for biological invasions (Taylor and Hastings 2005; Lewis et al. 2016). For example, healthy pine trees produce and exude resin to defend themselves against harmful insects, such as the mountain pine beetle (*Dendroctonus ponderosae* Hopkins). As long as only a few beetles attack a tree, they will die in the resin. When a large number of beetles attack a tree, the resin is insufficient to kill all of them. The tree succumbs and the beetles can reproduce (Powell and Bentz 2014). This is a classical setup for an Allee effect; see also Sect. 12.6.

In the presence of an Allee effect, mathematical analysis becomes more difficult. We already know from the nonspatial model in (2.22) and the steady-state analysis in Sect. 4.5 that the linearization at zero may not provide information about the existence of a positive steady state and that the eventual state of the population may depend on the initial condition. Similarly, in the question of invasions, none of the

© Springer Nature Switzerland AG 2019

F. Lutscher, *Integrodifference Equations in Spatial Ecology*, Interdisciplinary Applied Mathematics 49, https://doi.org/10.1007/978-3-030-29294-2_6

explicit calculations for spread with linear equations from the previous chapter carry over to the case with Allee effect, since the phenomenon is inherently nonlinear.

Throughout this section, we consider a strong Allee effect with a monotone growth function. As always, we can scale steady states to be $F(0) = 0$, $F(1) = 1$, as in (2.22). With a strong Allee effect, there is an *Allee threshold* $N_a \in (0, 1)$ with the property (see Fig. 2.2)

(A1) $F(N) < N$ for $N \in (0, N_a)$ and $F(N) > N$ for $N \in (N_a, 1)$.

For a piecewise-constant caricature Allee function, we can explicitly calculate conditions for spread as well as the corresponding asymptotic speeds (Kot et al. 1996). In Sect. 6.3, we present a simple criterion for the direction of the traveling wave (Wang et al. 2002). The last section is devoted to the more abstract theory of the existence of traveling waves (Lui 1983).

6.2   A Caricature Allee Function

For an analytically tractable example, we choose the piecewise-constant growth function (Kot et al. 1996)

$$F(N) = \begin{cases} 0, & N < N_a , \\ 1, & N \geq N_a , \end{cases} \tag{6.1}$$

with Allee threshold $N_a \in (0, 1)$ and carrying capacity equal to unity. The population density after the growth phase is either one or zero. As in the previous chapter, we consider a homogeneous landscape and a symmetric dispersal kernel of the form $K(x - y)$.

Because of the Allee effect, we expect the initial spatial extent of a population to determine whether the population will persist and spread or decline and retreat. We assume that the initial population exceeds the Allee threshold exactly on some bounded interval and calculate subsequent densities. Since the landscape is homogeneous and the growth function is binary, we may choose $N_0(x) = \chi_{[-x_0, x_0]}(x)$, the characteristic function of that interval, i.e., $N_0(x) = 1$ if $x \in [-x_0, x_0]$ and $N_0(x) = 0$ otherwise. Then $F(N_0) = N_0$. The density in the next generation is

$$N_1(x) = \int_{-\infty}^{\infty} K(x - y)\chi_{[-x_0, x_0]}(y)\mathrm{d}y = \int_{-x_0}^{x_0} K(x - y)\mathrm{d}y = \int_{x-x_0}^{x+x_0} K(y)\mathrm{d}y . \tag{6.2}$$

By symmetry, if $N_1(x) \geq N_a$ for some x, then $N_1(x) \geq N_a$ on some interval $[-x_1, x_1]$, where x_1 satisfies the implicit equation

$$\int_{x_1-x_0}^{x_1+x_0} K(y)\mathrm{d}y = N_a . \tag{6.3}$$

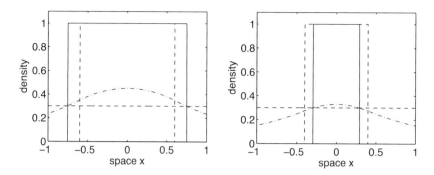

Fig. 6.1 Expansion (left) and retraction (right) with the caricature Allee function in (6.1). The dashed line is the density before dispersal, the dash-dot line is the profile after dispersal, and the solid line represents the density after applying the growth function. The horizontal dashed line indicates the Allee threshold of 0.3. We used the Laplace kernel with dispersal distance one.

After the subsequent growth phase, the population density will be $F(N_1(x)) = \chi_{[-x_1, x_1]}(x)$. Inductively, we obtain the extent x_{t+1} from x_t by solving

$$\int_{x_{t+1}-x_t}^{x_{t+1}+x_t} K(y)\mathrm{d}y = N_a .$$ (6.4)

We expect that if the initial spatial extent is small, it will shrink over time ($0 \leq x_{t+1} < x_t$) and the population will die out. If the initial extent is large enough, it will grow over time ($x_{t+1} > x_t$) and the population will spread. These two cases are illustrated in Fig. 6.1. We obtain the critical spatial extent where the population remains constant by setting $x_{t+1} = x_t = x_c$ or

$$\int_0^{2x_c} K(y)\mathrm{d}y = N_a .$$ (6.5)

Since the kernel is a symmetric probability density, the integral is bounded by $1/2$. Hence, we require $N_a < 1/2$; otherwise a population cannot persist or spread.

For the Laplace kernel in (2.27) with parameter a, the integral in (6.4) can be evaluated explicitly, but we have to distinguish two cases. When $x_{t+1} > x_t$, we find

$$\int_{x_{t+1}-x_t}^{x_{t+1}+x_t} K(y)\mathrm{d}y = e^{-ax_{t+1}} \sinh(ax_t) .$$ (6.6)

Hence, the spatial extent satisfies the difference equation

$$x_{t+1} = \frac{1}{a} \ln\left(\frac{\sinh(ax_t)}{N_a}\right).$$ (6.7)

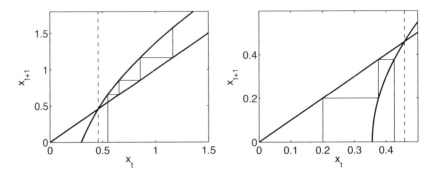

Fig. 6.2 Illustration of the recursions of spatial extent from one generation to the next with Laplace kernel (6.7). The plot on the left illustrates the case $x_{t+1} > x_t \geq x_c$; the plot on the right has the reversed inequalities. The mean dispersal distance is unity, and the Allee threshold is $N_a = 0.3$, so that $x_c \approx 0.4581$ (dashed vertical line).

When $x_{t+1} < x_t$, it is easier to write the backward iteration

$$x_t = \frac{1}{a} \ln\left(\frac{\cosh(ax_{t+1})}{1 - N_a}\right). \tag{6.8}$$

The critical value x_c from (6.5) is given by $2ax_c = -\ln(1 - 2N_a)$. We illustrate the cobweb for both of these iterations in Fig. 6.2 and the critical value as the vertical dashed line.

We can use the same approach to calculate the asymptotic spreading speed for the population. If the population spreads asymptotically with constant speed $c^* > 0$, then $x_{t+1} - x_t \to c^*$ and $x_{t+1} + x_t \to \infty$ as $t \to \infty$. From (6.4), we find c^* implicitly as

$$\int_{c^*}^{\infty} K(z)\mathrm{d}z = N_a \quad \text{or} \quad \int_0^{c^*} K(z)\mathrm{d}z = \frac{1}{2} - N_a , \tag{6.9}$$

where we used the symmetry of the kernel again. (Recall also that K is a probability density.)

As before, the necessary condition for spread ($c^* > 0$) is $N_a < 1/2$. For certain kernels, (6.9) can be solved for c^*. For the Gaussian (2.25) and Laplace (2.27) kernel we obtain

$$c^*_{\text{Gauss}} = \sqrt{2\sigma^2}\,\text{erf}^{-1}(1 - 2N_a) \quad \text{and} \quad c^*_{\text{Laplace}} = -\sqrt{\sigma^2/2}\,\ln(2N_a) , \tag{6.10}$$

respectively, where $\text{erf}(x)$ is the error function. We plot the speeds for these two kernels in Fig. 6.3 as a function of the variance (left plot). We see that the speed for the Gaussian kernel is lower than for the Laplace kernel when the Allee threshold

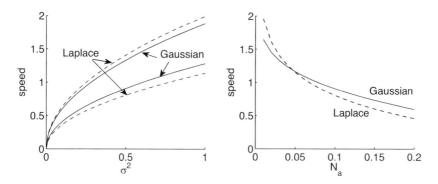

Fig. 6.3 Speeds of spread for the caricature Allee function (6.1) with Gaussian (solid) and Laplace (dashed) kernels. **Left plot:** Speed as a function of variance when $N_a = 0.1$ (lower curves) and $N_a = 0.03$ (upper curves). **Right plot:** The speed of spread as a function of N_a with $\sigma^2 = 0.5$.

is very small, but higher when N_a is large. It turns out that the value N_a^* where the two speeds are equal is independent of the variance. It is implicitly given by

$$2\mathrm{erf}^{-1}(1 - 2N_a^*) = -\ln(2N_a^*) \ . \tag{6.11}$$

We find the critical value numerically as $N_a^* \approx 0.0464$. Figure 6.3 also shows the speed as a function of N_a (right plot).

Several other kernels allow for an explicit calculation of c^* from (6.9). For the double Weibull kernel (see Table 3.1) we calculate

$$c^*_{\mathrm{Weibull}} = \theta \left(-\ln(2N_a) \right)^{1/k} \ . \tag{6.12}$$

For the Cauchy kernel (5.8), we find

$$c^*_{\mathrm{Cauchy}} = \beta \tan \left(\frac{\pi}{2}(1 - 2N_a) \right), \tag{6.13}$$

and for the exponential square root kernel from (5.37), we can use the Lambert W function again (see (5.28)) to find

$$c^*_{\mathrm{ExpRoot}} = \frac{1}{a^2} \left(-W_{-1}(-2N_a/e) - 1 \right)^2 . \tag{6.14}$$

We note that both heavy-tailed kernels admit a finite asymptotic spreading speed here because the Allee function ensures that the population occupies only a finite region after the growth phase. In general, however, heavy-tailed kernels can generate accelerating invasions, even with an Allee effect (Wang et al. 2002).

We can use a slight variation of the above reasoning to calculate the speed of a (monotone) traveling wave as well. Clearly, the traveling wave profile, $N^*(x)$, after the growth phase must have the form of a characteristic function, e.g., $F(N^*(x)) = \chi_{(-\infty, 0]}$. Then

$$N^*(x - c) = \int_{-\infty}^{\infty} K(x - y) F(N^*(y)) \mathrm{d}y = \int_{x}^{\infty} K(z) \mathrm{d}z = \frac{1}{2} - \int_{0}^{x} K(z) \mathrm{d}z .$$
(6.15)

After the subsequent growth phase, the profile will be the characteristic function on $(-\infty, c]$, where c is calculated from

$$\frac{1}{2} - \int_{0}^{c} K(z) \mathrm{d}z = N_a ,$$
(6.16)

which is the same as (6.9). When $N_a < 1/2$, then c is positive and the population advances; when $N_a > 1/2$, then c is negative and the population retreats. For $N_a = 1/2$, there is a constant profile with speed zero. This behavior is typical when an Allee effect is present, as we shall see in the next section.

6.3 The Direction of a Traveling Front

We saw that the speed of a traveling front in the IDE with the caricature Allee effect can have any sign; i.e., the front may invade or retreat or remain stationary. Somewhat surprisingly, the direction of the front depends only on the growth function and is independent of the dispersal kernel (as long as it is symmetric). This result by Wang et al. (2002) generalizes the corresponding, well-known result for a reaction-diffusion equation with strong Allee effect (Kot 2001). The proof in the discrete-time case is much more involved. Our exposition follows Wang et al. (2002).

Theorem 6.1 (Wang et al. 2002) *Consider the IDE $N_{t+1}(x) = (K * F(N_t))(x)$ with monotone growth function F and steady states $N = 1$ and $N = 0$. Assume that there is a monotone decreasing traveling front with speed c and profile $N(x)$, connecting the two states; see Fig. 6.4. Furthermore, assume that F and N are real analytic functions, that derivatives of any order of N vanish as $x \to \pm\infty$, and that the derivatives $d^i F(N(x))/dx^i$ are bounded uniformly in i. Then we have the following relation for the sign of the speed of the traveling front:*

$$\mathrm{sign}(c) = \mathrm{sign}\left(\int_{0}^{1} [F(N) - N] \mathrm{d}N\right).$$
(6.17)

The integral on the right is the signed area between $F(N)$ and N. The region where $F(N) - N$ is positive (negative) is indicated by a $+$ ($-$) sign in Fig. 6.4. The statement of the theorem does not require a (strong) Allee effect. However, if there is a weak or no Allee effect, then $F(N) > N$ for $0 < N < 1$, and so the integral on the right-hand side will be positive.

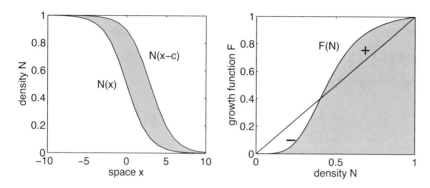

Fig. 6.4 Illustration for Theorem 6.1. **Left plot:** The difference between the traveling front profile in subsequent generations. **Right plot:** Growth function with strong Allee effect and illustration of the integral in (6.17).

Proof Since the front is decreasing, we have $c > 0$ if and only if $N(x) < N(x-c)$ for all x. Since F is monotone increasing and N is monotone decreasing, the derivative

$$\frac{dF}{dx} := \frac{d}{dx}(F(N(x))) \tag{6.18}$$

is negative. Hence, we find that $c > 0$ if and only if $[N(x) - N(x - c)]\frac{dF}{dx} > 0$ for all x. In Lemma 6.1, we show the integral equality

$$\int_{-\infty}^{\infty} [N(x) - N(x - c)]\frac{dF}{dx}dx = \int_{-\infty}^{\infty}[N - F(N)]\frac{dF}{dx}dx . \tag{6.19}$$

The transformation of variables $y = F(N(x))$ applied to the integral on the right results in

$$\int_{-\infty}^{\infty}[N - F(N)]\frac{dF}{dx}dx = \int_{0}^{1}[y - F^{-1}(y)]dy . \tag{6.20}$$

The graph of the function $y = F(N)$ partitions the unit square into the gray and white areas in the right plot in Fig. 6.4 so that

$$\int_{0}^{1} F(N)dN + \int_{0}^{1} F^{-1}(y)dy = 1 . \tag{6.21}$$

A similar argument applies to the function $y = N$ so that the expression in (6.20) can be written as

$$\int_{0}^{1}[y - F^{-1}(y)]dy = 1 - \int_{0}^{1} NdN - \left(1 - \int_{0}^{1} F(N)dN\right) = \int_{0}^{1}[F(N) - N]dN . \tag{6.22}$$

Altogether, we find that $c > 0$ if and only if the expression in (6.22) is positive. The same reasoning applies for $c < 0$ and $c = 0$. Hence, the theorem is proved. □

Lemma 6.1 *Under the conditions of Theorem 6.1, integral equality (6.19) holds.*

Proof We need to show the equality

$$\int_{-\infty}^{\infty} N(x-c)\frac{\mathrm{d}F}{\mathrm{d}x}\mathrm{d}x = \int_{-\infty}^{\infty} F(N)\frac{\mathrm{d}F}{\mathrm{d}x}\mathrm{d}x . \tag{6.23}$$

In the defining equation for the traveling front,

$$N(x-c) = \int_{-\infty}^{\infty} K(z)F(N(x-z))\mathrm{d}z , \tag{6.24}$$

we split $F(N(x-z))$ into its even and odd parts with respect to z, i.e.,

$$F_e(x,z) = \frac{1}{2}[F(N(x-z))+F(N(x+z))], \quad F_o(x,z) = \frac{1}{2}[F(N(x-z))-F(N(x+z))],$$

and obtain

$$N(x-c) = \int_{-\infty}^{\infty} K(z)[F_e(x,z)+F_o(x,z)]\mathrm{d}z = \int_{-\infty}^{\infty} K(z)F_e(x,z)\mathrm{d}z . \tag{6.25}$$

The last equality above arises since the integral of the product of an even function and an odd function is zero. Now we multiply the equality in (6.25) by $\mathrm{d}F/\mathrm{d}x$ and integrate. Since the integrand is of one sign, we use Tonelli's theorem to exchange the order of integration and obtain

$$\int_{-\infty}^{\infty} N(x-c)\frac{\mathrm{d}F}{\mathrm{d}x}\mathrm{d}x = \int_{-\infty}^{\infty}\int_{-\infty}^{\infty} K(z)F_e(x,z)\frac{\mathrm{d}F}{\mathrm{d}x}\mathrm{d}x\mathrm{d}z . \tag{6.26}$$

Next, we expand F_e in a power series around $z = 0$. Because the function is even, all derivatives of odd order vanish. The derivatives of even order at $z = 0$ are the same as the derivatives of $F(N(x))$. Using the fact that the kernel integrates to unity, we find that the integral above equals

$$\int_{-\infty}^{\infty} F(N(x))\frac{\mathrm{d}F}{\mathrm{d}x}\mathrm{d}x + \int_{-\infty}^{\infty}\int_{-\infty}^{\infty} K(z)\sum_{i=1}^{\infty}\frac{z^{2i}}{(2i)!}\left(\frac{\mathrm{d}^{2i}}{\mathrm{d}x^{2i}}F(N(x))\right)\frac{\mathrm{d}F}{\mathrm{d}x}\mathrm{d}x\mathrm{d}z . \tag{6.27}$$

It remains to show that the double integral vanishes. By Levi's theorem, we may interchange summation and the inner integration. Using integration by parts, we verify that each of the inner integrals vanishes as follows. For $i = 1$, we calculate

$$\int_{-\infty}^{\infty} \frac{d^2}{dx^2} F(N(x)) \frac{dF}{dx} dx = \frac{1}{2} \left(\frac{dF}{dx} \right)^2 \Big|_{-\infty}^{\infty} = 0 . \tag{6.28}$$

For $i = 2$, we apply integration by parts twice

$$\int_{-\infty}^{\infty} \frac{d^4}{dx^4} F(N(x)) \frac{dF}{dx} dx = \frac{d^3 F}{dx^3} \frac{dF}{dx} \Big|_{-\infty}^{\infty} - \int_{-\infty}^{\infty} \frac{d^3}{dx^3} F(N(x)) \frac{d^2 F}{dx^2} dx$$

$$= 0 - \frac{1}{2} \left(\frac{d^2 F}{dx^2} \right)^2 \Big|_{-\infty}^{\infty} = 0 . \tag{6.29}$$

Successively, each term in the infinite sum vanishes by repeated application of integration by parts. At this point, we have used the assumption that all derivatives of $N(x)$ vanish as $x \to \pm\infty$ and that all derivatives of F are (uniformly) bounded. □

6.4 General Theory

Explicit calculations for the spreading speed and traveling fronts in an IDE with Allee growth functions are rarely possible. Abstract results about spreading properties and traveling fronts, however, appear simultaneously with those mentioned in the previous chapter when there is no Allee effect (Weinberger 1982; Lui 1983). Clearly, there cannot be a formula analogous to (5.17) for a spreading speed based on the linearization at zero, but that formula can be used to bound the spreading speed even in the case with Allee effect (Lui 1983). Since a population may retreat and not advance, it is also clear that Definition 5.1 of the asymptotic spreading speed cannot apply in the presence of an Allee effect. In particular, the example in Sect. 6.2 shows that a locally introduced population with Allee effect may collapse below the Allee threshold in finite time, but the second inequality in (5.31) requires that the population remain above a positive threshold in the wake of the invasion front for all times. This difficulty is reflected in the formulation of the result below. The following theorem summarizes several aspects of the first published work on spreading speeds and traveling waves in IDEs with Allee effect growth function.

Theorem 6.2 (Lui 1983) *Consider the IDE $N_{t+1}(x) = (K * F(N_t))(x)$ where*

(i) *K is a continuous, symmetric probability distribution with finite moment-generating function;*
(ii) *there is a constant C such that $\int_x^{\infty} K(y)dy \leq CK(x)$ for large x;*
(iii) *F is continuously differentiable with $F(0) = 0 = F(1) - 1$, and (A1) holds;*
(iv) *$F'(N) \leq F'(N_a)$ for $N \in [0, 1]$; and*
(v) *$F'(0)N \leq F(N) \leq F'(1)(N - 1) + 1$ for $N \in [0, 1]$.*

Then the following statements hold.

1. *There exists an asymptotic spreading speed, c^*, in the following sense. If $N_0(x) = 0$ for $x > 0$ and $N_0(x) > N_a$ as $x \to -\infty$, then*

$$\limsup_{t \to \infty} \max_{x > (c^*+\epsilon)t} N_t(x) = 0 \quad \text{and} \quad \liminf_{t \to \infty} \min_{x < (c^*-\epsilon)t} N_t(x) = 1 .$$

2. *A monotone traveling wave can exist for at most one speed.*
3. *There exists $c^* \in \mathbb{R}$ and a family of monotone traveling waves with speed c^*.*

Lui's results are more general than we have stated here. The dispersal kernel can have some discontinuities, and it does not have to be symmetric. When the kernel is not symmetric, we obtain a spreading speed in each direction. The results are not more difficult to prove, but they are more tedious to state. We consider a particular form of asymmetry in Sect. 12.2. The condition on the moment-generating function may be relaxed as in the previous chapter, but some boundedness condition is necessary. When the kernel is heavy tailed, accelerating fronts do exist even with Allee effects (Wang et al. 2002, 2013).

Condition (*v*) requires the graph of F to be bounded between the tangent line at zero and the tangent line at one. Condition (*iv*) requires the slope of the growth function to be maximal at the Allee threshold. These conditions can be weakened (Pan and Zhang 2011). The two asymptotic requirements on the initial condition make it look "wave-like." Lui's original formulation is for compactly supported initial data and needs additional assumptions.

While there is no explicit formula for the speed in the presence of an Allee effect, Lui (1983) gives an upper bound of c^* as

$$c^* \leq \max_{s>0} \frac{1}{s} \ln \left(\max_{N>0} \frac{F(N)}{N} M(s) \right) . \tag{6.30}$$

This bound can be obtained by bounding the growth function F with a function that is monotone and concave down. Lui constructs such a function as

$$F^+(N) = \begin{cases} mN, & 0 \leq N \leq \tilde{N}, \\ 1, & N > \tilde{N}, \end{cases} \tag{6.31}$$

where $m = \max_{N>0} \frac{F(N)}{N}$ and $\tilde{N} = 1/m$ (Fig. 6.5, left panel). At the point $N = 1/m$ where F^+ is not differentiable, it can be "smoothed out" so that it still has the required properties. Alternatively, if F is monotone and concave down whenever it is above the diagonal, we can define F^+ as above for $N < \tilde{N}$ and $F^+ = F$ for $N \geq \tilde{N}$ (Fig. 6.5, right panel). In the latter case, the function is continuously differentiable. In both cases, F^+ is monotone and concave down. Hence, the spreading speed for the IDE with F^+ is linearly determined and given explicitly by the formula in (5.17), which becomes (6.30) for this example. Since $F \leq F^+$, the spreading speed with function F is bounded above by the spreading

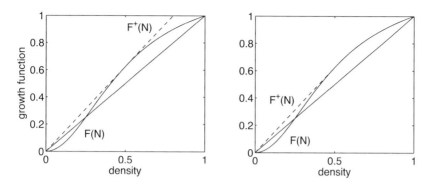

Fig. 6.5 Illustration of a function without Allee effect (F^+) bounding a function with Allee effect (F). The left panel shows the construction by Lui (1983), which has a corner. The right panel shows the alternative construction, which is smooth (see text).

speed with function F^+, as in Sect. 5.4. For the Allee function in (2.22) with $R > \gamma = 2$, we can explicitly calculate

$$m = \frac{R}{2\sqrt{R-1}} \qquad \text{and} \qquad \tilde{N} = \frac{1}{\sqrt{R-1}} .$$

An upper bound for the spreading speed is then given by

$$c^* \leq \max_{s>0} \frac{1}{s} \ln \left(\frac{R}{2\sqrt{R-1}} M(s) \right) . \tag{6.32}$$

To end this chapter, we relate the results about spreading speed to the observations about the existence of a positive steady state on a bounded domain from Sect. 4.5. There, we found that for a positive steady state to exist, we need more than just the existence of a stable positive state in the nonspatial model and a small variance of the dispersal kernel. In fact, the condition we found was independent of the variance of the dispersal kernel: it required the growth to be "strong enough." The threshold between existence and nonexistence of a positive steady state was given by $H(1) - 1/2 = H(0)$, where $H(N)$ is an antiderivative of F. For population spread with an Allee effect, formula (6.17) states that a traveling front invades only if

$$0 < \int_0^1 [F(N) - N]dN = H(1) - H(0) - 1/2 . \tag{6.33}$$

In other words, a steady state on a bounded domain can only exist if the speed of a traveling wave on the unbounded domain is positive. Numerically, we can observe that a locally introduced population on a bounded domain can spread in a front-like fashion to fill the domain and establish a positive steady state.

6.5   Further Reading

A comprehensive review of models for spatial spread with Allee effect can be found in Taylor and Hastings (2005) and more recent results in relation to biological invasions in Lewis et al. (2016). However, while there are many theoretical results, there are relatively few applications of the IDE with Allee effect to real ecosystems. We discuss some in Sect. 12.6.

Since the dynamics with strong Allee effect have two stable steady states, the equation is sometimes called the bistable equation. In contrast, dynamics of Beverton–Holt type have a single stable steady state and are called monostable. Accordingly, traveling fronts are sometimes called bistable fronts and monostable fronts, respectively. When the speed of a front is determined by the linearization at zero (monostable equations; see previous chapter), the front is referred to as a pulled front since the few individuals ahead of the front "pull" it along. In contrast, in the bistable equation, there has to be sufficient growth at higher density (see Theorem 6.1) to "push" the population forward. We sometimes speak of pushed fronts in that case. Bistable traveling fronts in reaction–diffusion equations have received considerably more attention than in IDEs, particularly in combustion problems. Early references can be found in Lui (1983); for recent results and extensions, see, e.g., Hamel (2016).

The original results by Lui (1983) were extended to multiple space dimensions by Creegan and Lui (1984). Later, Lui (1985) showed that solutions with compact initial data converge to a double-front profile. He also proved that solutions were trapped by translations of the traveling front. The existence and stability of clines, i.e., traveling fronts with speed zero, was shown in Lui (1986). More recently, Pan and Zhang (2011) showed the existence, uniqueness, and asymptotic stability of bistable traveling fronts for IDEs by a squeezing technique. Even more general results that include, e.g., spatially heterogeneous environments can be found in Fang and Zhao (2015). Similar, but independent results can be found in Coutinho and Fernandez (2004).

The theory for the monostable equation from Chap. 5 can be applied to prove the existence of a different kind of traveling waves in the bistable equation (Corollary after Proposition 3 in Lui 1983). We assume that F satisfies the conditions from Theorem 6.2. We define the function

$$G(N) = F(N + N_a) - N_a , \qquad N \in [0, 1 - N_a] , \qquad (6.34)$$

and the IDE $N_{t+1}(x) = (K * G(N_t))(x)$. Then $G(0) = 0 = G(1 - N_a) - N_a$, $G'(0) = F'(N_a) > 1$ and $G(N) \le G'(0)N$ on $[0, 1 - N_a]$. In other words, the IDE with growth function G satisfies all the conditions for the theory in Chap. 5. Hence, the bistable equation has monotone traveling waves that connect N_a with one, and their minimal speed is given by the formula in (5.17) with $R = G'(0)$. In Sect. 11.4, we will use the idea behind the construction of G to investigate spreading phenomena in the Ricker equation when the positive steady state is unstable.

Chapter 7
Modeling the Dispersal Process

Abstract In this chapter, we focus on the *process* of dispersal and use it to define dispersal kernels based on mechanistic principles. Many empirical studies examine how individuals move over a short period of time; a dispersal kernel describes where individuals are after a long period of time. By modeling the dispersal process, we can scale from short to long time. We model movement as a random walk and derive the governing partial differential equations. Depending on the form and duration of movement, we derive different dispersal kernels. We begin with uncorrelated random walks on the real line and fixed or exponentially distributed durations, which will lead us to the Gaussian and the Laplace kernel, respectively. For other durations, we obtain various other kernels. Finally, we consider random walks on bounded domains.

7.1 From Process to Outcome

Many empirical studies measure dispersal distances and produce histograms or parametric fits of their distribution, e.g., in release–recapture experiments. A typical example is the collection of wind-dispersed seeds of a single plant in an array of seed traps (Neubert et al. 1995). The results are empirical dispersal kernels that can be used directly in the IDE. The ease with which IDEs can accommodate different dispersal patterns makes them appealing to the ecological community and provides more versatility than reaction–diffusion models. For some species, however, the dispersal outcome is difficult to observe directly, e.g., when recapturing is rare. Instead, empiricists collect data on much smaller spatial and temporal scales, e.g., on the speed and direction of a moving individual over a few minutes to hours (Turchin 1998). The question then is how we can scale from the process (short time and distance) to the outcome (long time and distance). One way to make this connection is to model the dispersal process and solve the resulting equations to derive the dispersal kernel.

In this chapter, we model individual movement as a random walk and make some assumptions about the stopping times of this random walk. Depending on the type of random walk and the distribution of stopping times, we derive different

© Springer Nature Switzerland AG 2019   87
F. Lutscher, *Integrodifference Equations in Spatial Ecology*, Interdisciplinary
Applied Mathematics 49, https://doi.org/10.1007/978-3-030-29294-2_7

forms of dispersal kernels. This chapter follows the pioneering work by Neubert et al. (1995) and Van Kirk (1995). The Gaussian and the Laplace kernel can be derived from a reaction–diffusion equation for uncorrelated random walks with fixed and exponentially distributed stopping times, respectively. Other stopping times, described by *hazard functions*, will yield different kernels.

Having a movement model to derive a dispersal kernel will also allow us to consider dispersal processes in heterogeneous space. In this case, we cannot expect the kernel to depend on dispersal distance only, but rather on the specific initial and final location. Empirically measuring such kernels is extremely demanding since we would need to release individuals from every point and in each case measure their final density at every point. Instead, a process-based movement model will allow us to incorporate spatially varying dispersal behavior and then obtain the corresponding kernel.

7.2 Random Walks and the Dispersal Kernel

We describe the dispersal process as an uncorrelated random walk on a regular one-dimensional lattice with space steps Δ_x and time steps Δ_t; see Fig. 7.1 and Turchin (1998). During each time step, an individual may move one space step to the right with probability p^+ or to the left with probability p^-. With probability p_s, it ends its walk and settles. With the remaining probability, $1 - p^+ - p^- - p_s$, it does nothing. It cannot move farther than one space step per time step. Each individual moves independently of all other individuals.

To derive the *master equation* for this random walk, we denote by $u(t, x)\Delta_x$ the probability that the individual is in the interval of length Δ_x around x at time t. The master equation for u is the bookkeeping equation for all the choices that the individual has in its walk. An individual can be at some location x at time $t + \Delta_t$ if it was already there at time t and did not move, if it was one step to the left and moved to the right, or if it was to the right and moved to the left; see Fig. 7.1. In mathematical terms, we obtain the equation

$$u(t + \Delta_t, x) = p^+ u(t, x - \Delta_x) + p^- u(t, x + \Delta_x) + (1 - p^+ - p^- - p_s)u(t, x).$$

$$(7.1)$$

This master equation is suitable for simulation models but unwieldy for analytical purposes. We continue with the assumption that the time and space steps are small so that we can expand both sides of (7.1) in a Taylor series. For the left-hand side,

Fig. 7.1 Schematic illustration of the lattice and the movement probabilities for a simple random walk.

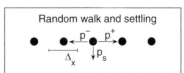

we find

$$u(t + \Delta_t, x) = u(t, x) + \Delta_t \frac{\partial}{\partial t} u(t, x) + O(\Delta_t^2), \tag{7.2}$$

whereas on the right-hand side, we find terms of the form

$$u(t, x + \Delta_x) = u(t, x) + \Delta_x \frac{\partial}{\partial x} u(t, x) + \frac{\Delta_x^2}{2} \frac{\partial^2}{\partial x^2} u(t, x) + O(\Delta_x^3). \tag{7.3}$$

Substituting these expansions into (7.1) and combining terms, we arrive at

$$\frac{\partial}{\partial t} u(t, x) = \frac{\Delta_x^2}{2\Delta_t}(p^- + p^+) \frac{\partial^2}{\partial x^2} u(t, x) + \frac{\Delta_x}{\Delta_t}(p^- - p^+) \frac{\partial}{\partial x} u(t, x) - \frac{p_s}{\Delta_t} u(t, x) + O(\Delta_t, \Delta_x^3). \tag{7.4}$$

We want to pass to a limit as $\Delta_x, \Delta_t \to 0$. Different limits result in different equations. The most commonly used and most successful is the *parabolic limit*, where Δ_x^2/Δ_t converges to a finite, positive number. (We chose to expand only to first order in t but to second order in x precisely because we had this limit in mind.) To carry out the parabolic limit, we need to assume that the difference between the left and right movement probability scales with the spatial step size ($p^- - p^+ \sim \Delta_x$) and that the settling probability scales with the time step ($p_s \sim \Delta_t$). Then we obtain the limiting reaction–diffusion–advection equation

$$\frac{\partial}{\partial t} u(t, x) = D \frac{\partial^2}{\partial x^2} u(t, x) - q \frac{\partial}{\partial x} u(t, x) - \alpha u(t, x), \tag{7.5}$$

where the motility, the movement bias (advection), and the settling rate are

$$D = \lim_{\Delta_x, \Delta_t \to 0} \frac{\Delta_x^2}{2\Delta_t}(p^- + p^+), \quad q = \lim_{\Delta_x, \Delta_t \to 0} \frac{\Delta_x}{\Delta_t}(p^+ - p^-), \quad \alpha = \lim_{\Delta_t \to 0} \frac{p_s}{\Delta_t}, \tag{7.6}$$

respectively. Clearly, D and α are positive, but q can be of any sign.

We let y be the initial location of the individual so that the initial condition is the Dirac delta distribution $u(0, x) = u(0, x; y) = \delta(x - y)$ (Sect. 5.2). Then (7.5) describes the probability density of the location of a single random walker as long as it is still moving. To obtain the dispersal kernel, we "add up" the probabilities that the walker stopped at location x. This distribution of settling locations in the limit of large times is given by

$$K(x, y) = \int_0^\infty \alpha u(t, x; y) dt. \tag{7.7}$$

The additional argument y in u indicates the initial location. We shall omit it if no confusion can arise.

In two space dimensions, the analogue of (7.5) is

$$\frac{\partial}{\partial t} u(t, \mathbf{x}) = D\nabla^2 u(t, \mathbf{x}) - \mathbf{q} \cdot \nabla u(t, \mathbf{x}) - \alpha u(t, \mathbf{x}), \tag{7.8}$$

where $\mathbf{q}$ is now the vector that points in the direction of the movement bias. The definition of K remains unchanged.

Depending on the assumptions about the settling rate, we obtain different kernels from this definition. Since space is homogeneous, we can assume that the starting point is $y = 0$ and use $K(x, y) = K(x - y, 0) = \widetilde{K}(x - y)$. We will drop the tilde when no confusion can arise.

The Laplace Kernel

In the first scenario, the settling rate $\alpha > 0$ is a constant; i.e., there is an equal probability per unit time that the individual settles. In other words, settling is a Poisson process with mean $1/\alpha$, which is the average time that an individual moves before it settles. We begin with an unbiased random walk so that $q = 0$. Equation (7.5) has the explicit solution

$$u(t, x) = \frac{e^{-\alpha t}}{\sqrt{4\pi Dt}} \exp\left(-\frac{x^2}{4Dt}\right). \tag{7.9}$$

In particular, $u \to 0$ uniformly in x as $t \to \infty$.

It is unwieldy to substitute the explicit solution into the definition of the kernel (7.7). Instead, we derive an equation for K directly. We integrate both sides of (7.5) over time from zero to infinity and find

$$u(\infty, x) - u(0, x) = \frac{D}{\alpha} \int_0^\infty \alpha \frac{\partial^2}{\partial x^2} u(t, x)\mathrm{d}t - \int_0^\infty \alpha u(t, x)\mathrm{d}t. \tag{7.10}$$

The last integral in this equation is exactly the expression for the kernel in (7.7). By the above, we have $u(\infty, 0) = 0$ and the initial condition $u(0, x) = \delta(x)$. For the first term on the right-hand side, we assume that we can interchange integration with respect to t and differentiation with respect to x. We obtain the equation for K as

$$-\delta(x) = \frac{D}{\alpha} K''(x) - K(x). \tag{7.11}$$

Hence, K is the Green's function of the differential operator $1 - (D/\alpha)(\partial^2/\partial x^2)$ and can be calculated by standard methods (Keener 2000). Using the asymptotic boundary conditions at infinity, we write

$$K(x) = \begin{cases} A \exp\left(-\sqrt{\alpha/D}\, x\right), & x > 0, \\ B \exp\left(\sqrt{\alpha/D}\, x\right), & x < 0. \end{cases} \tag{7.12}$$

Continuity of K (or symmetry) requires $A = B$. The easiest way to find the value of the remaining parameter is to use the integral condition (2.24). Alternatively, (7.11) requires that K' have a jump of size $-\alpha/D$ at zero. Formally, we integrate (7.11) over $(-\epsilon, \epsilon)$ and let $\epsilon \to 0$ to find

$$-1 = \frac{D}{\alpha}(K'(0^+) - K'(0^-)) = -2\frac{D}{\alpha}\sqrt{\frac{\alpha}{D}} A. \tag{7.13}$$

Hence, $A = \sqrt{\alpha/D}/2$, and the kernel is the Laplace kernel

$$K(x) = \frac{1}{2}\sqrt{\alpha/D} \exp\left(-\sqrt{\alpha/D}|x|\right). \tag{7.14}$$

The variance of K is $\sigma^2 = 2D/\alpha$; see (2.27).

From the two-dimensional random walk process with constant settling rate, we obtain the modified Bessel function

$$K(\mathbf{x}) = \frac{\alpha}{2\pi D} K_0\left(\sqrt{\frac{\alpha}{D}}\|\mathbf{x}\|\right) \tag{7.15}$$

with variance $\sigma^2 = 2D/\alpha$. Note that K_0 is unbounded at the origin. The one-dimensional Laplace kernel results as the marginal distribution along any line through the origin (Van Kirk 1995).

So far, we have assumed an unbiased random walk ($q = 0$), but movement in rivers and other advective environments is often biased. In one space dimension, the same steps as above can be carried out for $q > 0$ and lead to the asymmetric Laplace kernel (Van Kirk 1995; Lutscher et al. 2005)

$$K(x) = \begin{cases} A \exp(a_1 x), & x < 0, \\ A \exp(a_2 x), & x > 0, \end{cases} \quad A = \frac{a_1|a_2|}{a_1 + |a_2|}, \tag{7.16}$$

where $a_1 > 0 > a_2$ are given by $a_{1,2} = \frac{q \pm \sqrt{q^2 + 4\alpha D}}{2D}$. The variance and mean of this kernel are $\sigma^2 = 2D/\alpha + q^2/\alpha^2$ and $\mu = q/\alpha$, respectively (Lutscher et al. 2010).

We illustrate the effect of movement bias on the Laplace kernel in the left plot in Fig. 7.2. We will use this kernel to explore questions of persistence and spread in an advective environment in Sect. 12.2.

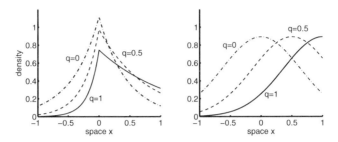

Fig. 7.2 Left plot: Asymmetric Laplace kernel resulting from the assumption of constant settling rate and movement bias. **Right plot:** Gaussian kernel resulting from the assumption of fixed time to settling and movement bias. Parameters are $D = 0.2$ and $\tau = 1/\alpha = 1$.

The Gaussian Kernel

In a second scenario, we assume that the settling rate is not constant but that an individual settles after exactly $\tau > 0$ time units have passed. Formally, we can write the time-dependent settling rate using the delta distribution: $\alpha = \alpha(t) = \delta(t - \tau)$. Then the dispersal kernel is simply the fundamental solution (7.9) of (7.5) with $\alpha = 0$ at time τ, or

$$K(x) = \int_0^\infty \delta(t - \tau)u(t, x)\mathrm{d}t = u(\tau, x) = \frac{1}{\sqrt{4\pi D\tau}} \exp\left(-\frac{(x - q\tau)^2}{4D\tau}\right),$$
(7.17)

with variance $\sigma^2 = 2D\tau$ and mean $\mu = q\tau$. In two space dimensions, we find the Gaussian kernel

$$K(\mathbf{x}) = \frac{1}{2\pi D\tau} \exp\left(-\frac{\|\mathbf{x} - \mathbf{q}\tau\|^2}{4D\tau}\right).$$
(7.18)

The marginal distribution of the two-dimensional kernel is the one-dimensional version.

To compare the Gaussian and the Laplace kernel, we choose the mean time before settling in the Laplace kernel to equal the exact time before settling in the Gaussian kernel, i.e., $1/\alpha = \tau$. The two kernels have the same mean, $\mu = q\tau = q\alpha$, but the variance of the Gaussian kernel is smaller than that of the Laplace kernel, and the difference increases as τ increases. The comparison in Fig. 7.2 shows that advection has a markedly different effect on the two kernels. Whereas the Gaussian kernel is merely shifted by the mean dispersal distance per lifetime, the Laplace kernel retains its maximum at zero but becomes increasingly asymmetric and skewed. This difference will have a big impact on persistence conditions and spreading speeds for river-dwelling organisms; see Sect. 12.2.

7.3   Straight Walks and Hazard Functions

The dispersal process constitutes a developmentally important part of the life cycle of many species. For example, a number of aquatic organisms with sessile adult stages produce larvae that need to mature in the water column before they have the physical ability to settle on the benthic floor and become sessile adults (Scheltema 1986). Such developmental processes can depend on temperature and other aspects. Accordingly, we would like to have a general, time-dependent settling rate $\alpha(t)$ in (7.5) and derive the corresponding dispersal kernel from the equations as above. We could even consider the movement parameters to depend on time as well. Unfortunately, explicit expressions for the dispersal kernel are not available in these cases.

As an alternative, we can replace the random walk with a straight walk at constant speed and obtain explicit examples (Neubert et al. 1995). When an individual moves at constant speed $\gamma > 0$, starting from $x = 0$ along the straight line $x > 0$, and stops moving at the rate $\alpha(t)$, then the density function $u(t, x)$ for this process satisfies the equation

$$\frac{\partial}{\partial t} u(t, x) + \gamma \frac{\partial}{\partial x} u(t, x) = -\alpha(t) u(t, x), \quad u(0, x) = \delta(x). \tag{7.19}$$

The equation can be solved by the method of characteristics to obtain

$$u(t, x) = \delta(x - \gamma t) \exp\left(-\int_0^t \alpha(\tau) d\tau\right). \tag{7.20}$$

The dispersal kernel for $x \geq 0$ is then

$$K(x) = \int_0^\infty \alpha(t) \delta(x - \gamma t) e^{-\int_0^t \alpha(\tau) d\tau} dt = \frac{1}{\gamma} \alpha\left(\frac{x}{\gamma}\right) \exp\left(-\int_0^{x/\gamma} \alpha(\tau) d\tau\right). \tag{7.21}$$

For symmetric dispersal from $x = 0$ in both directions on the real line, we divide the expression by two and replace x by $|x|$.

For the constant settling rate $\alpha(t) = \alpha$ we obtain the Laplace kernel, albeit with different parameters,

$$K(x) = \frac{1}{2} \frac{\alpha}{\gamma} \exp\left(-\frac{\alpha}{\gamma} |x|\right). \tag{7.22}$$

When the settling rate is a power function,

$$\alpha(t) = k \left(\frac{\gamma}{\theta}\right)^k t^{k-1}, \tag{7.23}$$

the resulting kernel is the double Weibull kernel

$$K(x) = \frac{k}{2\theta} \left|\frac{x}{\theta}\right|^{k-1} \exp\left(-|x/\theta|^k\right); \tag{7.24}$$

see Table 3.1 and Fig. 3.3. Similarly, the double gamma distribution from that table can be obtained with an appropriately chosen settling rate; see Table 1 in Neubert et al. (1995).

7.4 Ballistic Dispersal

Certain plants forcibly eject their seeds from their pods and send them in a ballistic trajectory before they land. For example, some geraniums (e.g., *Geranium maculatum*), phlox (*Phlox drummondii*), and touch-me-nots (*Impatiens capensis*) use this mechanism, which is also referred to as "explosive dispersal" (Beer and Swaine 1977; Stamp and Lucas 1983). This dispersal process can often be better modeled by the physical equations of motion than by a random walk (Neubert et al. 1995).

In the simplest case, the seed is launched from the ground with speed $\gamma > 0$ and angle $\theta \in [0, \pi]$. We consider the effects of gravity only and neglect air resistance. We denote by x and y the horizontal and vertical coordinates of the location of the seed. Then the equations of motion in the x- and y-direction are

$$\ddot{x} = 0, \quad \ddot{y} = -g, \tag{7.25}$$

where g is the gravitational constant and a dot indicates differentiation with respect to time. The initial conditions are $x(0) = 0$, $y(0) = 0$, $\dot{x}(0) = \gamma \cos(\theta)$, and $\dot{y}(0) = \gamma \sin(\theta)$. The solutions to these equations are

$$x(t) = \gamma \cos(\theta)t, \quad y(t) = \gamma \sin(\theta)t - \frac{g}{2}t^2. \tag{7.26}$$

Hence, a seed with initial angle θ and speed γ will land at location

$$x(\theta) = \frac{\gamma^2}{g} \sin(2\theta). \tag{7.27}$$

In particular, there is a maximum dispersal distance of γ^2/g.

We can calculate the corresponding dispersal kernel from the distribution of launch angles by a change of variables. We suppose that $f = f(\theta)$ is the distribution of launch angles, i.e., f is a nonnegative function on $[0, \pi]$ with integral equal to unity. Then the probability that a seed is launched into the interval $[x_1, x_2]$ equals the probability that the seed is launched with angles in $\theta \in [0, \pi]$, such that $x(\theta) \in [x_1, x_2]$, or

$$\int_{x_1}^{x_2} K(x)dx = \int_0^\pi f(\theta)d\theta . \tag{7.28}$$

The function $x = x(\theta)$ is not globally invertible. For nonnegative x, we find

$$\theta(x) = \begin{cases} \theta_1(x) = 0.5\arcsin(gx/\gamma^2), & \theta \in [0, \pi/4], \\ \theta_2(x) = \pi/2 - 0.5\arcsin(gx/\gamma^2), & \theta \in [\pi/4, \pi/2]. \end{cases} \tag{7.29}$$

For every $\bar{x} \in (0, \gamma^2/g)$, there is an interval $x_1 < \bar{x} < x_2$ with $x_i \in (0, \gamma^2/g)$. Then, by substitution, the formula in (7.28) yields

$$\int_{x_1}^{x_2} K(x)dx = \int_{x_1}^{x_2} f(\theta_1(x))\theta_1'(x)dx + \int_{x_2}^{x_1} f(\theta_2(x))\theta_2'(x)dx . \tag{7.30}$$

If the launch angles are uniformly distributed, we have $f(\theta) = 1/\pi$, and we calculate explicitly

$$K(x) = \frac{1}{\pi} \frac{1}{\sqrt{(\gamma^2/g)^2 - x^2}} , \qquad |x| < \frac{\gamma^2}{g} . \tag{7.31}$$

This ballistic dispersal kernel is compactly supported but it is not continuous at the boundary of its support; in fact, it has poles at this boundary (Fig. 7.3). Its integral, however, is finite and equals unity.

Fig. 7.3 Illustration of ballistic dispersal. **Top panel:** trajectories of seeds at different launch angles. **Bottom panel:** resulting dispersal kernel. Parameters are $\gamma^2 = g = 10$.

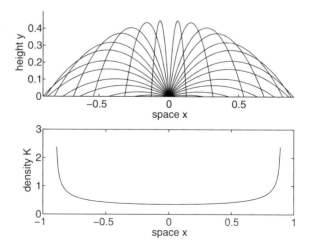

For increased realism, we can consider a positive initial height, include air resistance, and allow the angles to be nonuniformly distributed (Beer and Swaine 1977). When plants eject seeds from pods at different heights, we can integrate over the distribution of launch heights as well.

7.5 Random Walks on Bounded Domains

The previous examples of dispersal kernels were derived under the assumption of a homogeneous landscape, yet most landscapes are heterogeneous. An extreme form of heterogeneity is that of a single isolated patch surrounded by a hostile environment, as we encountered for the critical patch-size problem in Chap. 3.

Some dispersal processes are relatively unaffected by spatial heterogeneity. For example, wind-dispersed seeds may be transported from an island to the surrounding water without any special behavior at the edge. Wind patterns are, however, affected by topography, as can be the resulting seed dispersal process (Robbins 2004). Many other dispersal processes will strongly depend on habitat type. For example, many butterflies prefer meadows and will enter a surrounding forest only with very small probability. We present the derivation of kernels for movement on a single bounded patch by Van Kirk (1995) and Van Kirk and Lewis (1999). We generalize these ideas when we discuss persistence and spread in heterogeneous landscapes (Chap. 15).

Van Kirk (1995) assumed that inside the domain $[-L/2, L/2]$, an individual moves distance Δ_x left or right with probability $p^+ = p^- = 1/2$ per time increment Δ_t. At the boundary point $x = L/2$, the individual moves back into the domain with probability $p^- = 1/2$. Of the remaining probability, the individual leaves the domain with probability $c\Delta_x$ and stays at the boundary with probability $(1 - c\Delta_x)$; see Fig. 7.4. Here, $c \in [0, \infty)$ is the rate per unit length at which the individual leaves the domain. The master equation at the boundary takes the form

$$u(t + \Delta_t, L/2) = \frac{1}{2}u(t, L/2 - \Delta_x) + \frac{1 - c\Delta_x}{2}u(t, L/2). \qquad (7.32)$$

Expanding the terms and sorting by like powers, we obtain

$$\frac{\partial}{\partial t}u(t, L/2) = -\frac{\Delta_x}{2\Delta_t}\left(\frac{\partial}{\partial x}u(t, L/2) + cu(t, L/2)\right) + O(\Delta_t, \Delta_x^2). \qquad (7.33)$$

Fig. 7.4 Schematic illustration of the random walk at a boundary point.

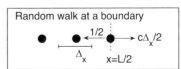

In the parabolic limit from (7.6), the fraction Δ_x / Δ_t becomes infinite. Therefore, the expression in parentheses has to equal zero. A similar argument applies at the boundary $x = -L/2$, with only the sign in the brackets changing. Hence, the movement model on the bounded interval is given by

$$\frac{\partial}{\partial t} u = D \frac{\partial^2}{\partial x^2} u - \alpha u , \quad x \in (-L/2, L/2) , \tag{7.34}$$

with boundary conditions

$$\frac{\partial}{\partial x} u(t, -L/2) = c u(t, -L/2) , \qquad \frac{\partial}{\partial x} u(t, L/2) = -c u(t, L/2) . \tag{7.35}$$

Before we calculate the dispersal kernel from this movement model, we simplify the notation by scaling space by L and time by α to obtain the equations

$$\frac{\partial}{\partial \tau} u(\tau, \xi) = \tilde{D}^2 \frac{\partial^2}{\partial \xi^2} u(\tau, \xi) - u(\tau, \xi) , \quad \xi \in (-1/2, 1/2) , \tag{7.36}$$

with $\alpha L^2 \tilde{D}^2 = D$ and boundary conditions

$$(1-p_e) \frac{\partial}{\partial \xi} u(\tau, -1/2) = p_e u(\tau, -1/2) , \quad (1-p_e) \frac{\partial}{\partial x} u(\tau, 1/2) = -p_e u(\tau, 1/2) , \tag{7.37}$$

where the dimensionless quantity $p_e = cL/(1 + cL)$ is the *edge permeability* (Van Kirk and Lewis 1999).

Following Van Kirk and Lewis (1999), we separate variables and represent the dispersal kernel as an infinite series. Separating variables for $u(\tau, \xi) = T(\tau) X(\xi)$ in (7.36) gives a Sturm–Liouville problem for X and a sequence of eigenvalues v_n^2 satisfying

$$T' = -(v_n^2 + 1) T , \qquad X'' = -\frac{v_n^2}{\tilde{D}^2} X , \tag{7.38}$$

with boundary conditions (7.37) for X. The general ansatz for X gives

$$X(\xi) = A \sin\left(\frac{v_n}{\tilde{D}} \xi\right) + B \cos\left(\frac{v_n}{\tilde{D}} \xi\right) , \tag{7.39}$$

and the boundary conditions result in the system

$$\left[(1-p_e) \frac{v_n}{\tilde{D}} \cos\left(\frac{v_n}{2\tilde{D}}\right) + p_e \sin\left(\frac{v_n}{2\tilde{D}}\right) \right] A + \left[p_e \cos\left(\frac{v_n}{2\tilde{D}}\right) - (1-p_e) \frac{v_n}{\tilde{D}} \sin\left(\frac{v_n}{2\tilde{D}}\right) \right] B = 0 ,$$

$$\left[(1-p_e) \frac{v_n}{\tilde{D}} \cos\left(\frac{v_n}{2\tilde{D}}\right) + p_e \sin\left(\frac{v_n}{2\tilde{D}}\right) \right] A - \left[p_e \cos\left(\frac{v_n}{2\tilde{D}}\right) - (1-p_e) \frac{v_n}{\tilde{D}} \sin\left(\frac{v_n}{2\tilde{D}}\right) \right] B = 0 .$$

This system has a solution only if one of A or B is zero. Accordingly, we obtain two sets of conditions that determine v_n as

$$p_e \cos\left(\frac{v_n}{2\tilde{D}}\right) = (1 - p_e)\frac{v_n}{\tilde{D}} \sin\left(\frac{v_n}{2\tilde{D}}\right) \qquad \text{if} \quad A = 0, \qquad (7.40)$$

$$p_e \sin\left(\frac{v_n}{2\tilde{D}}\right) = -(1 - p_e)\frac{v_n}{\tilde{D}} \cos\left(\frac{v_n}{2\tilde{D}}\right) \qquad \text{if} \quad B = 0. \qquad (7.41)$$

Explicit solutions are available when either $p_e = 1$ or $p_e = 0$; for intermediate cases, the transcendental equation has to be solved numerically. For $p_e = 1$, we find that $A = 0$ implies $v_n = (2n - 1)\pi \tilde{D}$ and $X_n(\xi) = \cos((2n - 1)\pi\xi)$, whereas $B = 0$ implies $v_n = 2n\pi \tilde{D}$ and $X_n(\xi) = \sin(2n\pi\xi)$. By standard Fourier theory, we can write the solution for u as

$$u(\tau, \xi) = \sum_{n \geq 1} C_n e^{-(v_n^2 + 1)\tau} \cos((2n - 1)\pi\xi) + S_n e^{-(v_n^2 + 1)\tau} \sin(2n\pi\xi), \qquad (7.42)$$

where, using the initial condition $u(0, \xi) = \delta(\xi - y)$, the constants are

$$C_n = 2\cos((2n - 1)\pi y), \qquad S_n = 2\sin(2n\pi y). \qquad (7.43)$$

Term-by-term integration gives the dispersal kernel (Van Kirk 1995)

$$K(\xi, y) = \sum_{n \geq 1} \frac{2}{(2n - 1)^2\pi^2\tilde{D}^2 + 1} \cos((2n - 1)\pi\xi) \cos((2n - 1)\pi y)$$
$$\qquad\qquad\qquad\qquad\qquad\qquad\qquad\qquad\qquad\qquad\qquad\qquad (7.44)$$
$$+ \frac{2}{4n^2\pi^2\tilde{D}^2 + 1} \sin(2n\pi\xi) \sin(2n\pi y).$$

We illustrate the kernel for hostile boundary conditions in Fig. 7.5; see also Figs. 1 and 2 in Van Kirk and Lewis (1999). We see that the kernel is *not* of the

Fig. 7.5 Dispersal kernels on the bounded domain $[-1/2, 1/2]$ with hostile boundary conditions ($p_e = 1$) and initial location $y = 0$ (blue) and $y = 0.3$ (black) and scaled diffusion constant $\tilde{D} = 0.1$ (solid) and $\tilde{D} = 0.2$ (dashed).

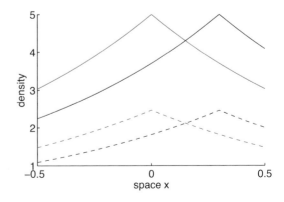

form $K(x - y)$ but depends on both the initial and the final location. With larger diffusion constant (dashed curves), individuals are more likely to get to the boundary before they settle, so that the probability of staying in the domain is decreased.

7.6 Multiple Dispersal Modes

Dispersal can be a multi-stage process involving different movement modes. For example, seeds can first be airborne and disperse with the wind, and once they settle on the ground they can be picked up by rodents and moved farther (Powell and Zimmermann 2004). Some stream insects, such as mayflies, drift downstream with the water flow in their larval stage, and then mature to winged adults and leave the water to fly and lay their eggs (Lutscher et al. 2010). Even within a river or stream, the flow speeds at different depths differ, and individuals (or particles) experience different movement when they change their vertical position (Lutscher and McCauley 2013). Individuals may also choose between different movement modes according to their internal state, e.g., they may switch between foraging and exploration (Tyson et al. 2011). The random-walk approach from Sect. 7.2 can and has been extended to several of these cases (Neubert et al. 1995; Skalski and Gilliam 2003; Lutscher et al. 2005).

If two or more dispersal processes happen consecutively, the derivations of the dispersal kernel and corresponding moment-generating function are straightforward extensions of the procedure presented earlier (Neubert et al. 1995). We consider two stages with diffusion coefficients D_i and drop-out rates α_i. The movement equations for an individual in stage $i = 1, 2$ are

$$
\begin{aligned}
\frac{\partial}{\partial t} u_1 &= D_1 \frac{\partial^2}{\partial x^2} u_1 - \alpha_1 u_1 , \\
\frac{\partial}{\partial t} u_2 &= D_2 \frac{\partial^2}{\partial x^2} u_2 - \alpha_2 u_2 + \alpha_1 u_1 ,
\end{aligned}
\tag{7.45}
$$

with initial conditions $u_1(0, x) = \delta(x - y)$ and $u_2(0, x) = 0$. We define kernels K_i as in (7.28) with α and u replaced by α_i and u_i, respectively. Since the landscape is homogeneous, we can set $y = 0$ and consider $K_i = K_i(x)$ as a function of a single variable only. By integration, we obtain the two coupled equations

$$
\begin{aligned}
-\delta(x) &= \frac{D_1}{\alpha_1} K_1''(x) - K_1(x) , \\
-K_1(x) &= \frac{D_2}{\alpha_2} K_2''(x) - K_2(x) .
\end{aligned}
\tag{7.46}
$$

We already know the Green's function of the first equation from the previous calculation as the Laplace kernel with parameter $a_1 = \sqrt{\alpha_1/D_1}$. Similarly, the

Green's function of the second equation is the Laplace kernel with parameter $a_2 = \sqrt{\alpha_2/D_2}$. Hence, the solution of the second equation in (7.46) is the convolution (Neubert et al. 1995)

$$
K_2(x) = \int_{-\infty}^{\infty} \frac{a_2}{2} e^{-a_2|x-y|} \frac{a_1}{2} e^{-a_1|y|} dy
$$

$$
= \begin{cases} \frac{a_1^2 a_2^2}{2(a_2^2-a_1^2)} \left(\frac{1}{a_1} e^{-a_1|x|} - \frac{1}{a_2} e^{-a_2|x|} \right), & a_1 \neq a_2, \\[2mm] \frac{a_1}{4}(1 + a_1|x|)e^{-a_1|x|}, & a_1 = a_2. \end{cases} \tag{7.47}
$$

This is our desired dispersal kernel. We illustrate the effect of the second dispersal stage in Fig. 7.6.

The easiest way to calculate the moment-generating function is to use the property of the exponential transform that the moment-generating function of a convolution is the product of the moment-generating functions. With this, we find

$$
M(s) = \frac{a_1^2}{(a_1^2 - s^2)} \frac{a_2^2}{(a_2^2 - s^2)} . \tag{7.48}
$$

In principle, more complicated movement equations can also be solved by this Green's function method, e.g., when individuals can settle from both modes, when they can switch back and forth between modes, and when there is advection (Lutscher et al. 2005).

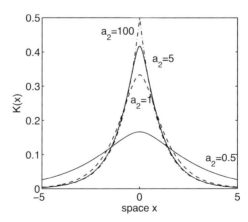

Fig. 7.6 Dispersal kernel (7.47) for the model of two subsequent dispersal stages. When the second stage is of short duration (α_2 large) and/or of little movement (D_2 small), then a_2 is large and the dispersal kernel is essentially the Laplace kernel of the first dispersal mode only (see dash-dot line for $a_2 = 100$). As a_2 decreases, the second mode becomes more important and individuals disperse considerably further.

When we are interested in calculating spread rates for linear models, we do not need to know the dispersal kernel but rather its moment-generating function; see (5.17). It can be obtained directly from the equation(s) for the kernel(s) via an exponential transform without calculating the kernel(s) explicitly. We illustrate the method using the preceding example.

When we multiply the first equation in (7.46) by $\exp(sx)$ and integrate with respect to x, we obtain

$$- \int_{-\infty}^{\infty} \delta(x) e^{sx} dx = a_1^2 \int_{-\infty}^{\infty} K_1''(x) e^{sx} dx - \int_{-\infty}^{\infty} K_1(x) e^{sx} dx . \qquad (7.49)$$

By the properties of the delta distribution, the term on the left side equals -1. By definition, the second term on the right becomes the moment-generating function, $M_1(s)$, of K_1; see (5.15). The first term on the right is more delicate since K'' is not continuous at zero. We split the integral as

$$\int_{-\infty}^{\infty} K''(x) e^{sx} dx = \int_{-\infty}^{-\epsilon} K''(x) e^{sx} dx + \int_{-\epsilon}^{\epsilon} K''(x) e^{sx} dx + \int_{\epsilon}^{\infty} K''(x) e^{sx} dx .$$
$$(7.50)$$

Since K decays exponentially to zero for $|x| \gg 0$, the improper integrals converge for small enough $|s|$. Integrating by parts twice, we find

$$\int_{\epsilon}^{\infty} K''(x) e^{sx} dx = -K'(\epsilon) e^{s\epsilon} + K(\epsilon) s e^{s\epsilon} + \int_{\epsilon}^{\infty} s^2 K(x) e^{sx} dx , \qquad (7.51)$$

and a similar expression for the integral from $-\infty$ to $-\epsilon$. For the middle integral, we substitute the equation to get

$$\int_{-\epsilon}^{\epsilon} K''(x) e^{sx} dx = a_1^2 \int_{-\epsilon}^{\epsilon} (-\delta(x) + K_1(x)) e^{sx} dx \to -a_1^2 \qquad (7.52)$$

as $\epsilon \to 0$, again by the properties of the delta distribution and since K_1 is continuous. Substituting back and taking the limit in (7.50) gives

$$\int_{-\infty}^{\infty} K''(x) e^{sx} dx = K(0^-) - K(0^+) + K'(0^-) - K'(0^+) - a_1^2 + s^2 M_1(s) . \qquad (7.53)$$

The first two terms on the right side cancel because K is continuous. The second two cancel with the fifth by the jump condition (7.13), so that only the last term remains. Hence, (7.49) turns into

$$-1 = a_1^2 s^2 M_1(s) - M_1(s) , \quad \text{or} \quad M_1(s) = \frac{a_1^2}{a_1^2 - s^2} . \qquad (7.54)$$

We have re-derived the expression for the moment-generating function of the Laplace kernel.

There is no discontinuity in the equation for K_2 in (7.46). We can take the exponential transform and integrate by parts twice to get

$$- M_1(s) = a_2^2 s^2 M_2(s) - M_2(s) \,. \tag{7.55}$$

Solving for M_2, we get the expression in (7.48) with $a_i^2 = \alpha_i / D_i$ as usual.

For the most general (linear) dispersal model of this kind with two modes (or states), we assume that individuals in mode i move randomly with diffusion coefficient D_i and advection q_i. They end their dispersal phase with rate α_i. In addition, they may switch from mode i to the other mode with rate β_i. Then the equations for the location of the individual in motion read (Lutscher et al. 2005)

$$\frac{\partial}{\partial t} u_i = D_i \frac{\partial^2}{\partial x^2} u_i - q_i \frac{\partial}{\partial x} u_i - (\alpha_i + \beta_i) u_i + \beta_j u_j \,, \quad i \neq j. \tag{7.56}$$

The initial conditions

$$u_1(0, x) = p\delta(x - y) \,, \qquad u_2(0, x) = (1 - p)\delta(x - y) \,, \tag{7.57}$$

indicate that an individual starts dispersing in mode 1 with probability p and in mode 2 with probability $1 - p$. The densities of the settling locations from state i are

$$K_i(x, y) = \int_0^\infty \alpha_i u_i(t, x) \mathrm{d}t \,, \tag{7.58}$$

and the desired dispersal kernel is the sum $K = K_1 + K_2$.

Without loss of generality, we can set $y = 0$ and consider $K = K(x)$ as a function of a single variable only. By integration, we obtain the two coupled equations

$$-p\delta(x) = \frac{D_1}{\alpha_1} K_1''(x) - \frac{q_1}{\alpha_1} K_1'(x) - \frac{\alpha_1 + \beta_1}{\alpha_1} K_1(x) + \frac{\beta_2}{\alpha_2} K_2(x) \,,$$

$$-(1 - p)\delta(x) = \frac{D_2}{\alpha_2} K_2''(x) - \frac{q_2}{\alpha_2} K_2'(x) - \frac{\alpha_2 + \beta_2}{\alpha_2} K_2(x) + \frac{\beta_1}{\alpha_1} K_1(x) \,. \tag{7.59}$$

We can proceed in the same way as when finding the Green's function. For $x \neq 0$, we can derive a single linear fourth-order equation for K_1, say, by repeated differentiation and substitution (Lutscher et al. 2005). To solve the fourth-order differential equation we have to find the roots of its a characteristic equation, which is a polynomial of order four. Once we have the roots of the characteristic polynomial, we write the solution in terms of the corresponding exponential functions, and use the conditions at infinity as well as at zero to determine the constants.

Using the preceding ideas, it is much simpler to find the moment-generating function directly. Taking the exponential transform of equations (7.59), we have

$$
\begin{aligned}
-p &= \frac{D_1}{\alpha_1} s^2 M_1(s) + \frac{q_1}{\alpha_1} s M_1(s) - \frac{\alpha_1 + \beta_1}{\alpha_1} M_1(s) + \frac{\beta_2}{\alpha_2} M_2(s) , \\
-(1 - p) &= \frac{D_2}{\alpha_2} s^2 M_2(s) + \frac{q_2}{\alpha_2} s M_2(s) - \frac{\alpha_2 + \beta_2}{\alpha_2} M_2(s) + \frac{\beta_1}{\alpha_1} M_1(s) .
\end{aligned}
\tag{7.60}
$$

The solution of this system is a simple linear algebra problem that does not require root finding. One can solve the second equation for M_2 and substitute into the first to find M_1 (and thereby also M_2) explicitly. The desired moment-generating function for the process is then $M = M_1 + M_2$.

7.7 Further Reading

Modeling dispersal, deriving dispersal kernels, and estimating parameters are active areas of research in their own right. The ideas presented here can be extended in several directions, and many different approaches are possible. The mechanistic underpinning of dispersal kernels may also help in selecting appropriate kernels. For example, if empirical move-length distributions are available and several parametric kernels produce comparably good fits, process-based criteria might help choose one kernel over another.

We assumed spatially and temporally constant movement rates, but the diffusion–settling equation can be formulated more generally for heterogeneous landscapes (Turchin 1998). Even though we cannot find an explicit expression for the resulting dispersal kernel, we can use the fact that the kernel results in the Green's function of a certain differential operator to prove results about the critical patch-size and the spreading speed. We return to this issue in Chap. 15.

Chesson and Lee (2005) developed an analogous theoretical framework to derive dispersal kernels in discrete landscapes, where individuals take steps on a finite-dimensional integer lattice.

Additional individual-level processes can be included in the basic approach. For example, to account for mortality during dispersal, we can include a death rate in the dispersal equation and calculate corresponding kernels (Van Kirk 1995); see Sect. 12.1. Another question is that of a finite time of dispersal. If individuals need to settle within a certain time in order to survive, as certain marine invertebrates do, then the settling rate, $\alpha(t)$, is zero when t is larger than this maximum time. As in the case with dispersal-related mortality, the corresponding kernels will not integrate to unity. The detailed consequences of these effects are future research questions.

The mechanistic approach also provides an avenue to understanding the emergence of heavy-tailed kernels. For example, the moment-generating function for the Weibull kernel (7.24) is defined only for $k \geq 1$, but the kernel is defined for

$0 < k < 1$ as well. Secondly, if instead of using random walks, we consider Lévy flights as the movement process, a constant settling rate will produce the Cauchy kernel (Lutscher et al. 2005). Yet another approach is to use fractional diffusion equations (anomalous or super-diffusion) instead of the reaction–diffusion equations to model movement. This approach can also result in heavy-tailed kernels (Baeumer et al. 2007). There is empirical evidence for heavy-tailed kernels or mixed kernels; see Sect. 12.5 (Bullock and Clarke 2000). Clark (1998) developed a two-parameter kernel that can interpolate between the Gaussian kernel and certain heavy-tailed kernels.

Other mechanistic models exist, in particular for seed dispersal. One of the most prominent models is the so-called Wald analytical long-distance dispersal (WALD) model (Skarpaas and Shea 2007). This model arises from fluid dynamics and predicts a distribution of dispersal distances r according to

$$\widetilde{K}(r) = \left(\frac{\lambda}{2\pi r^3}\right)^{1/2} \exp\left(-\frac{\lambda(r-\mu)^2}{2\mu^2 r}\right), \tag{7.61}$$

where μ is the mean and λ is the scale parameter. These two parameters depend on seed release height, terminal velocity, and wind speed data. The actual kernel is calculated by integrating over the release height and wind speed distributions (Skarpaas and Shea 2007). Marchetto et al. (2010) use the WALD model to estimate the effect of different spatial arrangements of plants on invasion speeds. Bullock et al. (2012) use this approach in combination with predicted wind-speed distributions to estimate spread rates under climate change. Caplat et al. (2012) use perturbation analysis in conjunction with the WALD model to find sensitivities of spread rates with respect to parameters.

Yet another aspect of dispersal aims to estimate the relative contributions of different sources to a target population. Klein et al. (2006) show that these results differ significantly between thin- and heavy-tailed dispersal kernels and thereby help explain the phenomenon of accelerating invasions. Finally, Snäll et al. (2007) propose a general statistical framework to analyze dispersal patterns and data in a unified way.

Chapter 8
Computational Aspects

Abstract In this chapter, we present some numerical recipes to illustrate the analytical results on IDEs obtained so far and to facilitate exploring the dynamics beyond what can be proved analytically. Numerical simulation of IDEs is typically simpler than for corresponding reaction–diffusion equations since only space but not time needs to be discretized. When the IDE is in convolution form, the simplest algorithm uses a discrete Fourier transform. In other cases, an explicit quadrature rule can be used. Eigenvalue problems for IDEs can be approximated by matrix eigenvalue problems and solved with existing routines. Explicit analytical results help calibrate numerical schemes.

8.1 Numerical Methods

Numerical methods to solve IDEs are indispensable for applications where the equations are more complex than the ones that we have investigated analytically so far. Complex dynamic phenomena, such as cyclic and chaotic behavior with the Ricker or logistic growth function (see Chap. 4), can easily appear in IDEs but are almost impossible to analyze without computational help. Similarly, studying extensions such as the inclusion of multiple interacting species or spatio-temporal heterogeneity in later chapters often requires computational approaches. The analytical results obtained so far are useful for checking the accuracy of the numerical methods before applying the methods to problems that are inaccessible to analysis. The numerical methods presented here are relatively simple and are presented without numerical error analysis or convergence estimates. They are nonetheless widely used to explore the dynamics of IDEs even in chaotic regimes. For more rigorous approaches, we refer to Day et al. (2004) and Day and Kalies (2013).

The models in Chaps. 3 and 4 deal with bounded domains, whereas those in Chaps. 5 and 6 consider infinite domains. For computational purposes, these infinite domains have to be truncated. It then becomes a question of the relative scales of dispersal distance and patch-size. As a rule of thumb, to study spreading phenomena, we choose the length of the domain to be large compared to the variance

© Springer Nature Switzerland AG 2019

F. Lutscher, *Integrodifference Equations in Spatial Ecology*, Interdisciplinary Applied Mathematics 49, https://doi.org/10.1007/978-3-030-29294-2_8

of the dispersal kernel and let the initial density be localized far away from the boundary (or at least one of the boundary points). To study persistence phenomena, we choose the length of the domain to be small compared to the variance of the dispersal kernel and let the initial condition be positive everywhere.

We discuss several aspects. We begin with the standard forward-iteration scheme based on the discrete Fourier transform (Powell 2009), which can be applied when the integral is of convolution type. Then we mention three other ways of computing the integral operation in an IDE. Next, we outline Nyström's method in connection with solving eigenvalue problems for IDEs (Zhou and Kot 2013). The code presented in this chapter uses MATLAB/OCTAVE, but the ideas and structure carry over easily to other programming languages. This chapter is not an introduction to programming in general or to MATLAB in particular but rather assumes that the reader is familiar with the basics of MATLAB or any other similarly capable programming language. The code discussed here is available for download from the repository https://integrodifference.frithjof.ca/.

8.2 Integration Via Fast Fourier Transform

Our first method of evaluating the integral and calculating the population density from one generation to the next is based on a discrete *fast Fourier transform* (FFT), a highly efficient numerical algorithm for evaluating convolution integrals (Brigham 2002). The application of the FFT and its inverse to IDEs and an implementation in MATLAB are beautifully described by Powell (2009). We follow his procedure here. The FFT method applies to convolutions, i.e., when the dispersal kernel depends only on the (signed) distance between the initial and the final location. Hence, we numerically solve the equation

$$N_{t+1}(x) = (K * F(N_t))(x) = \int K(x - y)F(N_t(y))\mathrm{d}y . \tag{8.1}$$

We have already seen the usefulness of the (scaled) Fourier transform (5.9) in Chap. 5. Here, we take the unscaled version

$$\widehat{N}(\omega) = \frac{1}{\sqrt{2\pi}} \int_{-\infty}^{\infty} N(x)\mathrm{e}^{\mathrm{i}\omega x}\mathrm{d}x , \tag{8.2}$$

which has the same useful property as (5.10) that it turns convolution into multiplication, i.e.,

$$\widehat{K * N}(\omega) = \frac{1}{2\pi} \widehat{K}(\omega) \cdot \widehat{N}(\omega) . \tag{8.3}$$

The inverse Fourier transform,

$$N(x) = \frac{1}{\sqrt{2\pi}} \int_{-\infty}^{\infty} \widehat{N}(\omega)e^{-i\omega x} d\omega, \tag{8.4}$$

which we could find explicitly only in a few special cases related to linear IDEs in Chap. 5, can then be calculated numerically.

The discrete Fourier transform and its inverse are defined for functions on some interval with a periodicity assumption that identifies the two endpoints of the interval. We divide the interval into 2^m subintervals of length Δx and get the discrete Fourier transform as

$$\widehat{N}(\omega) = \sum_{j=-2^{m-1}}^{j=2^{m-1}} N(j\Delta x)e^{2\pi i\omega j \Delta x}. \tag{8.5}$$

Choosing the number of subintervals to be a power of two allows us to use some efficient algorithms.

The discrete Fourier transform and its inverse are implemented in MATLAB with the commands fft and ifft, respectively (Powell 2009). The code in Table 8.1 defines the vector of space points (x), the Gaussian dispersal kernel (GAUSS), and the initial density (N0) before it calculates the density after one dispersal event (N1) and displays the two in the same plot. We point out a few details that need to be understood to get the commands to work correctly in our setting. First, to take advantage of integer arithmetic, the fft routine multiplies the values by the number of subintervals. Hence, we have to divide by that number in the end. Second, the fft routine works on the interval [0, 1] so that we have to scale by the domain length that we are interested in. The factor dx takes care of these two scalings. Finally, since the interval in our setting is naturally centered at $x = 0$, we have to shift the result using the command fftshift. We take the real part to avoid error messages concerning real-valued plots of complex-valued functions.

In Fig. 8.1, we observe that individuals that disperse out of the domain at the left end reenter the domain at the right end (dashed curve). This phenomenon arises because fft assumes periodicity. It is undesirable and even incorrect for our application. For example, in the critical patch-size problem, individuals that leave the domain should not return. A relatively simple remedy for this (usually) undesired effect is to enlarge the computational domain to twice the actual domain and to set the population density to zero outside the actual domain at every step (Andersen 1991). The code in Table 8.2 implements this remedy. It defines a vector (xx) of space points on $[-L, L]$ that is twice as long as needed and a vector (PAD) that sets the density to zero outside the domain of interest $[-L/2, L/2]$. This code also implements scaled Beverton–Holt population dynamics and tracks the location of the front (if one exists) as the most rightward point where the density exceeds the value 0.1 by the line

```
Front(k) = dx*( max( find( (N>0.1) ) ) )-2*L;
```

Table 8.1 MATLAB code using FFT to evaluate the convolution integral: a single dispersal event.

```
% Fast Fourier transform of the convolution integral
clear

% parameters
L = 10;             % the domain length
sig2 = 0.5;         % variance of the Gaussian kernel
a=0;     % center of the initial condition

% set-up of the spatial grid
np = 2^12;          % number of grid points; a power of 2
dx = L/np;          % length of the subintervals
x = linspace(-L/2,L/2-dx,np); % vector of grid points
% the point x=L/2 is the same as x=-L/2

% definition of the kernel and the initial condition
GAUSS = 1/sqrt(2*pi*sig2)*exp(-x.^2/(2*sig2));
N0 = exp(-(x-a).^2/0.1);

% Taking FFT
FGAUSS = fft(GAUSS);  FN0=(fft(N0));
% Multiplying and taking the inverse transform
N1 = dx*real( fftshift( ifft( FN0.*FGAUSS ) ) );

% displaying the result of one dispersal event.
plot(x,N0,'k','linewidth',2), hold on
plot(x,N1,'k','linewidth',2), hold on
xlabel('space x','FontSize',16)
ylabel('density','FontSize',16)
```

Fig. 8.1 Output of the code in Table 8.1 with initial condition centered at 0 (solid) and at −4 (dashed). The higher, narrower peaks are the initial condition; the lower, broader peaks are the density after one dispersal event. The FFT assumes periodicity, so that individuals that disperse out of the domain to the left enter the domain at the right.

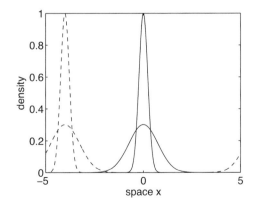

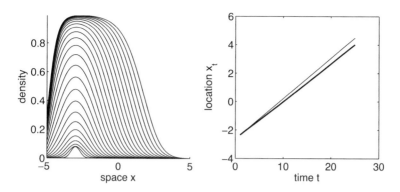

Fig. 8.2 Output of the code in Table 8.2. **Left plot:** The density evolves according to Beverton–Holt growth and Gaussian dispersal for 25 generations. **Right plot:** The actual location x_t where the population exceeds the threshold (thick line) remains slightly behind the theoretical prediction of the asymptotic spreading speed (thin line) but is almost parallel; i.e., the numerical simulation spreads at almost the asymptotic spreading speed.

This line should be removed when applying the code to study steady-state problems as in Chap. 4.

Figure 8.2 shows the output of the code in Table 8.2. The plot on the left illustrates how the density of the population grows and spreads. The plot on the right shows the numerically calculated location of the front and compares it with the theoretical asymptotic speed of a front starting from the same location. We see that the actual front moves slower than the asymptotic speed by about 5%. However, the asymptotic spreading speed is—as indicated by its name—an asymptotic quantity. When we simulate the dynamics for 70 generations on a suitably larger domain, the error drops below 3%. Unfortunately, running the code for 80 generations or more brings out instabilities of the numerical procedure that are inherent in the FFT approach. To avoid these instabilities, we can artificially set the density to zero when it is below some small threshold. For example, replacing the line that applies the padding with the command below allows us to run the simulations for longer and to obtain even better approximations for the spreading speed:

```
N = N.*PAD.*(N>0.000000001);
```

The FFT approach can also be applied to simulate IDEs in two spatial dimensions. The corresponding commands for the discrete Fourier transform in MATLAB are `fft2` and `ifft2`, respectively. The scaling has to occur in both spatial directions. Hence, if N and GAUSS are the two-dimensional density and dispersal kernel on a grid with cells of size `dx*dy`, then their convolution is calculated as

```
N = dx*dy*real( fftshift( ifft2( fft2(N).*fft2(GAUSS) ) ) );
```

Table 8.2 MATLAB code using FFT to evaluate the convolution integral: several dispersal and growth events.

```
% Simulating the IDE with scaled Beverton-Holt function
% Integral via Fast Fourier transform, padding

clear
L = 10;   R0 = 1.5; sig2 = 0.1;   Tsteps = 25;
np = 2^14;   dx = L/np;   x = linspace(-L/2,L/2-dx,np);
xx = linspace(-L, L-dx, 2*np);    % double domain for padding
PAD = (abs(xx)<=L/2); % padding outside [-L/2, L/2]
GAUSS = 1/sqrt(2*pi*sig2)*exp(-xx.^2/(2*sig2));
Ninit = exp(-(xx+3).^2/0.001);

% Iterating for Tsteps generations, using FFT
FGAUSS = fft(GAUSS);     % the Fourier transform
N = Ninit;
for t=1:Tsteps
    N = R0*N./(1+(R0-1)*N);     FN = fft(N);
    N = dx*real( fftshift( ifft( FN.*FGAUSS ) ) );
    N = N.*PAD;
    figure(1),   subplot(1,2,1), hold on
    plot(xx,N,'k'), axis([-L/2,L/2, 0, max(N)]), hold off
    xlabel('space x','FontSize',16')
    ylabel('density','FontSize',16')
    Front(t) = dx*(max(find((N>0.01))))-L;
end

figure(1), subplot(1,2,2)
plot(1:length(Front),Front,'k','linewidth',2), hold on
c=sqrt(2*sig2*log(R0));
plot(1:length(Front),Front(1)+c*(0:Tsteps-1),'k'), hold off
xlabel('time t','FontSize',16)
ylabel('location x_t','FontSize',16)
Theoretical_speed = c
Numerical_speed = (Front(Tsteps)-Front(Tsteps-10))/10
```

Figure 8.3 illustrates dispersal in two dimensions and growth. The corresponding code can be found in the repository mentioned above.

The accuracy of the discrete Fourier transform method increases as the number of grid points increases. The number 2^{14} used in the examples is usually a good compromise between speed and accuracy. The plots in this book are typically done with power 2^{16} or 2^{17}. For one-dimensional domains, this number of grid points still poses no problem for MATLAB on a standard PC or laptop computer. For two-dimensional domains, the limits are tighter. A spatial resolution of 2^8 grid points in each direction gives a matrix of 2^{16} entries, which is handled very easily. A resolution of 2^{10} or higher leads to a significant slowing of the program.

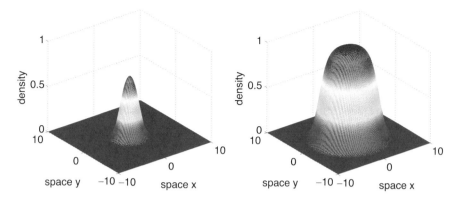

Fig. 8.3 Illustration of population growth and spread in a two-dimensional domain after 5 generations (left plot) and 25 generations (right plot). Dispersal follows the Gaussian kernel in (2.26) with $\sigma^2 = 0.1$, and growth is modeled by the scaled Beverton–Holt function with $R = 1.5$. The initial condition is one on $[-1, 1] \times [-0.1, 0.1]$ and zero elsewhere.

8.3 Alternative Methods of Numerical Iteration

When the integral in the IDE is not of convolution type, the Fourier transform method from the previous section cannot be used. Even if the integral is a convolution, certain applications prohibit the use of a large number of nodes in the spatial grid discretization. For example, Cobbold et al. (2005) perform a numerical bifurcation analysis of a host–parasitoid IDE model (see Sect. 14.3) with the help of AUTO (Doedel 1981). For that purpose, the number of grid points has to be much smaller than what the Fourier transform requires. They use direct integration via Simpson's rule and obtain very good accuracy with only 33 grid points. We briefly describe a few alternatives here. The repository contains the codes corresponding to those in Tables 8.1 and 8.2 with the alternatives implemented.

The MATLAB *Command* `conv`

MATLAB has the built-in command `conv` to carry out a discrete convolution that can be used instead of `fft` above. When multiplying two polynomials (given as sequences of coefficients with respect to the standard basis x^k), the sequence of coefficients of the product is the discrete convolution of the sequences of the factors. The same discrete convolution arises from discretizing the continuous convolution integral with an equidistant grid. However, since the degree of the product of two polynomials is the sum of the degrees of the factors, applying the command `conv` to two vectors leads to a vector whose length is the sum of the lengths of the two vectors minus one. To apply this command to simulate the IDE, we need to pick out the correct portion of the resulting vector that represents the population density on our domain of interest.

More specifically, if $N0$ and $GAUSS$ are vectors of length np as above, representing the density and kernel on the domain $[-L/2, L/2]$, then the command

```
N1long = dx * conv(N0,GAUSS);
```

produces a vector of length $2*np-1$. This vector represents the population density on the domain $[-L, L]$. The middle part, which represents the new population density on $[-L/2, L/2]$, is given by

```
N1 = N1long(np/2+1:np/2+np);
```

Selecting the middle part is the equivalent operation to the "padding" that had to be done in the FFT method (see above). The approximation for the spreading speed in the simple example for fft above appears to be of the same accuracy as for $conv$, but the program tends to be somewhat slower.

Nyström's Method

The idea of Nyström's method is to turn the integral operator in the IDE into a matrix multiplication. To that end, we discretize space with an equidistant grid and use the trapezoidal rule to approximate the integral. More specifically, we approximate the integral

$$\int_{-L/2}^{L/2} K(x, y)N(y)\mathrm{d}y \qquad (8.6)$$

with the matrix multiplication

$$\sum_j K_{ij} N_j \qquad (8.7)$$

for some appropriately chosen vector (N_j) and matrix (K_{ij}). We describe the method in slightly more detail when we discuss eigenvalue problems in Sect. 8.4. Here we give some sample code that generates the matrix entries G_{ij} for the Gaussian kernel:

```
for j=1:np
  G(1,j)=exp(-(x(1)-x(j))^2/(2*sig2))/sqrt(2*pi*sig2)/2;
  G(np,j)=exp(-(x(np)-x(j))^2/(2*sig2))/sqrt(2*pi*sig2)/2;
end

for i=2:np-1
  for j=1:np
    G(i,j)=exp(-(x(i)-x(j))^2/(2*sig2))/sqrt(2*pi*sig2);
  end
end
```

Integration is then approximated by the simple matrix multiplication

```
N1 = dx*G*N0;
```

The great advantage of this method over the previous two is that it can be applied to dispersal processes that are not given by a convolution. The program requires a lot fewer grid points than the Fourier transform to obtain similar accuracy, but it tends to be slower than the FFT approach. There do not seem to be any stability issues.

Simpson's Rule

When the kernel is a convolution kernel, the matrix generated in Nyström's method seems to store too much redundant information. The np^2 entries actually contain only $2*np$ different values. One can store the information required for the dispersal kernel in a vector of length $2*np$ and apply other quadrature rules as well. For example, we generate a vector GAUSSvec that evaluates the Gaussian kernel on an equidistant grid representing the interval $[-L, L]$ and a vector SIM with the weights of the formula for Simpson's rule, i.e.,

```
SIM = [1 4 2 4 2 ... 4 1];
```

Then we approximate the integration step by

```
for i = 1:np
    Nnew(i) = sum(SIM.*GAUSSvec(np-i+2:2*np-i+1).*N)*dx/3;
end
```

Since Simpson's rule has a higher accuracy than the trapezoidal rule, fewer space points are required to obtain comparable results to Nyström's method above. Implementing even higher accuracy quadrature rules to approximate the integral is straightforward from here.

8.4   Eigenvalues for Integral Operators

The previous two sections dealt with the time-dependent problem. Here, we consider the eigenvalue problem that appears in stability analysis, e.g., the critical patch-size problem in Chap. 3. We present one approach used by Zhou and Kot (2013). The goal is to find the dominant eigenvalue of the Fredholm integral equation

$$\lambda\phi = \int_{-L/2}^{L/2} \overline{K}(x, y)\phi(y)dy, \tag{8.8}$$

where the integral kernel $\overline{K}$ is the product of the dispersal kernel and the linearized growth rate $\overline{K}(x, y) = K(x, y)R(y)$.

We turn this problem into a finite-dimensional matrix eigenvalue problem by discretizing space (Nyström's method). We divide the interval into n subintervals of length $\Delta x = L/n$ with endpoints

$$y_j = -\frac{L}{2} + j\frac{L}{n}, \qquad j = 0, \ldots, n. \tag{8.9}$$

Then the trapezoidal rule approximates the integral in (8.8) as

$$\frac{\Delta x}{2} \sum_{j=0}^{n-1} \left[\overline{K}(x, y_j)\phi(y_j) + \overline{K}(x, y_{j+1})\phi(y_{j+1}) \right]. \tag{8.10}$$

We set $x_j = y_j$ and define $\phi_j = \phi(x_j)$. Then we obtain a discrete approximation to the eigenvalue problem as

$$\lambda\phi_i = \sum_{j=0}^{n} A_{ij}\phi_j, \tag{8.11}$$

with coefficients

$$A_{i0} = \frac{\Delta x}{2}\overline{K}(x_i, y_0),$$

$$A_{ij} = \Delta x \overline{K}(x_i, y_j), \qquad 1 \le j \le n-1,$$

$$A_{in} = \frac{\Delta x}{2}\overline{K}(x_i, y_n).$$

Hence, we have a finite-dimensional linear system that can be solved by standard methods implemented in many software packages.

Zhou and Kot (2013) point out two alternative ways to use this method. Suppose the goal is to find the persistence boundary with respect to some parameter, e.g., the domain length. Since the matrix entries A_{ij} depend continuously on model parameters, and since the eigenvalue depends continuously on the matrix entries, we can vary the parameter in question and include a root-finding algorithm to reach the threshold $\lambda = 1$. Alternatively, we can start by setting $\lambda = 1$ and use a root-finding algorithm to find conditions on the parameter in question that make the determinant of the system in (8.11) equal to zero. The latter approach requires some checking that the solution found corresponds indeed to the dominant eigenvalue and not to any other.

8.5 Further Reading

Powell (2009) gives a great introduction not only to the FFT method for simulating IDEs but also to some basic theory of IDEs, to some elementary MATLAB commands, and to further modeling ideas. He defines and illustrates a number of different kernels that include behavior other than simple random Gaussian dispersal. In addition, he introduces spatial heterogeneity as well as boundary behavior.

Haefner and Dugaw (2000) illustrate the computational advantage of using FFT for convolution integrals with a discrete-time model of quantitative genetics where the integration is over trait space rather than physical space.

When a two-dimensional domain is very large compared to the variance of the dispersal kernel, FFT may be computationally less efficient than discretization and direct computation of the convolution (Slone 2011). Different scales of spatial discretization have different effects on the accuracy of results (Bocedi et al. 2012), depending on the computational method used (Slone 2011). The authors caution against using coarse-grain grids and kernels since repeated convolution will amplify discretization errors, in particular when computing large-time dynamics. Slone (2011) develops correction equations.

Gilbert et al. (2017) develop a two-dimensional adaptive mesh solver for an IDE that models the spread of stage-structured populations on the scale of a country (Great Britain); this solver includes spatial heterogeneity. Their key idea is to use a fine mesh only around the invasion front and a coarse mesh far ahead and far behind the front.

Various numerical and approximation techniques for the eigenvalue problem are discussed by Zhou and Kot (2013) and Kot and Phillips (2015), in particular in the case of asymmetric kernels as they arise in the context of shifting-habitat models; see Sect. 12.3. For an in-depth treatment of numerical aspects of linear integral equations, see Kythe and Puri (2011).

Part II
Approximations and Applications

Chapter 9
Dispersal Success

Abstract In this chapter, we present techniques for approximating the steady-state profile and the dominant eigenvalue of an IDE on a bounded domain. The approximations are based on the idea that only partial information about dispersal may be available, corresponding to two different mark-recapture experiments. The approximations are surprisingly good when dispersal is symmetric but less reliable when dispersal is asymmetric.

9.1 Experiments and Dispersal Characteristics

A dispersal kernel summarizes the outcome of dispersal events as probabilities of going from any location to any other location. To find it empirically, we have to partition the landscape into small grid cells, release marked individuals from each cell, and record where they settle. As such, an enormous amount of experimental data is required to find $K(x, y)$. Even measuring dispersal distances under the assumption that the landscape is homogeneous can be extremely tedious. We can ask whether and how many useful insights can be obtained from an IDE when only certain characteristics of the dispersal kernel are known. The less information necessary to draw certain conclusions, the easier it is to obtain the required information.

We draw our inspiration from two kinds of experiments: the *point-release experiment* and the *area-release experiment* (described below). Each gives us only partial information about the dispersal kernel. Based on this partial information, we formally derive various approximations for the profile and stability of steady states of an IDE on bounded domains. Numerical experiments will reveal how accurate (or not) these approximations are.

In a *point-release experiment*, individuals are released from a single location and later recaptured (or not) within a certain (bounded) area of interest, Ω. If an individual is released many times from a location y or, equivalently, if many individuals are released simultaneously and move independently, the frequency of

© Springer Nature Switzerland AG 2019

F. Lutscher, *Integrodifference Equations in Spatial Ecology*, Interdisciplinary Applied Mathematics 49, https://doi.org/10.1007/978-3-030-29294-2_9

their recapture within Ω gives the *dispersal success function* (Van Kirk and Lewis 1997)

$$S(y) = \int_\Omega K(x, y) dx . \tag{9.1}$$

This function measures the probability that an individual disperser stays within the area of interest, Ω, in one dispersal phase. However, the exact settling location is unknown. Since dispersal success is a probability, we have

$$0 \le S(y) \le 1, \qquad y \in \Omega . \tag{9.2}$$

Typically, we expect an individual at the center of Ω to have a higher chance of staying within Ω than an individual near the boundary. The spatial average of the dispersal success function is the *average dispersal success* (Van Kirk and Lewis 1997)

$$\bar{S} = \frac{1}{|\Omega|} \int_\Omega S(y) dy = \frac{1}{|\Omega|} \int_\Omega \int_\Omega K(x, y) dx dy , \tag{9.3}$$

where $|\Omega| = \int_\Omega dx$ is the "volume" of Ω. We will frequently use the overbar to denote the spatial average of a function. The average dispersal success is a single number that integrates information about the movement process and the size and shape of the patch, as we will see in illustrations below.

The average dispersal success provides a tool to explore the questions raised at the end of Sect. 2.4: given a dispersal kernel and an area, what is the (spatially averaged) probability that an individual that starts the dispersal process in the area remains in the area after dispersal? In the next section, we shall see that the average dispersal success relates the spatially explicit IDE to some approximate, spatially implicit difference equation of the form (2.30).

In some areas of ecology, the term "dispersal success" refers to a disperser successfully reaching and settling in a patch *other* than its natal patch; see the introduction in Fagan and Lutscher (2006). The two notions should not be confused. One can extend the concept of average dispersal success to multi-patch systems (Lutscher and Lewis 2004) and thereby reconcile it with the more common ecological use of the term.

In an *area-release experiment*, individuals are released uniformly inside the bounded area of interest, Ω. Their density at the end of one dispersal period is the *redistribution function* (Lutscher and Lewis 2004)

$$U(x) = \int_\Omega K(x, y) dy . \tag{9.4}$$

Here, we know the final but not the initial location of an individual. This function is *not* a probability. For example, if individuals move preferentially to one particular

Fig. 9.1 Dispersal success function for the Laplace kernel with effective domain length $La = \hat{L} = 1$ (solid), $\hat{L} = 5$ (dashed), and $\hat{L} = 10$ (dash-dot). The corresponding values of the average dispersal success are $\bar{S} = 0.368, 0.801$, and 0.9, respectively. Figure 9.2 shows how $\bar{S}$ depends on the scaled domain length.

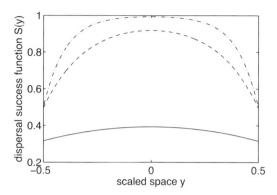

location within Ω, then the redistribution function will have a peak at that location, and the total density at that peak can exceed unity. Nonetheless, the spatial average of the redistribution function is also the average dispersal success, i.e., $\bar{U} = \bar{S}$. If dispersal is symmetric, i.e., $K(x, y) = K(y, x)$, then $S = U$.

The dispersal success function of the Laplace kernel (2.27) with mean dispersal distance $1/a$ on a bounded interval of length L can be computed explicitly as (Van Kirk and Lewis 1997)

$$S(y) = U(y) = 1 - e^{-\hat{L}/2} \cosh(ay) , \qquad (9.5)$$

and the average dispersal success is

$$\bar{S} = 1 - \frac{1 - e^{-\hat{L}}}{\hat{L}} , \qquad (9.6)$$

where $\hat{L} = La$ is the effective domain length; see Fig. 9.1.

A comparison of four different dispersal kernels shows how the average dispersal success integrates movement behavior and patch-size; see Fig. 9.2. The average dispersal success is an increasing function of domain length. It also increases as the probability of dispersing short distances increases. Accordingly, the average dispersal success has been used as a measure for persistence of a population (Lockwood et al. 2002). Under symmetric dispersal, long-distance dispersal events do not significantly affect the average dispersal success since it is computed only over a bounded domain. We discuss the case of biased dispersal in Chap. 12.2.

In the following two sections, we use these measures for dispersal success to approximate the spatial distribution at steady state as well as the persistence conditions. In particular, we give a more theoretic underpinning for the use of the average dispersal success as a persistence measure.

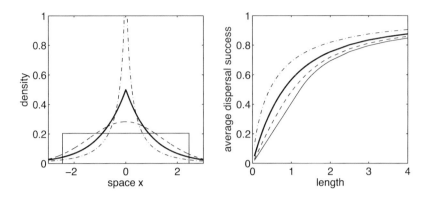

Fig. 9.2 Comparison of the average dispersal success (right plot) of four dispersal kernels (left plot). The exponential square root kernel (dash-dot) is the most concentrated at zero and has the highest average dispersal success; the top-hat kernel (solid, thin) is the least concentrated and has the lowest average dispersal success. The Gaussian kernel (dashed) and the Laplace kernel (solid, thick) are in between.

9.2 Dispersal Success Approximation of a Steady State

The dispersal characteristics $S(y)$, $U(x)$, and $\bar{S}$ are useful for approximating steady states and eigenvalues of IDEs since they emerge naturally from a Taylor series expansion. We present the dispersal success approximation of a positive steady state, originally given by Van Kirk and Lewis (1997) and extended by Lutscher and Lewis (2004).

We denote a positive steady state of (2.1) on the bounded domain Ω by N^* and its spatial average by $\bar{N}^*$, i.e.,

$$N^*(x) = \int_\Omega K(x, y) F(N^*(y)) \mathrm{d}y, \qquad \bar{N}^* = \frac{1}{|\Omega|} \int_\Omega N^*(x) \mathrm{d}x. \tag{9.7}$$

Assuming that N^* and F are sufficiently differentiable, we write the first-order Taylor polynomial for each $y \in \Omega$ as

$$F(N^*(y)) = F(\bar{N}^*) + F'(\hat{N}(y))(N^*(y) - \bar{N}^*) \tag{9.8}$$

for some $\hat{N}(y)$ between $\bar{N}^*$ and $N^*(y)$. Substituting this expression and (9.4) into (9.7), we obtain

$$N^*(x) = U(x) F(\bar{N}^*) + \int_\Omega K(x, y) F'(\hat{N}(y))(N^*(y) - \bar{N}^*) \mathrm{d}y. \tag{9.9}$$

To obtain an equation for $\bar{N}^*$, we average both sides and find

Fig. 9.3 Comparison of the dispersal success approximation (dashed) with the true steady state (solid). We used the Beverton–Holt updating function with $R = 3$ and symmetric dispersal according to the Laplace kernel with parameters $\hat{L} = La = 1, 2, 5$, and 10 (from bottom up). The corresponding values of the average dispersal success are $\bar{S} = 0.3679, 0.5677, 0.8013$, and 0.900. The approximation is more accurate for smaller values of $\hat{L}$.

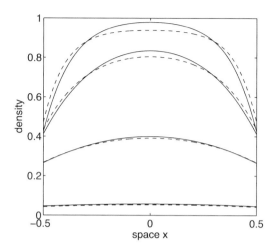

$$\bar{N}^* = \bar{S}F(\bar{N}^*) + \frac{1}{|\Omega|}\int_{\Omega} S(y)F'(\hat{N}(y))(N^*(y) - \bar{N}^*)dy\,. \qquad (9.10)$$

For the lowest-order approximation, we neglect the integral term and obtain equations for the approximation, $N_{\mathrm{ap}}(x)$, of the steady state and its spatial average as

$$N_{\mathrm{ap}}(x) = F(\bar{N}_{\mathrm{ap}})U(x) \quad \text{and} \quad \bar{N}_{\mathrm{ap}} = \bar{S}F(\bar{N}_{\mathrm{ap}})\,. \qquad (9.11)$$

For a symmetric kernel, the approximation becomes $N_{\mathrm{ap}}(x) = F(\bar{N}_{\mathrm{ap}})S(x)$, which is why it is called the *dispersal success approximation*.

We note that the first equation in (9.11) is exactly the steady-state equation for the very simple spatially implicit dispersal model (2.30) from Chap. 2.

Typically, we can only neglect the integral term in (9.11) when the factor $N^*(y) - \bar{N}^*$ is small. We expect that this difference could grow large at the boundary of the domain where $N^*(y)$ becomes small. However, near the boundary, the factor $S(y)$ is small. As a result, the approximation tends to be quite accurate even if the steady state is not close to a constant; see Fig. 9.3.

9.3 Dispersal Success Approximation of the Eigenvalue

The ideas from the previous section can also be applied to approximate the dominant eigenvalue and corresponding left and right eigenfunctions using the dispersal characteristics (Lutscher and Lewis 2004). We begin with the linear eigenvalue problem (setting $R = 1$) and scaled eigenfunction

$$\lambda\phi(x) = \int_{\Omega} K(x, y)\phi(y)dy\,, \qquad \bar{\phi} = \frac{1}{|\Omega|}\int_{\Omega}\phi(x)dx = 1\,. \qquad (9.12)$$

The same steps as for the steady-state approximation above lead to the approximations

$$\lambda_{ap} = \bar{S}, \qquad \phi_{ap}(x) = \frac{1}{S}U(x) \tag{9.13}$$

for the eigenvalue and the right eigenfunction. The analogous expression for the left eigenfunction (the eigenfunction of the adjoint operator), defined by

$$\lambda\psi(y) = \int_\Omega K(x, y)\psi(x)dx, \tag{9.14}$$

is $\psi_{ap}(y) = \frac{1}{S}S(y)$.

When the dispersal kernel is symmetric, the approximation systematically underestimates the true eigenvalue, as the following lemma by Fagan and Lutscher (2006) shows. Ecologically, the lemma says that the approximation is conservative in the sense that if a population can persist according to the approximation (i.e., $\lambda_{ap} > 1$), then it can persist in the full model (i.e., $\lambda > 1$).

Lemma 9.1 *Assume that K is symmetric. Then $\lambda_{ap} \leq \lambda$.*

Proof We choose the constant function $u(x) = \frac{1}{\sqrt{|\Omega|}}$ and apply the variational formula (3.43). By definition, u has norm one in the space of square-integrable functions on Ω. Then

$$\lambda \geq \int_\Omega \int_\Omega u(x)K(x, y)u(y)dydx = \frac{1}{|\Omega|} \int_\Omega \int_\Omega K(x, y)dydx = \bar{S} = \lambda_{ap}. \tag{9.15}$$

$\square$

The average dispersal success approximates the dominant eigenvalue, but we shall see later that this approximation is helpful only if the kernel is symmetric. However, the approximation $\lambda_{ap} = \bar{S}$ does not use the dispersal success and the redistribution functions. We present three ideas to improve the approximation by using some of this information.

Iterated Eigenvalue Approximation

Instead of the eigenvalue equation, we consider the iterated equation for the square of the eigenvalue (corresponding to two dispersal phases), given by

$$\lambda^2\phi(x) = \int_\Omega \int_\Omega K(x, y)K(y, z)\phi(z)dzdy. \tag{9.16}$$

Applying the same steps as above, we find the approximation

$$\lambda_{ap}^{(1)} = \sqrt{\frac{1}{|\Omega|} \int_\Omega S(y)U(y)dy} \,. \tag{9.17}$$

This expression can be written in a more insightful way in terms of the covariance between functions S and U, namely

$$\text{cov}(S, U) = \frac{1}{|\Omega|} \int_\Omega (S(y) - \bar{S})(U(y) - \bar{S})dy = \frac{1}{|\Omega|} \int_\Omega S(y)U(y)dy - \bar{S}^2. \tag{9.18}$$

Then the expression for $\lambda_{ap}^{(1)}$ becomes

$$\lambda_{ap}^{(1)} = \bar{S}\sqrt{1 + \frac{1}{\bar{S}^2}\text{cov}(S, U)} \,. \tag{9.19}$$

When the dispersal kernel is symmetric, the redistribution function, U, equals the dispersal success function, S, so that the covariance becomes the variance, which is positive. The same proof as in Lemma 9.1, applied to the iterated equation, shows that we have the same upper bound, i.e., $\lambda_{ap}^{(1)} \leq \lambda$, if dispersal is symmetric. In particular, for a symmetric kernel, we have $\lambda_{ap} \leq \lambda_{ap}^{(1)} \leq \lambda$, so that $\lambda_{ap}^{(1)}$ provides a better approximation.

The Power Method

The power method is an iterative scheme to approximate the dominant eigenvalue of a positive matrix; see, e.g., Zhou and Kot (2013). Given a positive matrix A, we choose a nonnegative vector ϕ_0 with $\|\phi_0\| = 1$. We define an iteration whose vectors are all of norm one as $\phi_{n+1} = A\phi_n/\|A\phi_n\|$. Then the scaling factors $\|A\phi_n\|$ converge to the dominant eigenvalue and ϕ_n to the corresponding eigenvector.

We apply the same scheme to the integral operator in (9.12). The constant function $\phi_0 = 1/|\Omega|$ on Ω has L^1-norm equal to unity. Taking this function as our initial function and applying the linear dispersal operator, we find

$$\phi_1 = \frac{\frac{1}{|\Omega|}\int_\Omega K(x, y)dy}{\int_\Omega \frac{1}{|\Omega|}\int_\Omega K(x, y)dydx} = \frac{\frac{1}{|\Omega|}U(x)}{\bar{S}} \,, \tag{9.20}$$

so that the scaling factor with respect to the L^1-norm is simply $\bar{S}$ and the "approximation" to the eigenfunction is $U(x)$. Hence, the dispersal success procedure from above is equivalent to the first step of the power method.

We cannot calculate the second iterate without using the dispersal kernel. But we can calculate the scaling factor in the second step using only the functions S and U. This scaling factor is given by

$$\int_\Omega \int_\Omega K(x,y) \frac{1}{\bar{S}|\Omega|} U(y) \mathrm{d}y \mathrm{d}x = \frac{1}{\bar{S}} \frac{1}{|\Omega|} \int S(y) U(y) \mathrm{d}y. \tag{9.21}$$

Hence, using (9.18), we get another approximation for λ as

$$\lambda_{\mathrm{ap}}^{(2)} = \bar{S} \left(1 + \frac{1}{\bar{S}^2} \mathrm{cov}(S,U) \right). \tag{9.22}$$

Since $\lambda_{\mathrm{ap}}^{(1)}$ is the geometric mean of λ_{ap} and $\lambda_{\mathrm{ap}}^{(2)}$, we have the following lemma.

Lemma 9.2 *If* $\mathrm{cov}(S,U) > 0$, *then* $\lambda_{\mathrm{ap}} \leq \lambda_{\mathrm{ap}}^{(1)} \leq \lambda_{\mathrm{ap}}^{(2)}$. *If* $\mathrm{cov}(S,U) < 0$, *then the inequalities are reversed.*

Weighted Averages

A completely different but ultimately equivalent approach to improving the dispersal success approximation of $\lambda \approx \bar{S}$ is presented by Reimer et al. (2016). They start with the biological interpretation of average dispersal success and reason that the spatial average in (9.3) could be weighted by the actual distribution of individuals in the domain instead of simply assuming a uniform distribution. The actual distribution, of course, is not known. Instead, they choose the (scaled) average dispersal success approximation of the steady state. In the symmetric case, they arrive at the expression

$$\frac{1}{|\Omega|} \int_\Omega \int_\Omega K(x,y) \frac{S(y)}{\bar{S}} \mathrm{d}x \mathrm{d}y = \frac{1}{\bar{S}} \frac{1}{|\Omega|} \int S^2(y) \mathrm{d}y, \tag{9.23}$$

which equals $\lambda_{\mathrm{ap}}^{(2)}$. In general, we replace the dispersal success function by the redistribution function; see (9.11). Reimer et al. (2016) also prove the estimate $\lambda_{\mathrm{ap}}^{(2)} > \lambda_{\mathrm{ap}}$ for the symmetric case, but the proof in the previous lemma via the expression in (9.22) is much simpler than theirs.

A natural question is then whether there are other weight functions for which the weighted average of the dispersal success function provides an even better approximation of the dominant eigenvalue. To simplify the notation, we use the inner product on the space of square-integrable functions, $L^2(\Omega)$, namely

$$\langle f, g \rangle = \int_\Omega f(x) g(x) \mathrm{d}x. \tag{9.24}$$

Then we can write the (weighted) average dispersal success as

$$\bar{S} = \frac{\langle S, 1 \rangle}{\langle 1, 1 \rangle} \quad \text{and} \quad \lambda_{\text{ap}}^{(2)} = \frac{\langle S, U \rangle}{\langle 1, U \rangle}, \tag{9.25}$$

where U is the redistribution function from (9.4).

Integrating the eigenvalue equation in (9.12) with respect to x, we find

$$\lambda = \frac{\langle S, \phi \rangle}{\langle 1, \phi \rangle}, \tag{9.26}$$

i.e., the true eigenvalue λ is the weighted average of the dispersal success function with the eigenfunction as weight. Hence, $\lambda_{\text{ap}}^{(2)}$ results by approximating the eigenfunction ϕ in (9.26) by U from (9.13).

The Laplace Kernel as an Example

To illustrate the difference between the various approximations and the true eigenvalue for the Laplace kernel, we calculate these quantities explicitly. From the explicit expression of the dispersal success function in (9.5), we find

$$\text{cov}(S, S) = \text{var}(S) = 1 - 2\frac{1 - e^{-\hat{L}}}{\hat{L}} + \frac{e^{-\hat{L}}}{2} + \frac{1 - e^{-2\hat{L}}}{2\hat{L}} - \bar{S}^2, \tag{9.27}$$

where, as before, $\hat{L} = aL$ is the effective domain length. In Fig. 9.4, we numerically compare the resulting quantities

$$\lambda_{\text{ap}} = \bar{S}, \quad \lambda_{\text{ap}}^{(1)} = \bar{S}\sqrt{1 + \text{var}(S)/\bar{S}^2}, \quad \lambda_{\text{ap}}^{(2)} = \bar{S}\left(1 + \text{var}(S)/\bar{S}^2\right). \tag{9.28}$$

Fig. 9.4 Comparison of the dominant eigenvalue with the three approximations for the (symmetric) Laplace kernel. The large plot shows the dominant eigenvalue λ (solid, thick), the average dispersal success approximation λ_{ap} (dash-dot), and the two approximations using the variance $\lambda_{\text{ap}}^{(1)}$ (solid, thin) and $\lambda_{\text{ap}}^{(2)}$ (dashed). The inset shows the relative errors, i.e., $(\lambda - \lambda_{\text{ap}})/\lambda$ and similarly for the others.

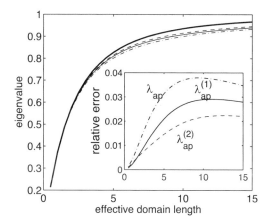

All three approximations underestimate the true value of the dominant eigenvalue. The relative error is highest for intermediate domain lengths and decreases for larger domains. Even the worst approximation is fairly accurate, lying within 5% of the true value, as the inset plot shows.

9.4 Application to Asymmetric Dispersal

When dispersal is asymmetric, the dispersal success function does not equal the redistribution function. The simple dispersal success approximation becomes less accurate. The improved approximations are somewhat better but also fail eventually when the asymmetry becomes too strong. We illustrate this behavior by using the asymmetric Laplace kernel (7.16) that arises from a biased random walk. We recall the definition

$$K(x) = \begin{cases} A \exp(a_1(x)), & x < 0, \\ A \exp(a_2(x)), & x > 0, \end{cases} \qquad A = \frac{a_1|a_2|}{a_1 + |a_2|}, \qquad (9.29)$$

where $a_1 > 0 > a_2$ are given in terms of diffusion rate D, advection speed q, and settling rate α as

$$a_i = \frac{1}{2D}\left(q \pm \sqrt{q^2 + 4\alpha D}\right), \qquad a_1 > |a_2|.$$

The corresponding eigenvalue problem is treated by Lutscher et al. (2005) in their Appendix F. As in Chap. 3, the integral equation can be turned into a (non-self-adjoint) Sturm–Liouville problem by repeated differentiation. A change of coordinates results in a self-adjoint problem for which we can find a transcendental equation that relates the parameters of the model to the eigenvalue λ as

$$\tan\left(\frac{(a_1 - a_2)L}{4}\sqrt{\frac{4Ra_1|a_2|}{\lambda(a_1 - a_2)^2} - 1}\right) = \sqrt{\frac{4Ra_1|a_2|}{\lambda(a_1 - a_2)^2} - 1}. \qquad (9.30)$$

Straightforward but tedious calculations give the dispersal success function and the redistribution function for this asymmetric kernel on the interval $[-L/2, L/2]$ as

$$S(y) = A\left[\frac{1}{a_1}\left(1 - e^{-\hat{L}_1/2}e^{-a_1 y}\right) + \frac{1}{a_2}\left(e^{-\hat{L}_2/2}e^{-a_2 y} - 1\right)\right] \qquad (9.31)$$

and

$$U(x) = A\left[\frac{1}{a_1}\left(1 - e^{-\hat{L}_1/2}e^{a_1 x}\right) + \frac{1}{a_2}\left(e^{\hat{L}_2/2}e^{a_2 x} - 1\right)\right], \qquad (9.32)$$

where $\hat{L}_i = a_i L$. The average dispersal success is given by

$$\bar{S} = 1 - \frac{a_1}{a_2 - a_1} \frac{1 - e^{\hat{L}_2}}{\hat{L}_2} - \frac{a_2}{a_2 - a_1} \frac{1 - e^{\hat{L}_1}}{\hat{L}_1} . \qquad (9.33)$$

Accordingly, we calculate the three approximations for the dominant eigenvalue and plot the relative errors in Fig. 9.5. The relative errors of $\lambda_{\mathrm{ap}}^{(1)}$ and $\lambda_{\mathrm{ap}}^{(2)}$ are smaller than that of λ_{ap}, but the errors of both increase as q increases and the kernel becomes more asymmetric. The assumption that the dominant eigenfunction is close to its spatial average is increasingly inaccurate, as we see in Fig. 9.6. The peak of the redistribution function moves in the direction of the bias as q increases, since an

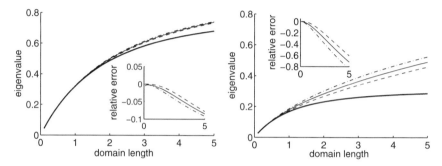

Fig. 9.5 Absolute and relative (inset) errors of the three approximations to the dominant eigenvalue for the asymmetric Laplace kernel (9.29). Line styles are as in Fig. 9.4. Parameters of the movement model for the dispersal kernel are $D = 1$, $\alpha = 1$, $q = 1$ (left), and $q = 3$ (right).

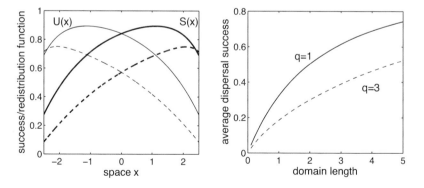

Fig. 9.6 Illustrating dispersal success and redistribution for the asymmetric Laplace kernel (9.29). **Left:** Dispersal success function from (9.31) (thick) and redistribution function from (9.32) (thin). **Right:** Average dispersal success from (9.33). Parameters are $L = 5$ (left plot), $q = 1$ (solid) and $q = 3$ (dashed). Other parameters are $D = 1$ and $\alpha = 1$. The corresponding coefficient values are $a_1 = 1.618$, $a_2 = -0.618$ (solid), and $a_1 = 3.3028$, $a_2 = -0.3028$ (dashed).

Fig. 9.7 Comparison of the dispersal success approximation (dashed) with the true steady state (solid). Dispersal according to the asymmetric Laplace kernel (9.29) and Beverton–Holt updating function with $R = 10$. Other parameters are as in Fig. 9.6, with $q = 1$ (top curves) and $q = 3$ (bottom curves).

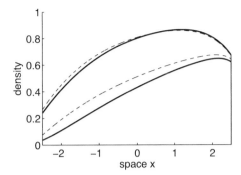

individual tends to settle in the direction of the bias from its original location. The peak of the dispersal success function moves in the opposite direction for the same reason.

Finally, we illustrate the dispersal success approximation for the steady state of an IDE with asymmetric dispersal. We choose the scaled Beverton–Holt growth function and the asymmetric Laplace kernel. Comparing the average dispersal success and the true steady-state profile shows that the approximation is reasonably good when q is small but becomes increasingly inaccurate when q is large, so that the asymmetry of the kernel is more pronounced; see Fig. 9.7.

9.5   Further Reading

Latore et al. (1998) evaluate the density function at a single point (midpoint) in the domain to estimate the critical patch-size directly rather than approximating the dominant eigenvalue.

The study of the dispersal success approximation began with the steady-state approximation for symmetric kernels by Van Kirk and Lewis (1997). The extension to asymmetric kernels and eigenvalues was initiated by Lutscher and Lewis (2004). They also use the average dispersal success to obtain approximations for bifurcations from the positive steady state, e.g., when the growth function is the Ricker function. They apply the technique to a two-stage model of juveniles and adults; see Chap. 13. Other applications to interacting species have been equally successful (Cobbold et al. 2005). The corresponding theory for reaction–diffusion equations was developed by Cobbold and Lutscher (2014).

The average dispersal success can also be defined for landscapes consisting of several spatially separated patches. For every *pair* of patches, Ω_i, Ω_j, say, we define the probability that an individual from patch j moves to patch i in one dispersal period, i.e.,

$$S_{ij} = \frac{1}{|\Omega_j|} \int_{\Omega_j} \int_{\Omega_i} K(x, y) \mathrm{d}x \mathrm{d}y . \qquad (9.34)$$

The result is a matrix of average dispersal success values that can be used to turn a spatially explicit IDE on several patches into a spatially implicit matrix equation; see Lutscher and Lewis (2004) for details.

A completely different approach of analytical approximations is taken by Kot and Phillips (2015) and applied to IDEs in more detail by Rinnan (2017). These approaches are based on classical and recent mathematical results in eigenvalue approximation for matrices, and in particular on "geometric symmetrization" (Kot and Phillips (2015) and references therein) and not on ecological ideas. The geometric symmetrization of the integral operator

$$\phi \mapsto \int_{-L/2}^{L/2} K(x, y)\phi(y)dy$$

is the operator

$$\phi \mapsto \int_{-L/2}^{L/2} \sqrt{K(x, y)K(y, x)}\phi(y)dy \,.$$

If K is symmetric, i.e., $K(x, y) = K(y, x)$, then the geometric symmetrization is identical to the original operator. The key insight is that the spectral radius of the geometric symmetrization is a lower bound for the spectral radius of the original operator, at least in the finite-dimensional matrix case (Kot and Phillips (2015) and references therein).

Rinnan (2017) then defines the "geometric success function" and its average as

$$G(y) = \int_{\Omega} \sqrt{K(x, y)K(y, x)}dx \,, \quad GS = \frac{1}{|\Omega|} \int_{\Omega} G(y)dy \,.$$

A priori, there is no reason to believe that GS is a good approximation for the dominant eigenvalue of the original IDE since the geometric symmetrization describes a process different from the original equation. However, Rinnan (2017) shows via numerical experiments that GS (and various other approximations derived from iterations of it) is a very good approximation for the dominant eigenvalue that we are looking for.

A heuristic argument for the success of the method rests on Taylor series expansion and constitutes a fitting conclusion to this chapter. We write

$$K(x, y) = \sqrt{K(x, y)K(x, y)} = \sqrt{K(x, y)K(y, x) + K(x, y)(K(x, y) - K(y, x))} \,.$$

Then the Taylor series expansion $\sqrt{x + \epsilon} \approx \sqrt{x} + \epsilon/(2\sqrt{x})$ gives

$$K(x, y) \approx \sqrt{K(x, y)K(y, x)} + \frac{1}{2}\sqrt{\frac{K(x, y)}{K(y, x)}}(K(x, y) - K(y, x)) \,.$$

Hence, geometric symmetrization can be seen as a low-order symmetric approximation to the kernel. It turns out that the symmetrization is also helpful for finding good approximations to the steady-state profiles (Rinnan 2017). Finding a rigorous connection between the original IDE and its symmetrization is still an open problem.

Chapter 10
Approximations for Spread

Abstract Continuing in the spirit of the previous chapter, we now aim to approximate characteristics of spread to reduce data requirements or simplify numerical
effort. First, we find the *kurtosis approximation* for the asymptotic spreading speed
and discuss its usefulness and limitations. Next, we use an exponential transform
method to solve the linear IDE and employ the *saddle-point approximation* to obtain
an approximate shape for the population distribution for large times. Both methods
are based on Taylor series expansion.

10.1 Approximating the Speed

One motivation for approximating the speed of spread of a population modeled by
an IDE stems from an observation in Fig. 5.4. The plot illustrates that the rates of
spread as a function of variance are similar for different kernels, given that their tails
are exponentially bounded. Furthermore, the (small) differences in speeds between
the different kernels seem to follow a predictable pattern. We begin by applying a
Taylor series expansion to the dispersion relation for the speed of a traveling wave.
The lowest-order term gives the speed as a function of the variance of the dispersal
kernel and is exact for the Gaussian kernel. Including the next term in the expansion
results in the *kurtosis approximation* for the spread speed.

All results in this chapter are based on the linear IDE

$$N_{t+1}(x) = R \int_{-\infty}^{\infty} K(x-y)N_t(y)\mathrm{d}x = R(K * N_t)(x). \tag{10.1}$$

With the localized initial condition $N_0(x) = \delta(x)$, its solution is

$$N_t(x) = R^t K^{*t}(x). \tag{10.2}$$

Under certain conditions, the spread rate for the linear equation equals the spread
rate for a corresponding nonlinear equation; see Chap. 5.

© Springer Nature Switzerland AG 2019

F. Lutscher, *Integrodifference Equations in Spatial Ecology*, Interdisciplinary
Applied Mathematics 49, https://doi.org/10.1007/978-3-030-29294-2_10

When the dispersal kernel, K, has zero mean and finite variance, σ^2, then the central limit theorem (CLT) states that its t-fold convolution, K^{*t}, approaches a Gaussian distribution with zero mean and variance $t\sigma^2$. Since the speed of spread is an asymptotic quantity (as $t \to \infty$), we could be tempted to think that the speed of spread with kernel K is the same as that for the Gaussian kernel with equal variance, namely

$$c_G = \sqrt{2\sigma^2 \ln(R)} \,, \tag{10.3}$$

as in (5.7). We call this value the Gaussian approximation. From an empirical point of view, if the variance of dispersal distances is the only information available, then c_G is the best and only guess for the speed of spread.

The plots in Fig. 5.4 show that the CLT cannot be applied to determine the asymptotic speed. In particular, if the dispersal kernel is the top-hat kernel (see Table 3.1), then the dispersal distance per generation is limited to the length of the support of the kernel. Therefore, the spread rate for a fixed variance has to be bounded independently of the growth rate. However, c_G in (10.3) increases without bound in R. On the other hand, the right plot in Fig. 5.4 indicates that the speeds as a function of σ^2 do not differ "too much" between the three kernels. In the following, we first explain why reasoning with the CLT does not lead to the correct result and then show how a better approximation can be obtained. The material presented here can be found in Lutscher (2007).

Rate of Convergence

It is, of course, true that the iterated convolutions for each of the kernels used in Fig. 5.4 converge to the Gaussian kernel, as illustrated in Fig. 10.1 (left plot); however, the convergence is slow in the tails of the kernels (right plot). It turns out that this convergence is simply too slow, compared to population growth, to yield the correct spreading speed.

To explain the rate of convergence, we compare the iterates $N_t = R^t K^{*t}$ for a kernel K with mean μ and variance σ^2 to those for the Gaussian kernel, $G = G(x; \mu, \sigma^2)$, with the same mean and variance:

$$\sup_x |R^t K^{*t}(x) - R^t G^{*t}(x; \mu, \sigma^2)| = R^t \sup_x |K^{*t}(x) - G(x; t\mu, t\sigma^2)|$$

$$= \frac{R^t}{\sigma\sqrt{t}} \sup_y |\sigma\sqrt{t}\, K^{*t}(\sigma\sqrt{t}\, y + t\mu) - G(y; 0, 1)| \tag{10.4}$$

$$\sim \frac{R^t}{\sigma\sqrt{t}} \frac{\text{const.}}{\sigma\sqrt{t}} = \frac{R^t}{\sigma^2 t} \text{const.} \tag{10.5}$$

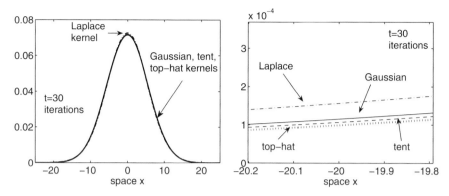

Fig. 10.1 Plot of K^{*30} for the Gaussian, Laplace, tent, and top-hat kernels to illustrate (slow) convergence of the CLT. **Left:** Only with the Laplace kernel and only near zero is there a noticeable difference at this scale. **Right:** At $x = -20$, the absolute differences are very small (note the scale on the y-axis) but the relative differences are clearly visible. Figure adapted from Lutscher (2007).

Equation (10.4) is obtained by standardizing the distributions via the change of variables $y = (x - t\mu)/(\sigma\sqrt{t})$ and by multiplying by $\sigma\sqrt{t}$. The rate of convergence of the normalized distribution is on the order of $1/(\sigma\sqrt{t})$ (see, e.g., Theorem 10, Chap. VII, in Petrov (1975)), which leads to (10.5).

The exponential in the numerator grows much faster in t than the linear term in the denominator. Therefore, this expression grows large with increasing t. Thus, as t increases, the approximation becomes worse or is not applicable at all.

To be precise, the above reasoning follows the *local limit theorem* and not the CLT, which deals with convergence of the cumulative density functions. To fully understand what happens in the tails of the distribution for iterated convolutions, we apply the theory of large deviations. This theory states that the mass of the t th convolution outside an interval $[-Ct, Ct]$ for any constant $C > 0$ will be exponentially small. Hence, the quantities that we are trying to compare are much smaller than their difference decays; see Lutscher (2007) for details.

The Kurtosis Approximation

The kurtosis approximation can be derived from formula (5.17) for the speed of a traveling exponential profile, i.e.,

$$c(s) = \frac{1}{s}\ln(RM(s)),\tag{10.6}$$

by repeated application of Taylor series expansion around $s = 0$. When the kernel is symmetric with zero mean, the moments of odd order vanish and the expansion of the moment-generating function up to order four in s is given by

Fig. 10.2 The exact
spreading speeds (solid) and
the kurtosis approximations
(dashed) for the kernels from
Table 10.1. The
approximation is highly
accurate for the top-hat
(bottom) and tent (second
from bottom) kernels. It
overestimates the speed for
the Laplace kernel (top), but
the relative error is much
smaller than that made by the
Gaussian approximation. The
growth rate is $R = 3$.

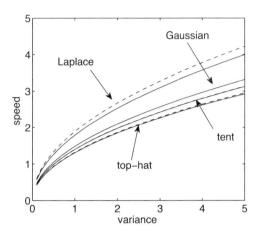

Table 10.1 Summary of dispersal kernels and their characteristic quantities.

Kernel	Gaussian	Laplace	Top-hat	Tent
Variance	σ^2	$\frac{2}{a^2}$	$\frac{\beta^2}{3}$	$\frac{\eta^2}{6}$
Excess kurtosis	0	3	$-6/5$	$-3/5$
$M(s)$	$\exp(\frac{\sigma^2 s^2}{2})$	$\frac{1}{1-s^2/a^2}$	$\frac{\sinh(s\beta)}{s\beta}$	$\frac{2(\cosh(\eta s)-1)}{\eta^2 s^2}$

The excess kurtosis is defined in (10.9); the moment-generating function is denoted by $M(s)$. When the excess kurtosis is positive (negative), we say that the kernel is leptokurtic (platykurtic)

$$M(s) = 1 + \frac{\sigma^2}{2}s^2 + \frac{\mu_4}{24}s^4, \quad \text{with} \quad \mu_4 = \int_{-\infty}^{\infty} x^4 K(x)\,dx. \tag{10.7}$$

The logarithm in (10.6) expanded to the same order gives

$$\ln(RM(s)) = \ln(R) + \frac{\sigma^2}{2}s^2 + \left(\frac{\mu_4}{24} - \frac{\sigma^4}{8}\right)s^4 = \ln(R) + \frac{\sigma^2}{2}s^2 + \epsilon\sigma^4 s^4, \tag{10.8}$$

where

$$\gamma_2 = \mu_4/\sigma^4 - 3 \tag{10.9}$$

is the *excess kurtosis* and $\epsilon = \gamma_2/24$. For all the kernels in Fig. 10.2, this number is small; see Table 10.1. We therefore choose ϵ as our small parameter and expand terms up to order one in ϵ.

The expression for the dispersion relation becomes

$$c(s) = \frac{\ln(R)}{s} + \frac{\sigma^2}{2}s + \epsilon\sigma^4 s^3. \tag{10.10}$$

To find its minimum, we set its derivative,

$$c'(s) = -\frac{\ln(R)}{s^2} + \frac{\sigma^2}{2} + 3\epsilon\sigma^4 s^2 , \tag{10.11}$$

equal to zero. The resulting equation is bi-quadratic. The first-order expansion for the zero is

$$s^2 = \frac{1}{\sigma^2}\left[2\ln(R) - 24\epsilon\ln(R)^2\right] . \tag{10.12}$$

Similarly, based on (10.10), we need to expand $1/s$, s, and s^3. We find

$$s = \sqrt{\frac{2\ln(R)}{\sigma^2}} - 12\epsilon\sqrt{\frac{\ln(R)^3}{2\sigma^2}} , \qquad \frac{1}{s} = \sqrt{\frac{\sigma^2}{2\ln(R)}} + 6\epsilon\sqrt{\frac{\sigma^2\ln(R)}{2}}, \tag{10.13}$$

and

$$s^3 = \frac{2\ln(R)}{\sigma^2}\sqrt{\frac{2\ln(R)}{\sigma^2}} . \tag{10.14}$$

Since the cubic term in (10.10) is multiplied by ϵ, we only need to calculate its zero-order expansion. Substituting these expressions into $c(s)$ and rearranging gives the kurtosis approximation

$$c(\gamma_2) = c_G\left(1 + \frac{\gamma_2}{12}\ln(R)\right) . \tag{10.15}$$

The excess kurtosis of the Laplace kernel is positive, whereas, for the tent and the top-hat kernels, it is negative. Accordingly, the kurtosis approximation predicts that the spread speed for the Laplace kernel is higher than c_G; for the other two kernels, it is lower. Figure 10.2 compares the exact spreading speeds and the kurtosis approximation.

Application to a Family of Kernels

We close this section by studying a family of dispersal kernels that interpolates smoothly between the Laplace and the Gaussian kernel. The marginal distribution of the *gamma-binomial distribution* (Tufto et al. 2005) with variance σ^2 and shape parameter α is given by

$$K_{GB}(x; \sigma, \alpha) = 2\frac{(\alpha/2)^{(\alpha/2+1/4)}}{\Gamma(\alpha)\sqrt{\pi}\sigma}\left|\frac{x}{\sigma}\right|^{\alpha-1/2} K_{1/2-\alpha}\left(\sqrt{2\alpha}\left|\frac{x}{\sigma}\right|\right) , \tag{10.16}$$

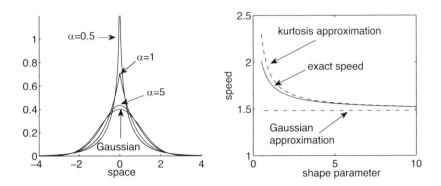

Fig. 10.3 **Left:** The shape of the gamma-binomial kernel interpolates between the Laplace kernel (for $\alpha = 1$) and the Gaussian kernel ($\alpha \to \infty$). **Right:** The kurtosis approximation captures the effect of the shape parameter on the rate of spread. Parameters are $R = 3$, $\sigma^2 = 1$. Right plot adapted from Lutscher (2007).

where K_γ is the Bessel function of the second kind of order γ; see Fig. 10.3. The moment-generating function is

$$M(s) = \left(1 - \frac{\sigma^2 s^2}{2\alpha}\right)^{-\alpha}, \qquad (10.17)$$

and the excess kurtosis is $\gamma_2 = 3/\alpha$. For $\alpha = 1$, this kernel equals the Laplace kernel, and as $\alpha \to \infty$, it approaches the Gaussian kernel. Since the variance of this kernel is independent of the shape parameter, the Gaussian approximation is independent of the shape parameter. The kurtosis approximation, however, reflects the shape and provides a good approximation for the exact speed; see Fig. 10.3.

10.2 Approximating the Shape

We now present a method for approximating not only the speed but also the evolution of the full spatial distribution of a population that is initialized at a single location. Such an approximation is particularly important for large times, because numerical simulations of (10.1) are prone to instabilities, as we have seen in Chap. 8. We use an exponential transform method similar to (5.9) to solve the linear IDE in (10.1) and employ the *saddle-point approximation* to obtain an approximate shape of the population distribution for large times. At the heart of this method is, again, a quadratic approximation. The approximate but explicit expression for the population density can be used to calculate an approximate spread rate. In the case of the Gaussian kernel, it gives the exact solution. The application of the saddle-point approximation and the method of steepest descent go back to Radcliffe and Rass (1997) and were extended later and publicized by Kot and Neubert (2008). For a recent application to a system of IDEs modeling the spread of a tree disease (white pine blister rust) caused by the rust fungus *Cronartium ribicola*, see Leung and Kot (2015).

Exponential Transform and the Gaussian Kernel

The *exponential transform* of a function N is defined as

$$\tilde{N}(s) = \int_{-\infty}^{\infty} N(x)e^{sx}\,\mathrm{d}x\,. \tag{10.18}$$

The exponential transform of the dispersal kernel is the moment-generating function $\tilde{K}(s) = M(s)$; see (5.15). Just like the Fourier transform, the exponential transform turns convolution into multiplication, so that IDE (10.1) becomes

$$\tilde{N}_{t+1}(s) = RM(s)\tilde{N}_t(s)\,. \tag{10.19}$$

The solution of (10.19) with initial condition $N_0(x) = n_0\delta(x)$ is then given by

$$\tilde{N}(s) = n_0 R^t M^t(s)\,. \tag{10.20}$$

The exponential transform can be inverted by a contour integral, so that the solution of IDE (10.1) is given by

$$N_t(x) = \frac{n_0 R^t}{2\pi \mathrm{i}} \int_{b-\mathrm{i}\infty}^{b+\mathrm{i}\infty} M^t(s)e^{-sx}\,\mathrm{d}s\,, \tag{10.21}$$

where b is chosen such that the line of integration is within the strip of convergence of the transform. The challenge is to calculate this inverse transform. The easiest example is for the Gaussian kernel.

The moment-generating function is $M(s) = \exp(\sigma^2 s^2/2)$, so that the integral becomes

$$\int_{b-\mathrm{i}\infty}^{b+\mathrm{i}\infty} \exp\left(t\left(\frac{\sigma^2 s^2}{2} - \frac{x}{t}s\right)\right)\mathrm{d}s = \int_{b-\mathrm{i}\infty}^{b+\mathrm{i}\infty} e^{tf(s)}\,\mathrm{d}s\,. \tag{10.22}$$

We expand the quadratic function f around its critical point $s_0 = x/(\sigma^2 t)$ and get

$$f(s) = f(s_0) + \frac{f''(s_0)}{2}(s - s_0)^2 = -\frac{x^2}{2\sigma^2 t^2} + \frac{\sigma^2}{2}(s - s_0)^2\,. \tag{10.23}$$

Substituting this expression, we simplify the integral in (10.22) to

$$
e^{-\frac{x^2}{2\sigma^2 t}} \int_{b-i\infty}^{b+i\infty} \exp\left(\frac{t\sigma^2}{2}(s-s_0)^2\right) ds .
$$

(10.24)

We can still choose b, and the choice $b = s_0$ makes the integration relatively easy by a change of variables since the exponent becomes real and nonpositive. We find

$$
\int_{s_0-i\infty}^{s_0+i\infty} e^{\frac{t\sigma^2}{2}(s-s_0)^2} ds = i \int_{-\infty}^{+\infty} e^{-\frac{t\sigma^2}{2}y^2} dy = i\sqrt{\frac{2\pi}{\sigma^2 t}} .
$$

(10.25)

Putting everything together, we find the explicit solution

$$
N_t(x) = n_0 R^t \frac{1}{\sqrt{2\pi\sigma^2 t}} e^{-\frac{x^2}{2\sigma^2 t}} .
$$

(10.26)

For kernels other than the Gaussian kernel, function f in the exponent in (10.22) is not a quadratic. The main idea in what follows is to use the quadratic approximation and continue with the same steps. Before we illustrate this procedure, we take a look at what the method has to do with saddle points and steepest descent.

Saddle Points and the Method of Steepest Descent

We consider the integral in (10.22) for an arbitrary holomorphic function $f(s) = f_1(x, y) + i f_2(x, y)$, where $s = x + iy$. The integrand becomes

$$
e^{tf(s)} = e^{tf_1(x,y)+itf_2(x,y)} .
$$

(10.27)

The main idea is to choose a path of integration that concentrates most of the contributions near a single point. Since the modulus of the integrand depends on the real part of f only, we aim to find a critical point of the real part, f_1.

Differentiating with respect to x gives

$$
f'(s) = f_{1,x}(x, y) + i f_{2,x}(x, y) ,
$$

(10.28)

where the second subscript indicates partial differentiation. Therefore, $f'(s) = 0$ if and only if $f_{1,x}(x, y) = 0 = f_{2,x}(x, y)$. By the Cauchy–Riemann equations, $f_{1,x} = f_{2,y}$ and $f_{2,x} = -f_{1,y}$. Therefore, these equalities imply $f_{1,y} = 0$. Hence, s is a critical point of f if and only if (x, y) is a critical point of f_1. We compute the Hessian, using the Cauchy–Riemann equations again, and find

$$
H(x, y) = \begin{vmatrix} f_{1,xx} & f_{1,xy} \\ f_{1,yx} & f_{1,yy} \end{vmatrix} = f_{1,xx} f_{1,yy} - f_{1,xy}^2 = -f_{1,xx}^2 - f_{2,xx}^2 \le 0 .
$$

(10.29)

Hence, s is a nondegenerate critical point of f if and only if (x, y) is a saddle point of f_1.

Ideally, then, we want to choose a path of integration that follows the steepest descent of f_1 near the saddle point. At the same time, we want to keep the imaginary part, f_2, constant so that the absolute value of the integrand does not oscillate around the path of steepest descent. Fortunately, we can reach both of these goals simultaneously, due to the Cauchy–Riemann equations. The path of steepest descent follows the gradient of f_1. A level set of f_2 is perpendicular to the gradient of f_2. By the Cauchy–Riemann equations, the gradients of f_1 and f_2 are perpendicular so that the path of steepest descent is automatically a level set for the imaginary part. We only have to be careful to pick the correct one of the two directions of the gradient of f_1.

At a critical point when $f'(s_0) = 0$, we write

$$f(s) - f(s_0) = \frac{f''(s_0)}{2}(s - s_0)^2 + \text{h.o.t.} \tag{10.30}$$

We want to choose a path so that the first term on the right-hand side is real and negative. In general, this choice requires careful consideration of the (complex) term $f''(s_0)$ (Kot and Neubert 2008). We will, however, only need the case where s_0 and $f''(s_0)$ are real and $f''(s_0)$ is positive. Then the choice of the steepest path is simple in that the real part of s equals s_0, so that $s - s_0$ is purely imaginary and its square is negative.

We return to the expression in (10.21). We write the integrand as

$$M^t(s)e^{-sx} = e^{t\left[\ln M(s) - \frac{x}{t}s\right]} = e^{tf(s)}. \tag{10.31}$$

The condition for a critical (saddle) point becomes

$$\frac{M'(s_0)}{M(s_0)} = \frac{x}{t}. \tag{10.32}$$

Typically, there will be a unique, real root of this equation (Kot and Neubert 2008). At the critical point, the second derivative is

$$f''(s_0) = \frac{d^2}{ds^2}\ln(M(s)) = \frac{M''(s_0)M(s_0) - M'^2(s_0)}{M^2(s_0)} = \frac{M''(s_0)}{M(s_0)} - \frac{x^2}{t^2}. \tag{10.33}$$

Since the moment-generating function is concave up, its logarithm has the same property, so this second derivative is positive. We obtain the expression

$$\int_{b-i\infty}^{b+i\infty} e^{tf(s)}ds \approx e^{tf(s_0)}\int_{b-i\infty}^{b+i\infty} e^{tf''(s_0)(s-s_0)^2/2}ds = iM^t(s_0)e^{-s_0x}\sqrt{\frac{2\pi}{tf''(s_0)}}.$$

Altogether, the approximation to the spatial profile from (10.21) becomes

$$N_t(x) \approx \frac{n_0 R^t M^t(s_0) e^{-s_0 x}}{\sqrt{2\pi t f''(s_0)}} .$$
(10.34)

We illustrate this method and the excellent fit for the Laplace kernel in the next section; more examples can be found in Kot and Neubert (2008).

The Laplace Kernel

For the Laplace kernel (2.27) with parameter $a = \sqrt{2/\sigma^2}$, the inverse transform can be found explicitly in terms of Bessel functions (Kot and Neubert 2008). The exact expression is

$$N_t(x) = n_0 R^t \frac{a^{t+1/2}}{2^{t-1/2}\Gamma(t)\sqrt{\pi}} |x|^{t-1/2} K_{t-1/2}(a|x|) ,$$
(10.35)

where $K_\nu(x)$ is the modified Bessel function of the second kind of order ν; see left plot in Fig. 10.1.

The moment-generating function and its derivatives are

$$M(s) = \frac{a^2}{a^2 - s^2}, \quad M'(s) = M(s)\frac{2s}{a^2 - s^2}, \quad M''(s) = M(s)\frac{2a^2 + 6s^2}{(a^2 - s^2)^2} .$$
(10.36)

The critical value, s_0, is a root of the quadratic equation $s^2 + \frac{2t}{x}s - a^2 = 0$. For each $t > 0$ and $x \neq 0$, we can choose a solution $-a < s_0 < a$ as

$$s_0 = \begin{cases} -\frac{t}{x} - \sqrt{\frac{t^2}{x^2} + a^2}, & x < 0, \\ -\frac{t}{x} + \sqrt{\frac{t^2}{x^2} + a^2}, & x > 0. \end{cases}$$
(10.37)

When $x = 0$, we choose $s_0 = 0$. The second derivative, $f''(s_0) = 2\frac{a^2 + s^2}{(a^2 - s^2)^2} > 0$, is positive. Substituting these expressions into (10.34), we find the approximation

$$N_t(x) \approx \frac{n_0 R^t (a^2 - s_0^2)}{2\sqrt{t\pi(a^2 + s_0^2)}} \left(\frac{a^2}{a^2 - s_0^2}\right)^t e^{-s_0 x} .$$
(10.38)

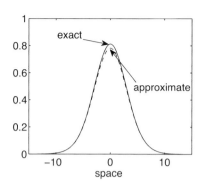 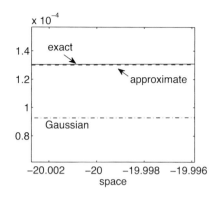

Fig. 10.4 Comparison of the exact solution and saddle-point approximation for (10.20). **Left:** The approximation underestimates the exact solution slightly near the peak but provides an excellent fit elsewhere ($R = 1.2$, $\sigma^2 = 1$, $t = 10$). **Right:** The approximation is much better than the corresponding Gaussian approximation (see previous section) in the tails of the distribution; compare Fig. 10.1 ($R = 1$, $\sigma^2 = 1$, $t = 30$).

Figure 10.4 shows that this approximation is highly accurate, not only near the peak of the distribution, but also in the tail, where the convergence to the Gaussian distribution is slow; compare Fig. 10.1.

Chapter 11
The Shape of Spatial Spread

Abstract In previous chapters, we studied the speed of spread and the existence of traveling waves. In this chapter, we focus on the shape of traveling-wave profiles and more general patterns of spatial spread. We first provide an approximation scheme, based on asymptotic expansion, for the shape of a monotone wave. Then we explore the existence of nonmonotone waves as well as more complicated patterns of spread when the growth function has a stable two-cycle. We generalize the notion of the asymptotic spreading speed and discuss dynamic stabilization.

11.1 Monotone Versus Nonmonotone Scenarios

In the study of biological invasions, the speed of spatial spread of the invading organism is arguably the most important quantity. The theory in Chaps. 5 and 6 guarantees the existence of such a speed and, in some cases, also provides a simple formula for it, but leaves open many questions about the shape of an invasion front. When the growth function is monotone, the results in Chap. 5 also establish the existence of monotone traveling waves. Information about the steepness of the wave, say, would tell us how fast the invading population grows at any fixed location. When the growth function is not monotone, we would like to know more about the shape of the profile behind the initial invasion front; see Fig. 5.6. Will the profile converge to the positive steady state, and if so how fast? But many more scenarios could arise. Since the positive steady state of the nonspatial Ricker model may be unstable, there is no reason to believe that a traveling profile in the spatial model with Ricker dynamics—if it exists—would settle at that positive state. If it does not, then what are the dynamics of the population in the wake of the invasion front? Would a profile oscillate in ways comparable to those that we have seen in Sect. 4.4? In this chapter, we present various approaches to answer some of these questions.

The first part of this chapter is based on the analysis of traveling waves. We assume that there is a traveling-wave solution of speed c, i.e., a profile N that satisfies

© Springer Nature Switzerland AG 2019 145
F. Lutscher, *Integrodifference Equations in Spatial Ecology*, Interdisciplinary
Applied Mathematics 49, https://doi.org/10.1007/978-3-030-29294-2_11

$$N(x+c) = Q[N](x) = \int_{-\infty}^{\infty} K(x-y)F(N(y))dy \,. \tag{11.1}$$

In this notation, the profile travels to the left when $c > 0$, so that we impose the boundary conditions $N(-\infty) = 0$ and $N(\infty) = 1$. The profile travels to the right when $c < 0$ and the boundary conditions are interchanged. When the dispersal kernel is the Laplace or the exponential kernel, we can reduce this equation to a second- or first-order delay equation, which we can study via linearization and asymptotic expansion. This material was originally developed by Kot (1992).

In the second part, we consider the case where the positive steady state of the nonspatial growth function is unstable and there is a stable two-cycle. We study this case by considering the second-iterate operator $N_{t+2} = Q \circ Q[N_t]$. Some aspects of the traveling-wave analysis from the first part of this chapter can be extended to this case. More important, the theory from Chap. 5 can be adapted to yield the existence of *generalized spreading speeds* and various forms of traveling profiles that appear after the initial invasion front and travel more slowly. This material originates in the thesis of Bourgeois (2016). The existence of two or more distinct traveling profiles at different speeds leads to *stacked waves* and the phenomenon of dynamic stabilization.

11.2   Asymptotic Expansion of Monotone Traveling Waves

When the growth function is of Beverton–Holt type (monotone, concave down) and the dispersal kernel is exponentially bounded, then there exist monotone traveling-wave solutions (11.1) of the IDE for every $c \geq c^*$; see Chap. 5, Theorem 5.2. We obtain a unique (leftward-moving) solution if we fix the density at one point, e.g., $N(0) = 1/2$.

When the dispersal kernel is the Laplace kernel (2.27), we can use repeated differentiation as in (3.7) to turn equation (11.1) into the second-order delay differential equation

$$N''(x+c) = a^2[N(x+c) - F(N(x))] \,. \tag{11.2}$$

A constant solution of this equation satisfies $N = F(N)$; i.e., it is a fixed point of the nonspatial dynamics. By our assumptions on F, there are exactly two such fixed points, namely $N = 0$ and $N = 1$, the asymptotic values of the traveling wave. We are looking for a solution that connects these two points.

To apply asymptotic expansion, we need to identify a small parameter in the equation. We rescale space by setting $x = zc$ and define $\tilde{N}(x/c) = N(x)$. Using the chain rule and dropping the tilde, we find the equation

$$\frac{1}{(ac)^2}N''(z+1) + F(N(z)) - N(z+1) = 0 \,. \tag{11.3}$$

We now consider $\epsilon = 1/(ac)^2$ as our small parameter. We know that the smallest traveling-wave speed $\hat{c} = c^*$ is an increasing function of the growth rate at low density $F'(0)$ for fixed $a > 0$. Hence, we can make ϵ small by making $F'(0)$ large.

We expand the solution $N(z)$ of (11.3) in a perturbation series

$$N(z) = N^{(0)}(z) + \epsilon N^{(1)}(z) + \epsilon^2 N^{(2)}(z) + \dots . \tag{11.4}$$

Inserting this expansion into (11.3), we obtain an infinite system of equations. The two lowest-order equations are

$$N^{(0)}(z+1) = F(N^{(0)}(z)), \tag{11.5}$$

$$N^{(1)}(z+1) = F'(N^{(0)}(z))N^{(1)}(z) + [N^{(0)}]''(z+1). \tag{11.6}$$

The lowest-order equation looks like the simple nonspatial equation. The difference is that the equation here is meant to hold for a continuous variable $z \in \mathbb{R}$ and not just for a discrete set $z \in \mathbb{N}$. Fortunately, we have an explicit solution of (11.5) if F is the scaled Beverton–Holt function (2.13), namely

$$N^{(0)}(z) = \frac{R^z}{1 + R^z} ; \tag{11.7}$$

see (2.16). This solution satisfies the two boundary conditions, $N^{(0)}(-\infty) = 0$ and $N^{(0)}(\infty) = 1$, and the condition at zero, $N^{(0)}(0) = 1/2$. It also defines a continuous function on $\mathbb{R}$. Hence, we have found an approximation to lowest order.

Comparing this lowest-order approximation to numerical simulations of the traveling front, we see that the approximation is very good even when ϵ is on the order of unity. The left plot in Fig. 11.1 shows that the two curves are virtually

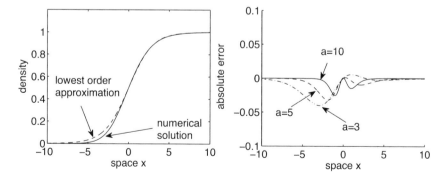

Fig. 11.1 Comparison of the numerically obtained shape of a traveling front and its lowest-order approximation. **Left:** The two densities. **Right:** The absolute error, $N(x) - N^{(0)}(x)$. The growth function is the Beverton–Holt function with $R = 1.25$. The dispersal kernel is the Laplace kernel with $a = 5$ (left plot) and values of a as indicated (right plot).

indistinguishable and that the absolute error, defined as $N(x) - N^{(0)}(x)$ (right plot), is small for $\epsilon = 1.0139$. As discussed above, ϵ decreases as R increases, so that the approximation should become even better. As a decreases and dispersal distances increase, the error increases in size and in spatial extent.

Kot (1992) proceeds to solve the equation for $N^{(1)}$ in (11.6) explicitly, but the computations are lengthy and the gain is relatively small since the zero-order approximation is already good. Instead, we consider a different example where the calculations of the higher-order term are somewhat simpler and provide greater improvement.

Asymptotics for the Exponential Kernel

The exponential kernel

$$K(x) = \begin{cases} a \exp(ax), & x \leq 0, \\ 0, & x > 0, \end{cases} \tag{11.8}$$

is the (scaled) "left half" of the Laplace kernel and allows leftward spread only. The moment-generating function is

$$M(s) = \frac{a}{a - s}. \tag{11.9}$$

The parametric representation of c according to (5.22) can be written as (compare (5.26))

$$ac = 1 + \bar{s}, \qquad R = \frac{e^{\bar{s}}}{1 + \bar{s}}, \tag{11.10}$$

with $\bar{s} = -\frac{s}{a-s}$. The defining equation for the traveling front with the exponential kernel turns into

$$N(x + c) = a \int_x^\infty e^{a(x-y)} F(N(y)) \mathrm{d}y. \tag{11.11}$$

Differentiating once, we obtain the first-order delay differential equation

$$N'(x + c) + a[F(N(x)) - N(x + c)] = 0. \tag{11.12}$$

Similar to the procedure for the Laplace kernel, we scale space by setting $x = zc$, introduce the small parameter $\epsilon = 1/(ac)$, and expand $N(z)$ in a perturbation series. The two lowest-order equations are

Fig. 11.2 Comparison of the numerically obtained shape of a traveling front and its two lowest-order approximations. The growth function is the Beverton–Holt function with $R = 1.5$. The dispersal kernel is the exponential kernel with $a = 5$.

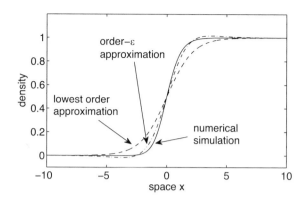

$$N^{(0)}(z + 1) = F(N^{(0)}(z)),\tag{11.13}$$

$$N^{(1)}(z + 1) = F'(N^{(0)}(z))N^{(1)}(z) + [N^{(0)}]'(z + 1).\tag{11.14}$$

The equation for $N^{(0)}$ is the same as (11.5), so that we already have its solution in (11.7). The plot in Fig. 11.2 shows that this approximation is not as close as for the Laplace kernel in the previous section. To improve the approximation, we calculate the term of order ϵ.

Substituting the expressions for $N^{(0)}$ and $F'(N)$ into (11.14), we arrive at

$$N^{(1)}(z + 1) = \frac{R(1 + R^z)^2}{(1 + R^{z+1})^2} N^{(1)}(z) + \ln(R)\frac{R^{z+1}}{(1 + R^{z+1})^2}\tag{11.15}$$

or, after rearranging,

$$\frac{(1 + R^{z+1})^2}{R^{z+1}} N^{(1)}(z + 1) = \frac{(1 + R^z)^2}{R^z} N^{(1)}(z) + \ln(R).\tag{11.16}$$

Hence, the expression $U(z) = \frac{(1+R^z)^2}{R^z} N^{(1)}(z)$ satisfies the simple recursion relation

$$U(z + 1) = U(z) + \ln(R).\tag{11.17}$$

To find its solution, we need an initial condition. Since $N^{(0)}(0) = N(0) = 1/2$, the requirement is $N^{(1)}(0) = 0 = U(0)$. Therefore, the solution for U is $U(z) = z \ln(R)$. As before, this solution is initially only determined for $z \in \mathbb{N}$, but we notice that it is a continuous function for all $z \in \mathbb{R}$.

The resulting approximation for $N(x)$ in the original parameters is

$$N(x) = \frac{R^{x/c}}{1 + R^{x/c}} + \frac{1}{ac}\frac{x}{c}\frac{\ln(R)R^{x/c}}{(1 + R^{x/c})^2} + O(\epsilon^2).\tag{11.18}$$

The improvement of this approximation over the previous one is obvious in Fig. 11.2.

The approximation procedure presented here requires that we find an explicit solution of the zero-order equation analytically. Numerical methods would only allow us to calculate approximations defined on the integers. In particular, we cannot easily use the method to study the shape of solutions when the growth function is not monotone. In the following section, we take a linearization-based approach to distinguish between monotone and nonmonotone solutions.

11.3   Traveling Waves in the Phase Plane

We return to Eq. (11.2), but now we treat it as a dynamical system in the phase plane. We turn the second-order equation into a pair of first-order equations and let $y = x + c$ to obtain

$$N'(y) = n(y), \qquad n'(y) = a^2[N(y) - F(N(y - c))]. \tag{11.19}$$

We point out that since the equation has a delay, the phase space is really infinite dimensional and the representation in the "phase plane" is somewhat misleading. For example, solutions of this system can cross in the phase plane. Nonetheless, the visualization in two dimensions turns out to be helpful.

The steady states of system (11.19) are of the form $(N^*, 0)$, where N^* is a solution of $F(N) = N$. A traveling wave of the IDE is a heteroclinic connection between the steady states $(0, 0)$ and $(1, 0)$. A necessary condition for such a connection to exist is that $(0, 0)$ be unstable and $(1, 0)$ have at least one stable direction. A monotone front requires that the eigenvalue at $(1, 0)$ be real and negative, whereas a front with damped spatial oscillations at $(1, 0)$ requires it to have nonzero imaginary and negative real part.

Figure 11.3 shows a monotone and a nonmonotone front for the IDE with Ricker dynamics together with their phase-plane representations. In the nonspatial Ricker model from (2.19), we have $F'(1) = 1 - r$. Solutions approach one in a monotone way when $0 < r \leq 1$ and in an oscillatory way when $1 < r < 2$. When $r > 2$, stability is lost, and cyclic or chaotic behavior appears. Accordingly, the lowest-order approximation to the traveling wave from (11.5) predicts monotone traveling fronts when $0 < r \leq 1$ and nonmonotone fronts for some $1 < r < 2$. Dispersal appears to have a dampening effect since Fig. 11.3 shows a monotone wave for some $r > 1$.

To study the stability of the two steady states in the phase plane, we linearize the traveling-wave equation. It is easier to work with (11.3). The scaling does not affect the sign of the eigenvalue. We find the transcendental eigenvalue problem

$$1 - \frac{\lambda^2}{a^2 c^2} = F'(N^*)e^{-\lambda}. \tag{11.20}$$

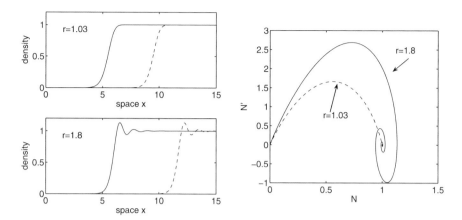

Fig. 11.3 Left: Traveling fronts with Ricker dynamics and a Laplace kernel can show a monotone profile ($r = 1.03$, top) or damped spatial oscillations ($r = 1.8$, bottom). The two profiles (solid and dashed lines) are taken 10 generations apart. **Right:** Phase-plane representations of the invasion fronts. The dispersal parameter is $a = 6$.

Alternatively, we linearize (11.19) and scale the eigenvalue; see Chap. 4 in Bourgeois (2016). We ask for conditions such that the parabola on the left intersects the exponential on the right. We distinguish three cases.

Case 1: $F'(N^*) > 1$

If $F'(N^*) > 1$, the parabola on the left-hand side of (11.20) and the exponential on the right cannot intersect for $\lambda < 0$. There are two positive intersections when c is large enough but none when c is small. We illustrate this case in Fig. 11.4, top left panel. The thick solid curve represents the right-hand side of (11.20); the thin solid (dashed) curve represents the left-hand side for small (large) c. The critical case, in which the curves are tangent, is given by the pair of equations

$$1 - \frac{\lambda^2}{a^2c^2} = F'(N^*)e^{-\lambda} \quad \text{and} \quad \frac{2\lambda}{a^2c^2} = F'(N^*)e^{-\lambda}. \tag{11.21}$$

These two equations lead precisely to the defining equations for the minimal speed of a traveling front with λ in place of $\tilde{s}$; see (5.26). This situation may arise at the zero state where $F'(0) = R_0 > 1$. Hence, the zero state in the phase plane is unstable precisely when the speed is at least the minimal speed. We have recovered the result from Theorem 5.1 that there can be no biologically meaningful traveling waves for $c < \hat{c}$.

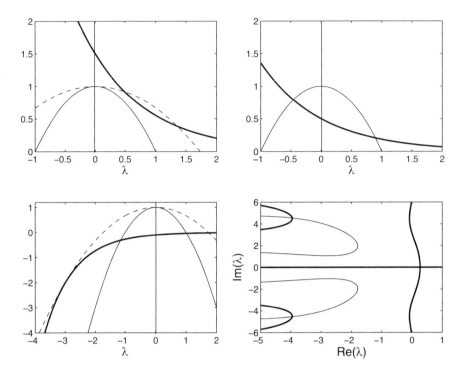

Fig. 11.4 Illustrating the potential roots of the eigenvalue problem in (11.20) as intersections of curves. **Top left:** When $F'(0) > 1$, there is a critical value for c for which a positive eigenvalue exists. **Top right:** When $0 < F'(1) < 1$, there is always one positive and one negative real eigenvalue. **Bottom left:** When $F'(1) < 0$, there is one positive eigenvalue, and there can be up to two negative eigenvalues. **Bottom right:** When $F'(1) < 0$ and there are no negative eigenvalues, there can be complex eigenvalues with negative real part. The thick (thin) curves indicate where the first (second) condition in (11.22) is satisfied. The intersection points indicate the eigenvalues. They all have negative real parts.

Case 2: $0 \leq F'(N^*) < 1$

If $0 \leq F'(N^*) < 1$, there is exactly one positive and one negative solution of (11.20); see Fig. 11.4, top right panel. As before, the thick (thin) curve represents the right-hand (left-hand) side. This situation arises at the positive steady state with the Beverton–Holt function but also with the Ricker function when $0 < r < 1$. The necessary conditions for the existence of a monotone traveling wave are satisfied.

Case 3: $F'(N^*) < 0$

When the slope of the updating function at the positive steady state is negative, (11.20) always has one positive solution. When c and/or $|F'(N^*)|$ are small, there are two negative solutions; when they are large, there is no negative solution; see Fig. 11.4, bottom left panel. As usual, the thick curve represents the exponential in (11.20); the thin solid (dashed) curve represents the quadratic with small (large) c. There are two negative intersections for small c. The dashed curve shows the critical case where the curves are tangent, which leads to the pair of equations (11.21).

When there are no real negative solutions, we consider complex-valued solutions. We write $\lambda = \alpha + i\beta$ and split the equation in (11.20) into real and imaginary parts, namely

$$1 - \frac{\alpha^2 - \beta^2}{a^2 c^2} = F'(N^*)e^{-\alpha}\cos(\beta), \qquad \frac{2\alpha\beta}{a^2 c^2} = -F'(N^*)e^{-\alpha}\sin(\beta). \qquad (11.22)$$

The bottom right panel in Fig. 11.4 illustrates the solution curves of these equations in the complex plane. Their intersection points correspond to eigenvalues.

We return to the case of the Ricker function in Fig. 11.3. When $r = 1.03$, we have $F'(1) = -0.03 < 0$. As $|F'(N^*)|$ is small, we have two negative eigenvalues as in the bottom left panel of Fig. 11.4. Hence, there can be a monotone traveling wave, as observed in Fig. 11.3. When $r = 1.8$, we have $F'(1) = -0.8 < 0$. As $|F'(N^*)|$ is now much larger, we have no negative eigenvalues but complex eigenvalues with negative real parts, as in the bottom right panel of Fig. 11.4. Hence, there cannot be a monotone traveling wave, but there is a wave with decaying oscillations around one, as observed in Fig. 11.3. For the existence of traveling waves, see Sect. 5.4.

11.4 Invasion Dynamics with a Two-Cycle

So far, we have assumed that the positive state $N^* = 1$ is stable for the growth dynamics. Without this assumption, there is no reason to believe that a traveling wave could exist, since the dynamic behavior in the wake of a wave is often determined by the stable steady states for the nonspatial equation. What can we expect if there is no stable steady state? Kot (1992) finds a "traveling two-cycle" in numerical simulations; see also Kot (2003). We illustrate this kind of solution to motivate the subsequent analysis.

The Ricker map, $F(N) = N\exp(r(1 - N))$, has a globally stable two-cycle for $r = 2.2$, with densities $0 < n_- < 1 < n_+$, satisfying $F(F(n_-)) = F(n_+) = n_-$. The invasion dynamics of the spatial model are shown in the left plot in Fig. 11.5. At the invasion front, the density increases from zero, overshoots the state $N = 1$, and then shows damped spatial oscillations to that state in the wake. Somewhere behind the invasion front, a second front alternates between increasing to n_+ in even generations (dashed) and decreasing to n_- in odd generations (solid). The right plot in Fig. 11.5 reveals that the first front with the damped oscillations travels faster than the alternating second front that connects to n_- and n_+, respectively. Hence, the term "traveling two-cycles" (Kot 1992) could be misleading because this is not a

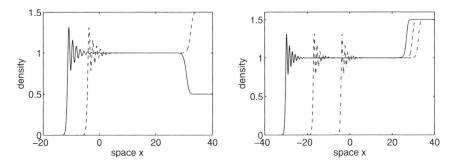

Fig. 11.5 Invasion dynamics with a stable two-cycle. **Left:** The profile of the front at odd generations (solid) connects with the lower value of the nonspatial two-cycle, n_-, whereas the profile at even generations (dashed) connects with the higher values, n_+. The two profiles are 11 generations apart. **Right:** Plotting the profile 20 generations apart shows that the nonmonotone first front travels faster than the monotone second front. Parameters are $r = 2.2$ for the Ricker growth function and $a = 6$ for the Laplace kernel. The initial condition is a step function.

Fig. 11.6 Plot of the second-iterate map of the Ricker function with $r = 2.2$.

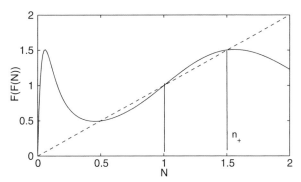

single profile traveling at a constant speed but rather two different objects traveling at different speeds.

Few authors have considered this problem. Theorem 5.3 guarantees the existence of a spreading speed even when the positive state is unstable. Hsu and Zhao (2008), Li et al. (2009), and Yu and Yuan (2012) prove the existence of a traveling-wave profile under some conditions, but the shape of the profile behind the initial front is unclear. Bourgeois (2016) provides a more detailed study of the spatial invasion dynamics with an unstable fixed point; see also Bourgeois et al. (2018) and Bourgeois et al. (2019). We present some of these results here.

To gain some preliminary insight into the second front, we consider the iteration of two generations. For the nonspatial Ricker model with $r = 2.2$, the map $N_{t+2} = F(F(N_t))$ has four fixed points. The fixed points of F and each of the points of the two-cycle of F are fixed points of $F \circ F$; see Fig. 11.6. The points zero and one are unstable, whereas $n_\pm$ are stable. The corresponding two-generation IDE is

$$N_{t+2}(x) = Q \circ Q[N_t](x) = \int K(x - y) F\left(\int K(y - z) F(N_t(z)) \mathrm{d}z \right) \mathrm{d}y .$$

$$(11.23)$$

We are looking for a monotone front that connects $N = 1$ with $N = n_+$ or $N = n_-$. Such a front satisfies the equation

$$N(x + 2c) = \int_{-\infty}^{\infty} K(x - y)F\left(\int_{-\infty}^{\infty} K(y - z)F(N(z))dz\right)dy. \quad (11.24)$$

We write the speed as $2c$ because we consider two generations in the equation. We study its behavior near $N = 1$ by linearization. The linearized equation is

$$n(x + 2c) = F'(1)^2 \int \int K(x - y)K(y - z)n(z)dy\,dz. \quad (11.25)$$

Taking an exponential transform or substituting an exponential profile leads to the dispersion relation

$$e^{2sc} = (F'(1))^2 M^2(s), \quad (11.26)$$

where $M(s)$ is the moment-generating function of K as usual. After taking square roots, we find the minimal speed at which a profile in the linearized equation can travel as

$$\hat{c}_1 = \frac{1}{s}\ln(|F'(1)|M(s)). \quad (11.27)$$

For the Ricker function, we have $F'(0) = e^r$ and $F'(1) = 1 - r$, so that $F'(0) > |F'(1)|$ for all $r > 0$. In particular, the slowest speed for a front in the linearized equation at zero is always larger than for the linearized equation at one. For the parameters in Fig. 11.5, the speed of the initial front is $c_0 = 0.67$, whereas the speed of the second front is $c_1 = 0.14$. The plateau region at level $N = 1$ between the first and second fronts grows by the difference $c_0 - c_1$ per generation. In particular, even though the state $N = 1$ is unstable for the nonspatial dynamics, it appears stable for a potentially long time in the spatial model. This same phenomenon, but in a two-component reaction–diffusion model, has been termed dynamic stabilization by Malchow et al. (2008).

11.5 Generalized Spreading Speeds

To study the speed of spread of the second front that we observe in the simulations in Fig. 11.5, we generalize the definition of the asymptotic spreading speed in (5.31) to allow for the state ahead of the front to be positive (Bourgeois 2016; Bourgeois et al. 2018). The construction is a slight extension of the one by Weinberger (1982).

We consider an operator $\widetilde{Q}$ that acts on continuous functions on $\mathbb{R}$. We assume that there are two constants $0 \le \pi_0 < \pi_1 < \infty$ such that $\widetilde{Q}[\pi_i] = \pi_i$ for the corresponding constant functions. We denote the set of all continuous functions with values in $[\pi_0, \pi_1]$ as $\mathscr{C}_{[\pi_0,\pi_1]}$ and define a sequence of functions by the iteration $N_{t+1} = \widetilde{Q}[N_t]$.

Definition 11.1 The value $c^*_{(\pi_0,\pi_1)}$ is called the *generalized asymptotic spreading speed* of $\widetilde{Q}$ from π_0 to π_1 if the following conditions hold:

(i) For any $N_0 \in C_{[\pi_0,\pi_1]}$ such that $N_0 - \pi_0$ has compact support,

$$\lim_{t\to\infty} \sup_{|x|\geq ct} N_t(x) = \pi_0 \text{ for all } c > c^*_{(\pi_0,\pi_1)} . \tag{11.28}$$

(ii) For any $N_0 \in C_{[\pi_0,\pi_1]}$ such that $N_0 - \pi_0 \not\equiv 0$,

$$\lim_{t\to\infty} \inf_{|x|\leq ct} N_t(x) = \pi_1 \text{ for all } c \in (0, c^*_{(\pi_0,\pi_1)}) . \tag{11.29}$$

For $\pi_0 = 0$ and $\pi_1 = 1$, this definition agrees with Definition 5.1; i.e., $c^*_{(0,1)} = c^*$. If $\pi_0 = 0$, we can always achieve $\pi_1 = 1$ by rescaling.

The existence of a generalized spreading speed and associated traveling waves for operator $\widetilde{Q}$, satisfying appropriate conditions, follows from Theorem 5.1.

Theorem 11.1 *Assume that operator $\widetilde{Q}$ acts on the space of continuous functions $C_{[\pi_0,\pi_1]}$ as follows:*

(i) *Translation invariance:* $\widetilde{Q}[N(\cdot - a)](x) = \widetilde{Q}[N](x - a)$.
(ii) *Invariance on $C_{[\pi_0,\pi_1]}$:* $N \in C_{[\pi_0,\pi_1]} \Rightarrow \widetilde{Q}[N] \in C_{[\pi_0,\pi_1]}$.
(iii) *Fixed points:* $\widetilde{Q}[\pi_0] = \pi_0$, $\widetilde{Q}[\pi_1] = \pi_1$, $\widetilde{Q}[\alpha] > \alpha$ *for* $\alpha \in (\pi_0, \pi_1)$.
(iv) *Monotonicity:* $\pi_0 \leq N \leq \tilde{N} \leq \pi_1 \Rightarrow \widetilde{Q}[N] \leq \widetilde{Q}[\tilde{N}]$.
(v) *Continuity: If* $\{f_t\} \subset C_{[\pi_0,\pi_1]}$ *and* $f_t \to f$ *uniformly on compact subsets of* $\mathbb{R}$, *then* $\widetilde{Q}[f_t] \to \widetilde{Q}[f]$ *pointwise as* $t \to \infty$.
(vi) *Compactness: Every sequence* $\{f_j\}$ *in* $C_{[\pi_0,\pi_1]}$ *has a subsequence* $\{f_{j_i}\}$ *such that* $\{\widetilde{Q}[f_{j_i}]\}$ *converges uniformly on every bounded subset of* $\mathbb{R}$.

Then there exists a generalized spreading speed, $c^*_{(\pi_0,\pi_1)}$, *for* $\widetilde{Q}$ *from* π_0 *to* π_1. *For all* $c \geq c^*_{(\pi_0,\pi_1)}$, *there exists a monotone traveling-wave solution* $W(x - ct)$ *with* $W(-\infty) = \pi_1$ *and* $W(\infty) = \pi_0$.

Proof We construct an operator Q on $C_{[0,\pi_1-\pi_0]}$ as $Q[f] = \widetilde{Q}[f + \pi_0] - \pi_0$. This operator inherits all the qualitative properties from $\widetilde{Q}$, shifted to the interval $[0, \pi_1 - \pi_0]$. Hence, it satisfies the assumptions of Theorem 5.1, which guarantees the existence of a spreading speed and traveling waves.   □

We want to apply this theorem to the second-iterate operator

$$\widetilde{Q}[N](x) = Q \circ Q[N](x) = \int K(x - y) F\left(\int K(y - z) F(N(z)) dz\right) dy .$$

$$\tag{11.30}$$

The following theorem gives conditions on F for the existence of a generalized spreading speed and monotone traveling waves (Bourgeois 2016; Bourgeois et al. 2018).

Theorem 11.2 *Assume that K is continuous and that its moment-generating function is bounded for at least one nonzero value. Let F be a growth function that satisfies the following conditions:*

(i)   F is bounded and continuously differentiable.
(ii) F has exactly one stable two-cycle, i.e., there exist $n_\pm$ such that $0 < n_- < 1 < n_+$, $F(n_-) = n_+$ and $F(n_+) = n_-$, and all nonnegative initial conditions converge to this two-cycle under the map $N_{t+1} = F(N_t)$.
(iii) $N = 1$ is the only fixed point of F on the interval $[n_-, n_+]$.
(iv) $F'(1) < -1$.
(v)   F is nonincreasing on the interval $[n_-, n_+]$.

Then, there exists a spreading speed $c^*_{(1,n_+)}$ for the operator $\widetilde{Q}$ in (11.30) from one to n^+. Furthermore, for every $c \geq c^*_{(1,n_+)}$, there exists a monotone traveling-wave profile $W(x - ct)$ with $W(-\infty) = n_+$ and $W(\infty) = 0$.

Proof Translation invariance, continuity, and compactness of $\widetilde{Q}$ follow from the corresponding properties of Q. Function F maps the interval $[1, n_+]$ into $[n_-, 1]$, and vice versa. Hence, if $N \in [1, n_+]$, then $Q[N] \in [n_-, 1]$ and also $Q(Q[N]) \in [1, n_+]$. Therefore, $C_{[1,n_+]}$ is invariant under $\widetilde{Q}$. Since $F(1) = 1$ and $F(F(n_+)) = n_+$, we have $\widetilde{Q}(1) = 1$ and $\widetilde{Q}(n_+) = n_+$ in the sense of constant functions. From (iv) we have $(F \circ F)'(1) > 1$, and hence $F(F(\alpha)) > \alpha$ for some $\alpha > 1$. Since there is no fixed point between one and n_+, we must have $F(F(\alpha)) > \alpha$ for $\alpha \in (1, n_+)$. The same relation holds for constant functions under $\widetilde{Q}$. To show monotonicity, assume that $1 \leq N(x) \leq \tilde{N}(x) \leq n_+$. Then by (v), we have $1 \geq F(N(x)) \geq F(\tilde{N}(x)) \geq n_-$ and hence also $1 \geq Q[N] \geq Q[\tilde{N}] \geq n_-$. But then we repeat the argument, since F maps $[n_-, 1]$ into $[1, n_+]$, and we obtain $1 \leq Q(Q[N]) \leq Q(Q[\tilde{N}]) \leq n_+$. □

The monotonicity assumption (v) is satisfied for the Ricker function with $2 < r < 2.2565$ and for the logistic function with $2 < r < 2.2362$, respectively. For larger values of r, function F is not monotone in the interval $[n_-, n_+]$. Under certain conditions, the existence of a generalized spreading speed for $\widetilde{Q}$ can still be shown. The key idea is to construct operators $\widetilde{Q}^\pm$ that bound $\widetilde{Q}$, just as we constructed operators $Q^\pm$ that bounded Q for overcompensatory dynamics in Sect. 5.4. Details can be found in the thesis by Bourgeois (2016) and in Bourgeois et al. (2018).

Theorem 11.2 and its generalizations to nonmonotone dynamics have an immediate analogue on the interval $[n_-, 1]$. By applying the one-step operator Q to any solution of the two-step operator $\widetilde{Q}$, we obtain traveling waves that connect one to n_-.

The theorem gives the existence of a spreading speed, but no simple formula to calculate it. In particular, it is not known whether the linear conjecture holds, i.e., whether the generalized spreading speed $c^*_{(1,n_+)}$ is determined by the linearization at 0. Based on numerical simulations, Bourgeois (2016) conjectures that the speed is linearly determined and that $c^*_{(1,n_+)} = \hat{c}_1$ from (11.27).

11.6 Further Reading

The dynamics of the nonspatial Ricker map range from monotone or nonmonotone convergence to a unique positive steady state to chaotic dynamics. We only discussed spreading phenomena in a few cases of a stable steady state or a stable two-cycle. The ideas presented here can be and have been applied to higher-order cycles. For example, Bourgeois (2016) uses the fourth power of operator Q to study spreading dynamics with four-cycles; see also the discussion in Bourgeois et al. (2018). Seemingly chaotic behavior in the wake of an invasion can be observed in numerical simulations; see Andersen (1991) and Li et al. (2009).

Bourgeois et al. (2018) do not address the stability of the traveling profiles in the case where the growth function has a stable two-cycle. They use step functions as initial conditions in their simulations. The resulting shapes (e.g., Fig. 11.5) appear to be stable in some sense since they were obtained by two independent numerical schemes. On the other hand, Li et al. (2009) prove the existence of traveling waves for operator Q and find a spatially oscillating profile by numerical fixed-point iteration (see Fig. 4 in their paper). It is unclear whether this profile is stable for the dynamics of the IDE.

Many more phenomena can occur with overcompensatory dynamics. For example, since $\tilde{Q}$ has the three positive fixed points, $n_- < 1 < n_+$, and since n_- and n_+ are stable, there could be a bistable front connecting n_- to n_+. From Theorem 6.2, we know that a bistable front of operator Q can and does exist for only one speed c^*. The techniques in Lui (1983) unfortunately do not apply for proving uniqueness of the speed in this case. However, Bourgeois et al. (2018) show that if the speed is unique, then it has to be zero, i.e., the wave is a standing wave. Numerical simulations suggest the existence of a standing wave; see Fig. 9 in Bourgeois et al. (2018).

We can extend the phase-plane approach from Sect. 11.3 to study spreading phenomena and dynamic stabilization as the growth parameter in the Ricker function increases so that the nonspatial dynamics exhibit oscillations. There is evidence for a Hopf bifurcation as the intersection points of the thin and thick curves in the bottom right panel in Fig. 11.4 move across the imaginary line to have positive real parts; see Fig. 4.19 by Bourgeois (2016) for an illustration and Bourgeois et al. (2019) for more details.

The theory and examples in this chapter all assumed that the growth dynamics have no Allee effect. Schreiber (2003) studies several nonspatial models with Allee effect and Ricker dynamics and finds that overcompensation and a strong Allee effect could lead to complex extinction dynamics. Sullivan et al. (2017) combine Allee dynamics and overcompensation with spatial spread and find pulsating traveling waves. Otto (2017) in his PhD thesis also studies an IDE that includes an Allee effect and overcompensation. He finds a novel type of solution that neither spreads nor retreats but stays in place with compact support in a homogeneous environment.

Chapter 12
Applications

Abstract The theory studied so far has considered somewhat idealized population dynamics. In this chapter, we present various extensions of this theory to include more realistic conditions and several applications of scalar IDEs to real biological systems. Quite naturally, as soon as we want to model any particular scenario with the simple IDE in (2.1), we find that the model may need to be adjusted in various ways to more accurately describe biological reality. The first example deals with dispersal-induced mortality. Next, we consider the effects of biased dispersal in rivers and along coastlines in the context of the drift paradox. Closely related is the third topic of moving-habitat models, where we incorporate certain aspects of climate change into the equations. We then take a closer look at populations where individuals differ in their dispersal behavior: in one extreme, some individuals could be immobile (sessile); in the other extreme, some individuals may disperse much farther than the majority. We discuss the latter aspects in the context of Reid's paradox of rapid tree migration. We take a closer look at how to model Allee effects. Finally, we present some applications to two-dimensional domains.

12.1 Dispersal-Induced Mortality

All the dispersal kernels that we have considered so far have the property

$$\int_{-\infty}^{\infty} K(x, y)\mathrm{d}x = 1 \quad \text{for all } y, \tag{12.1}$$

which reflects the assumption that there is no mortality during dispersal. On an infinite landscape, every individual will find some new location. This assumption is unrealistic since dispersal is generally risky. For example, wind-dispersed seeds may land in unsuitable habitats. Dispersing animals may face dangers from predation or human activity, or they may simply require more energy to move. It is therefore

© Springer Nature Switzerland AG 2019

F. Lutscher, *Integrodifference Equations in Spatial Ecology*, Interdisciplinary Applied Mathematics 49, https://doi.org/10.1007/978-3-030-29294-2_12

more realistic to assume that the dispersal success (see Chap. 9) can be less than one, i.e.,

$$\int_{-\infty}^{\infty} K(x, y)dx = S(y) \le 1 \quad \text{for all} \quad y. \tag{12.2}$$

A relatively simple scaling allows us to keep using the theory developed in previous chapters. Namely, we set

$$\widetilde{K}(x, y) = \frac{1}{S(y)} K(x, y). \tag{12.3}$$

This modified kernel has property (12.1). In a homogeneous landscape, $S(y) = \bar{S}$ is a constant. To compensate for the scaling, we multiply the growth function by $\bar{S}$, i.e., we replace $F(N_t(y))$ by $\bar{S}F(N_t(y))$. As a result, the minimal requirement for population persistence is $\bar{S}F'(0) > 1$.

Sometimes, however, we are interested in how the detailed mechanisms of movement and mortality affect population dynamics, such as the speed of range expansion. As the climate changes, the length of the dispersal period during an individual's life cycle might increase or decrease. For example, butterflies disperse only when the temperature is above a certain threshold. When the average length of the dispersal phase increases, the average distance dispersed should also increase, and the population should spread faster. On the other hand, if mortality is associated with dispersal, then fewer individuals will survive the dispersal phase. This mechanism should decrease the spreading speed. What then is the combined effect of these two mechanisms on population range expansion?

The dispersal process with mortality rate $\beta > 0$ can be modeled by the reaction–diffusion equation (see Chap. 7)

$$\frac{\partial}{\partial t}u(t, x) = D\frac{\partial^2}{\partial x^2}u(t, x) - \alpha u(t, x) - \beta u(t, x), \quad u(0, x; y) = \delta(x - y). \tag{12.4}$$

Since only surviving individuals produce offspring, the dispersal kernel is still given by

$$K(x, y) = \int_0^{\infty} \alpha u(t, x; y)dt. \tag{12.5}$$

When all individuals disperse for exactly τ time units, the resulting dispersal kernel is the modified Gaussian kernel

$$K(x, 0) = \frac{e^{-\beta\tau}}{\sqrt{4\pi D\tau}} \exp\left(-\frac{x^2}{4D\tau}\right). \tag{12.6}$$

When individuals settle at a constant rate α, we obtain the modified Laplace kernel (Van Kirk 1995)

$$K(x, 0) = \frac{1}{2} \frac{\alpha}{\alpha + \beta} \sqrt{\frac{\alpha + \beta}{D}} \exp\left(-\sqrt{\frac{\alpha + \beta}{D}} |x|\right). \tag{12.7}$$

We note that mortality enters the modified Gaussian kernel only through the scaling factor, whereas it also affects the exponential decay rate in the modified Laplace kernel.

We compare how the speed of spatial spread depends on the mean dispersal time for these two kernels. We assume that the growth function has no Allee effect and that the spreading speed is linearly determined (see Chap. 5) from the equation

$$N_{t+1}(x) = R \int_{-\infty}^{\infty} K(x - y) N_t(y) dy. \tag{12.8}$$

To make the two kernels comparable, we set $\alpha = 1/\tau$ so that the mean dispersal time for surviving individuals is the same for both scenarios. The values of the dispersal success from (12.2) turn out to be

$$S_G(y) = \bar{S}_G = e^{-\beta \tau} \quad \text{and} \quad S_L(y) = \bar{S}_L = \frac{\alpha}{\alpha + \beta} = \frac{1}{1 + \beta \tau}, \tag{12.9}$$

respectively, for the modified Gaussian and Laplace kernels. Accordingly, the minimal growth rates for the population to persist and spread are

$$R > e^{\beta \tau} \quad \text{and} \quad R > 1 + \beta \tau, \tag{12.10}$$

respectively, for the two kernels. For fixed mean dispersal time τ, the minimum growth rate for the modified Gaussian kernel is greater than for the modified Laplace kernel. Conversely, for fixed growth rate R, the maximum average dispersal time is larger for the modified Laplace kernel than for the modified Gaussian kernel.

The variance of the modified Gaussian kernel, $\sigma^2 = 2D\tau$, grows linearly with τ. The variance of the modified Laplace kernel, $\sigma^2 = 2D\tau/(1 + \beta \tau)$, also grows with τ but remains bounded. How do these relationships affect the spreading speed?

For the modified Gaussian kernel, we adapt the explicit expression in (5.20) and find

$$c^*(\tau) = \sqrt{2\sigma^2 \ln(Re^{-\beta \tau})} = \sqrt{4D\tau(\ln(R) - \beta \tau)}. \tag{12.11}$$

In particular, the speed is a hump-shaped function of τ with a maximum at $\tau^* = \ln(R)/(2\beta)$. The maximum speed is $c^*(\tau^*) = \ln(R)\sqrt{D}/\beta$. When individuals disperse only for a short time, they do not move far, so the spread rate is slow. When they disperse for a long time, mortality takes a toll and the spread rate is, again, slow. Intermediate dispersal times give the highest speed of spread; see Fig. 12.1.

Fig. 12.1 The speed of
spread in the presence of
dispersal-induced mortality is
a hump-shaped function of
the mean dispersal time. With
the modified Gaussian kernel
(thick curves), the population
spreads only as long as
$\beta\tau < \ln(R)$. For the modified
Laplace kernel (thin curves),
this threshold is at
$\beta\tau = R - 1 > \ln(R)$.
Parameters are $R = 3$,
$D = 1$, and $\beta = 1$ (solid) or
$\beta = 0.7$ (dashed).

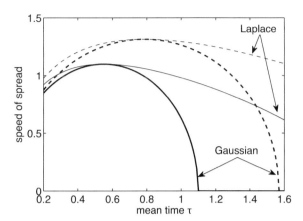

For the modified Laplace kernel, we apply formula (5.17), i.e.,

$$c^*(\tau) = \min_{s>0} \frac{1}{s} \ln\left(\frac{R}{1 + \beta\tau - D\tau s^2}\right), \tag{12.12}$$

to evaluate the speed numerically. As with the modified Gaussian kernel, the
speed for the modified Laplace kernel is hump shaped; see Fig. 12.1. Also, quite
predictably, the speed is no smaller than that of the modified Gaussian kernel.
Somewhat surprisingly, the maximum possible speed seems to be the same for the
two kernels and seems to occur for the same mean dispersal time.

We can prove this observation analytically by using the explicit expression for
the spreading speed for the Laplace kernel from (5.28). In terms of the -1-branch
of the Lambert W function, the speed of spread is given by

$$\hat{c}(\tau) = -\sqrt{\frac{D\tau}{1+\beta\tau}}\, W_{-1}(z) \sqrt{\frac{2}{W_{-1}(z)} + 1}, \quad \text{where} \quad z = \frac{-2(1+\beta\tau)}{R\,e^2}. \tag{12.13}$$

Differentiating this expression with respect to τ and using the identity $W_{-1}(z) = zW'_{-1}(z)(1 + W_{-1}(z))$, we find

$$\frac{d}{d\tau}\hat{c}(\tau) = \frac{\sqrt{\frac{D\tau}{1+\beta\tau}}\sqrt{\frac{2}{W_{-1}}+1}}{1+\beta\tau}\, W_{-1}(z) \left(\frac{\beta}{(1 + W_{-1})(2 + W_{-1})} - \frac{\beta}{1 + W_{-1}} - \frac{1}{2\tau}\right),$$

where $W_{-1} = W_{-1}(z)$. We require $W_{-1} \neq -1, -2$. Under this assumption, the
derivative is zero if and only if W_{-1} satisfies the quadratic equation

$$W^2 + (3 + 2\beta\tau)W + 2\beta\tau + 2 = 0. \tag{12.14}$$

The two solutions are $W = -1$, which we excluded above, and $W = -2 - 2\beta\tau$. Hence, we need to find the value of τ that satisfies

$$W_{-1}\left(\frac{-2 - 2\beta\tau}{R\,e^2}\right) = -2 - 2\beta\tau . \tag{12.15}$$

We recall that the defining equation for the Lambert W function is $W(y e^y) = y$ for real numbers y. Setting $y = -2 - 2\beta\tau$ and rewriting $R\,e^2 = e^{2 + \ln(R)}$, we finally find the solution $\tau^* = \ln(R)/\beta$, which agrees with the value of τ^* that we found above for the Gaussian kernel.

12.2 Biased Dispersal: Streams and Rivers

Many environments are characterized by external forces that bias individual movement in one direction. A prominent example is rivers, where water flow transports individuals downstream. Coastlines with longshore current are another example, and many more examples exist (Lutscher et al. 2010). How can populations persist in a particular river reach if individuals are constantly at risk of being washed out to downstream reaches? In river ecology, this question is known as the drift paradox. Together with the corresponding questions in other systems with biased (or asymmetric) dispersal, it has generated some fascinating mathematical modeling and analysis.

We apply the concepts of critical patch-size and spreading speeds to some simple models for a single population in a stream environment to address this question. It turns out that the definition of persistence requires some subtle distinction between local and global persistence, and that the definition of spreading speed has to include direction. The theory presented here stems from Byers and Pringle (2006) and Lutscher et al. (2010).

We study the linear IDE

$$N_{t+1}(x) = R \int_{-\infty}^{\infty} K(x - y) N_t(y)\,dy \tag{12.16}$$

to find persistence conditions and invasion speeds for the kernels that we derived from the random-walk model with bias in Chap. 7. We begin with the Gaussian kernel with variance σ^2 and nonzero mean μ; see (7.17):

$$K(x - y) = G(x - y; \mu, \sigma^2) = \frac{1}{\sqrt{2\pi\sigma^2}} \exp\left(-\frac{(x - y - \mu)^2}{2\sigma^2}\right). \tag{12.17}$$

With initial condition $N_0(x) = \delta(x)$, (12.16)–(12.17) have the solution

$$N_t(x) = R^t G(x; \mu t, \sigma^2 t) = \frac{R^t}{\sqrt{2\pi\sigma^2 t}} \exp\left(-\frac{(x - \mu t)^2}{2\sigma^2 t}\right), \tag{12.18}$$

since the convolution of two Gaussian distributions is again a Gaussian distribution with added means and variances. The total population size, $\int N_t(x)dx$, grows when $R > 1$. In fact,

$$\int_{-\infty}^{\infty} N_t(x)dx = R^t \int_{-\infty}^{\infty} G(x; \mu t, \sigma^2 t)dx = R^t. \tag{12.19}$$

At any particular point, however, the situation is quite different. Since the landscape is homogeneous, we may choose $x = 0$ and calculate

$$N_t(0) = \frac{R^t}{\sqrt{2\pi\sigma^2 t}} e^{-\frac{(-\mu t)^2}{2\sigma^2 t}} = \frac{1}{\sqrt{2\pi\sigma^2 t}} \left(R e^{-\frac{\mu^2}{2\sigma^2}}\right)^t. \tag{12.20}$$

Hence, $N_t(0)$ grows if and only if

$$R > R^* = \exp\left(\frac{\mu}{2\sigma^2}\right). \tag{12.21}$$

Therefore, population persistence at $x = 0$ and, in fact, at any given point requires a sufficiently large growth rate or, equivalently, a sufficiently small bias, namely $|\mu| < \mu^* = \sqrt{2\sigma^2 \ln(R)}$.

Figure 12.2 illustrates how a population with $1 < R < R^*$ (left plot) fails to persist at any given point and is swept downstream even though the total

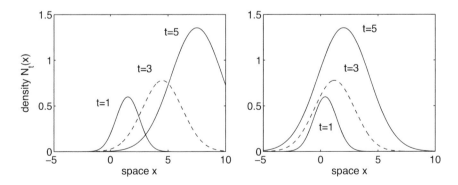

Fig. 12.2 Illustration of population density with drift according to (12.18). **Left plot:** The population grows but cannot persist at any point. **Right plot:** The population can persist at every point and spread upstream. Parameters are $R = 1.5$, $\sigma^2 = 1$, and $\mu = 1.5$ (left) versus $\mu = 0.4$ (right). The bias threshold is $\mu^* \approx 0.59$.

population grows over time. When $R > R^*$ (right plot), the population density increases at every point and expands in both directions. This observation suggests that persistence at a point could be characterized by the population being able to expand upstream, i.e., against the direction of the movement bias. It turns out that the threshold for directional bias, μ^*, is exactly the asymptotic speed, c^*, at which the population would spread in the absence of bias. We formalize this idea.

When movement is biased, there will be two spreading speeds in the system, one in the direction of the bias (c^+) and one against it (c^-) (Lutscher et al. 2005). These two speeds can still be characterized by the minimal speed of a traveling wave in the linearized equation. The ansatz

$$N_t(x) = N(x - ct) = \exp(-s(x - ct)) \tag{12.22}$$

models a profile traveling to the right if $c > 0$ and to the left if $c < 0$. The profile is decreasing if $s > 0$ and increasing if $s < 0$. If we assume that the bias is toward the right (downstream), then a profile with $s > 0$ and $c > 0$ indicates the downstream spread of a population. A profile with $s < 0$ and $c < 0$ indicates an upstream-spreading profile, while $s < 0$ and $c > 0$ indicates a downstream-retreating profile.

The dispersion relation for the traveling wave is the same as before, namely $e^{sc} = RM(s)$, where M is the moment-generating function of the kernel. The minimal rightward (downstream) speed, c^+, is given by

$$c^+ = \min_{s>0} c(s) = \min_{s>0} \frac{1}{s} \ln(RM(s)), \tag{12.23}$$

whereas the minimal leftward (upstream) speed, c^-, is given by

$$c^- = \max_{s<0} c(s) = \max_{s<0} \frac{1}{s} \ln(RM(s)) = -\min_{s>0} \frac{1}{s} \ln(RM(-s)). \tag{12.24}$$

If K is symmetric, then so is M, which implies that $c^+ = -c^-$.

The upstream speed is negative (so that the population spreads upstream) if $RM(s) > 1$ for all s where M is defined. But if $RM(s) < 1$ for some $s < 0$, then $c^- > 0$, so that the population retreats downstream; see Fig. 12.3. The critical value between spread and retreat is $c^- = \max_{s<0} c(s) = c(s^*) = 0$, which arises precisely when $RM(s^*) = 1$ and $M'(s^*) = 0$. Hence, the condition for upstream spread is (Lutscher et al. 2010)

$$R > R^* = \frac{1}{M(s^*)} = \frac{1}{\min M}. \tag{12.25}$$

This quantity can be calculated much more easily than the minimum of $c(s)$ in general. We give three examples.

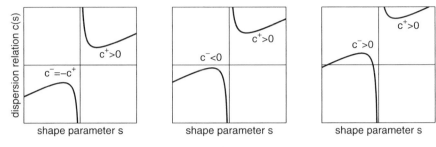

Fig. 12.3 Illustration of the dispersion relation for a symmetric (left plot) and two asymmetric (middle and right plots) moment-generating functions. The spreading speed in the direction of the bias (c^+) is the minimum of the curve for positive s, whereas the speed in the opposite direction (c^-) is the maximum over negative s. In all cases, $c^+ > 0$. In the symmetric case, $c^- = -c^+$. In the middle plot, $c^- < 0$; in the right plot, $c^- > 0$. The plots result from the Gaussian kernel with variance 1 and mean 0 (left plot), 0.5 (middle), and 1 (right).

The Gaussian Kernel

We continue the above example with the Gaussian kernel and its moment-generating function

$$M(s) = \exp\left(\frac{\sigma^2 s^2}{2} + \mu s\right).$$ (12.26)

Its minimum occurs at $s^* = -\mu/\sigma^2$ and equals $M(s^*) = \exp(\frac{\mu^2}{2\sigma^2})$. Hence, the threshold $R^* = 1/M(s^*)$ for upstream invasion is the same as the threshold for persistence at $x = 0$ found in (12.21).

The Shifted Laplace Kernel

Next, we choose a shifted Laplace kernel with mean μ and variance σ^2, namely

$$K(x) = \frac{1}{\sqrt{2\sigma^2}} \exp\left(-\frac{|x - \mu|}{\sqrt{\sigma^2/2}}\right).$$ (12.27)

The minimum of its moment-generating function, $M(s) = e^{\mu s}(1 - \frac{\sigma^2 s^2}{2})^{-1}$, occurs at

$$s^* = \frac{1}{\mu}\left(1 - \sqrt{1 + \frac{2\mu^2}{\sigma^2}}\right),$$ (12.28)

which gives a persistence threshold of

$$R^* = \frac{\sigma^2}{\mu^2}\left(\sqrt{1 + \frac{2\mu^2}{\sigma^2}} - 1\right)\exp\left(\sqrt{1 + \frac{2\mu^2}{\sigma^2}} - 1\right). \qquad (12.29)$$

The Asymmetric Laplace Kernel

We return to the asymmetric Laplace kernel derived from the biased random walk in (7.16). Its moment-generating function is (with $a_1 > 0 > a_2$)

$$M(s) = \frac{a_1 a_2}{(a_1 + s)(a_2 + s)}, \qquad s \in (-a_1, -a_2). \qquad (12.30)$$

The minimum of M occurs at $s^* = -(a_1 + a_2)/2$ and is given by (Lutscher et al. 2005)

$$M(s^*) = \frac{4a_1|a_2|}{(a_1 - a_2)^2}. \qquad (12.31)$$

Instead of relating parameters back to the random walk parameters D, q, and α in (7.16), we introduce the mean and variance of the asymmetric Laplace kernel, namely

$$\mu = \frac{1}{a_1} + \frac{1}{a_2} = \frac{a_1 + a_2}{a_1 a_2} \quad \text{and} \quad \sigma^2 = \frac{1}{a_1^2} + \frac{1}{a_2^2} = \frac{a_1^2 + a_2^2}{a_1^2 a_2^2}. \qquad (12.32)$$

We can express the critical growth rate in terms of μ^2/σ^2, as we did in the previous two examples, namely (Lutscher et al. 2010)

$$R^* = \frac{(a_1 - a_2)^2}{4a_1|a_2|} = \frac{1 - \frac{\mu^2}{2\sigma^2}}{1 - \frac{\mu^2}{\sigma^2}}. \qquad (12.33)$$

Figure 12.4 shows the three different kernels (left plot) and compares the threshold values R^* for upstream invasion (right plot). The dimensionless quantity μ^2/σ^2 corresponds to the inverse squared of the coefficient of variation. The threshold values R^* in the case of the shifted Gaussian and Laplace kernels differ only a little, but the upstream invasion condition for the asymmetric Laplace kernel is substantially more restrictive.

For the asymmetric Laplace kernel, we can relate the critical threshold for upstream invasion to the critical patch-size for persistence on a bounded domain. The calculation of the critical patch-size for the asymmetric Laplace kernel follows

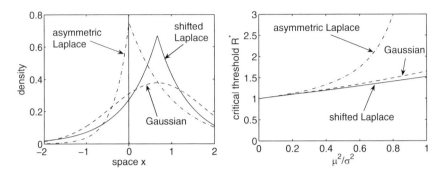

Fig. 12.4 Left: Illustrating the three kernels with movement bias. Parameter values are $\mu = 2/3$ and $\sigma^2 = 10/9$. **Right:** Threshold value R^* for upstream invasion according to (12.25) as a function of μ^2/σ^2. Plots adapted from Lutscher et al. (2010).

the same steps as for the symmetric kernel in Chap. 3 (Lutscher et al. 2005). We obtain the general relation between parameters and eigenvalues in (9.30). Setting $\lambda = 1$ in that relation, we obtain an explicit expression for the critical patch-size by solving for L as

$$
L^* = \frac{4 \arctan\left(1 \Big/ \sqrt{\frac{4Ra_1|a_2|}{(a_1-a_2)^2} - 1}\,\right)}{(a_1 - a_2)\Big/\sqrt{\frac{4Ra_1|a_2|}{(a_1-a_2)^2} - 1}} = \frac{4 \arctan\left(1\Big/\sqrt{RM(s^*) - 1}\right)}{(a_1 - a_2)\Big/\sqrt{RM(s^*) - 1}} . \tag{12.34}
$$

The critical patch-size approaches infinity as R approaches $1/M(s^*)$. In other words, the population can spread upstream if and only if there is a finite critical patch-size. This observation and similar results for continuous-time models can be found in Lutscher et al. (2005, 2010). Pringle et al. (2009) study an extension to structured populations and non-Gaussian kernels, such as the double Gaussian and others.

Lutscher et al. (2010) also study how various assumptions of movement patterns in streams relate to different shapes of dispersal kernels for stream insects with winged adult stages. For example, if larvae simply drift downstream and settle at a constant rate, their movement is described by a (one-sided) negative exponential kernel, as in (11.8). Once they emerge as adults and fly to lay eggs, oviposition could be modeled by a Laplace kernel. The resulting kernel from one generation to the next is then the convolution of these two kernels. Lutscher et al. (2010) discuss several more possible patterns. In all cases, the resulting kernel between generations is a convolution of two or more kernels that describe each dispersal stage. The critical patch-size for these kernels cannot be calculated explicitly. The threshold value R^*, on the other hand, can be computed relatively easily since the moment-generating function of the kernel is simply the product of the moment-generating functions of each stage. Vasilyeva et al. (2016) derive and analyze a mixed continuous-discrete model for stream populations that can incorporate more detail (in terms of population dynamics or even stream characteristics). Its linearization at zero reduces to a linear IDE as studied here.

Kanary et al. (2014) apply and extend these ideas to the spread of the European green crab (*Carcinus maenas L.*) along the Atlantic coast of Canada; see Sect. 12.4. In particular, they study how the interaction between two different genotypes, introduced in two different locations, would shape the relative densities of the two. Marculis and Lui (2015) formulate an age-structured model for the same species and study how a parasite as a potential biological control agent could spread in these crabs. Gagnon et al. (2015) derive a much more complex model for the dispersal of the green algae *Codium fragile* along the east coast of North America. They simulate an individual-based model that includes buoyancy, oxygen levels, daylight, water drift, and other aspects to estimate a dispersal kernel and then calculate the spread rates in different directions.

Stover et al. (2014) explore how population heterogeneity with respect to dispersal behavior affects population persistence and spread in the face of advection. We briefly touch on these questions in Sect. 12.5. Lewis et al. (2018) study the effect of biased dispersal on the dynamics of evolutionarily neutral variation in a population; see also Chap. 17.

12.3 Moving-Habitat Models

Most species have a preferred or an optimal temperature range within which they can successfully survive and reproduce. When individuals are exposed to temperatures outside that range, higher or lower, they suffer increased mortality and decreased reproductive output. Climate change scenarios predict that the spatial locations of a given mean annual temperature will shift toward higher altitude or latitude. To remain within their optimal temperature range, individuals will have to move in space and follow their shifting habitats. Zhou and Kot (2011) formulate the first IDE model to explore which combination of dispersal ability and growth function would allow a population to track their optimal habitat conditions and persist in the face of climate change. They parameterize their model for Fender's blue butterfly (*I. icarioides fenderi*). We present some of their ideas and results here.

We assume that the suitable habitat is an interval of length L that moves to the right at some constant speed $c > 0$. Inside this habitat, the population grows according to some growth function F; outside the habitat, the environment is hostile. Individuals move according to some dispersal kernel $K = K(x - y)$. The equation for the population density is (Zhou and Kot 2011)

$$N_{t+1}(x) = \int_{-L/2+ct}^{L/2+ct} K(x - y) F(N_t(y)) \, dy \, . \tag{12.35}$$

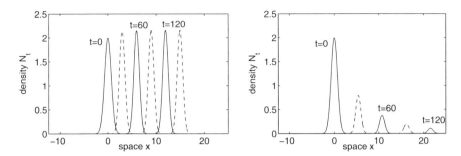

Fig. 12.5 A population in a moving habitat may persist if the habitat moves slowly ($c = 0.1$, left plot) or go extinct if the habitat moves fast ($c = 0.18$, right plot), according to model (12.35) with Beverton–Holt growth function (2.13) and Gaussian dispersal kernel. Parameters are $R = 1.7$, $K = 50$, $\sigma^2 = 0.25$, and $L = 1$. The suitable habitat is initially centered at $x = 0$. The initial density is a Gaussian, also centered at $x = 0$. Densities are plotted, alternating between solid and dashed curves, every 30 generations.

Since the habitat is moving, we cannot expect a steady-state solution. Instead, the simulations in Fig. 12.5 show two scenarios, depending on parameter values. For small values of c, the initial population distribution moves to the right, following the suitable habitat, and seems to stabilize at a "traveling pulse" solution (left plot). For large values of c, the initial distribution also travels to the right but continually decreases in height and seems to approach zero (right plot). Hence, we expect there to be some critical speed for habitat shift, below which the population can track its optimal range and above which it cannot. We aim to find this threshold.

A (rightward) traveling pulse with speed $c > 0$ is a solution of the form

$$N_t(x) = P(x - ct),\tag{12.36}$$

with the property that $P(\pm\infty) = 0$. Substituting this pulse solution into the IDE above, we find the equation

$$P(x - c(t + 1)) = \int_{-L/2+ct}^{L/2+ct} K(x - y)F(P(y - ct))\mathrm{d}y.\tag{12.37}$$

The change of variables $\bar{x} = x - c(t + 1)$ and $\bar{y} = y - ct$ results in

$$P(\bar{x}) = \int_{-L/2}^{L/2} K(\bar{x} + c - \bar{y})F(P(\bar{y}))\mathrm{d}\bar{y}.\tag{12.38}$$

The suitable habitat is now fixed in time at $[-L/2, L/2]$. Instead, the dispersal kernel is shifted by the term $+c$ in its argument. The situation is reminiscent of and closely related to the model with biased dispersal in the previous section. We want to find conditions for which (12.38) has a stable positive solution. Since this problem is a steady-state problem, we apply the theory from Chap. 4. We assume that growth function F is monotone and has no Allee effect, linearize at the trivial steady state, and obtain the existence of a stable positive steady state if the trivial state is unstable.

The eigenvalue problem for the stability of the trivial state is

$$\lambda \phi(\bar{x}) = R \int_{-L/2}^{L/2} K(\bar{x} - (\bar{y} - c)) \phi(\bar{y}) d\bar{y}, \tag{12.39}$$

where $R = F'(0)$. Unfortunately, the reduction of the integral equation with the Laplace kernel to a differential equation (see Chap. 3) does not carry over to the situation here when $c \neq 0$ since no appropriate boundary conditions can be derived. Therefore, most studies of moving-habitat models are based on numerical evaluation or analytical approximation of the eigenvalue. An exception is the separable cosine kernel from (3.24).

We write the cosine kernel in the form

$$K(x) = \frac{1}{4\ell} \cos\left(\frac{x}{2\ell}\right) \quad \text{for} \quad |x| \le \ell\pi \tag{12.40}$$

and $K(x) = 0$ for $|x| > \ell\pi$. To ensure that the kernel is positive on the domain of integration, we require

$$\ell\pi > L \quad \text{and} \quad 0 < c < \ell\pi - L. \tag{12.41}$$

Just as in (3.24), this kernel is separable, and the integral eigenvalue problem (12.39) turns into an eigenvalue problem of the 2×2 matrix (compare (3.29))

$$\frac{R}{4} \begin{bmatrix} \left[\frac{L}{2\ell} + \sin\left(\frac{L}{2\ell}\right)\right]\cos\left(\frac{c}{2\ell}\right) & \left[\frac{L}{2\ell} - \sin\left(\frac{L}{2\ell}\right)\right]\sin\left(\frac{c}{2\ell}\right) \\ \left[-\frac{L}{2\ell} - \sin\left(\frac{L}{2\ell}\right)\right]\sin\left(\frac{c}{2\ell}\right) & \left[\frac{L}{2\ell} - \sin\left(\frac{L}{2\ell}\right)\right]\cos\left(\frac{c}{2\ell}\right) \end{bmatrix}. \tag{12.42}$$

The stability boundary of the trivial state of IDE (12.38) is given when the dominant eigenvalue of the linearization (12.39) equals one. Substituting $\lambda = 1$ into the characteristic polynomial of the above matrix gives us the defining equation for the stability boundary as

$$1 - \frac{RL}{4\ell}\cos\left(\frac{c}{2\ell}\right) + \frac{R^2}{16}\left(\frac{L^2}{4\ell^2} - \sin^2\left(\frac{L}{2\ell}\right)\right) = 0. \tag{12.43}$$

To ensure that $\lambda = 1$ is indeed the dominant eigenvalue of the matrix, we require

$$\frac{R^2}{16}\left(\frac{L^2}{4\ell^2} - \sin^2\left(\frac{L}{2\ell}\right)\right) < 1. \tag{12.44}$$

Fig. 12.6 Critical shift speed according to (12.45) for the cosine kernel. The population can persist in the moving habitat if the shift speed is slower than the critical speed. The parameter is $\ell = 1$.

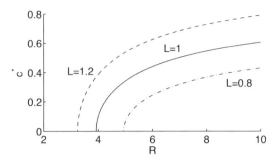

We can solve this equation for speed c and obtain the critical shift speed

$$c^* = \frac{L}{2\ell} \cos^{-1} \left(\frac{16 - R^2 \left(\frac{L^2}{4\ell^2} - \sin\left(\frac{L}{2\ell}\right)\right)}{8R\frac{L}{2\ell}} \right). \tag{12.45}$$

This formula is valid as long as the restrictions in (12.41) and (12.44) hold. If the habitat shifts faster than this critical speed, the population will go extinct. If the habitat shifts more slowly, then the population will persist and a positive traveling pulse will form. The critical speed in (12.45) depends on two quantities: the population growth rate, R, and a measure of the size of the habitat relative to the dispersal distance, L/ℓ. Figure 12.6 shows that the critical shift speed increases with each of these parameters.

For kernels where no explicit calculation of the eigenvalue is possible, Zhou and Kot (2013) present several ways to approximate the dominant eigenvalue in (12.39) and find corresponding approximations of the critical shift speed. Kot and Phillips (2015) apply results on eigenvalue approximation to derive bounds for spreading speeds in moving-habitat models. Rinnan (2017) uses some of their ideas, including symmetrization, to calculate approximate persistence conditions; see Chap. 9. Following Zhou and Kot (2013), we present one approach based on polynomial approximation.

If we expand a dispersal kernel in terms of the (orthogonal) Legendre polynomials on the interval $[-L/2, L/2]$, we obtain an infinite matrix eigenvalue problem; see Chap. 3. We obtain approximations by truncating this infinite matrix at some finite number. Since the lowest-order Legendre polynomial is a constant, truncating at the lowest order is equivalent to the dispersal success approximation of the eigenvalue in (9.13), i.e.,

$$\lambda \approx \frac{R}{L} \int_{-L/2}^{L/2} \int_{-L/2}^{L/2} K(\bar{x} + c - \bar{y}) d\bar{y} d\bar{x}. \tag{12.46}$$

We now assume that the kernel is differentiable and expand the term under the integral to second order to find

$$\lambda \approx \frac{R}{L} \int_{-L/2}^{L/2} \int_{-L/2}^{L/2} \left(K(\bar{x}) + K'(\bar{x})(c - \bar{y}) + \frac{K''(\bar{x})}{2}(c - \bar{y})^2 \right) d\bar{y}d\bar{x}.$$

(12.47)

The integrals can be evaluated particularly easily if the kernel is symmetric, i.e., $K(x) = K(-x)$. We find

$$\lambda \approx R \int_{-L/2}^{L/2} K(\bar{x})d\bar{x} + RK'(L/2) \left(c^2 L + \frac{L^2}{12} \right).$$

(12.48)

When the kernel is not symmetric, the expression is more complicated but still quadratic in c (Zhou and Kot 2013). We can solve for an approximation of the critical shift speed, c^*, by setting $\lambda = 1$. We find

$$(c^*)^2 \approx \left(\frac{1}{R} - \int_{-L/2}^{L/2} K(\bar{x})d\bar{x} \right) \left(K'\left(\frac{L}{2}\right) \right)^{-1} - \frac{L^2}{12}.$$

(12.49)

There are two critical shift speeds, one in the positive and one in the negative direction. The direction in which the habitat shifts does not matter. The two have the same magnitude since we assumed the kernel to be symmetric.

The Gaussian kernel with zero mean is differentiable and symmetric and hence satisfies the assumptions. The above formula applied to the Gaussian kernel with variance σ^2 gives the expression

$$c^* \approx \pm \sqrt{\frac{2\sigma^3 \sqrt{2\pi} \left[R \operatorname{erf}(\sqrt{2}L/(4\sigma)) - 1 \right]}{R L \exp(-L^2/(8\sigma^2))} - \frac{L^2}{12}},$$

(12.50)

where $\operatorname{erf}(\cdot)$ denotes the error function. For the parameters in Fig. 12.5, i.e., $L = 1$, $R = 1.7$, and $\sigma^2 = 0.25$, we find $c^* \approx \pm 0.12$, which is between the faster and the slower speed chosen in the figure.

Moving-habitat models have also been studied in the framework of reaction–diffusion equations, even prior to the work in IDEs. We refer to the references in Zhou and Kot (2011) and Harsch et al. (2017) for this literature. Here, we concentrate on IDEs. Fuller et al. (2015) include harvesting in the model by Zhou and Kot (2011) and explore how the combination of a shifting climate and exploitation by humans affects population dynamics. They parameterize their model for black rockfish (*Sebastes melanops*). Harsch et al. (2014) extend many of these ideas to species with a more complex life cycle; see Chap. 13. They obtain systems of IDEs and study the resulting eigenvalue problems for persistence numerically. They parameterize models for four different plant species and calculate the sensitivity and elasticity of the critical speed to model parameters. Bouhours and Lewis (2016) study population persistence in a moving-habitat model where habitat size and population growth rate vary stochastically in time; see Chap. 16. Santini et al. (2016) ask how well terrestrial mammals might be able to track climate. They generate virtual species with realistic life histories and compare the stage-structured IDE models from Harsch et al. (2014) with individual-based simulation models. Lewis et al. (2018) study how a moving habitat affects the dynamics of neutral genetic variation within a population. Li et al. (2016a) consider a different spatial setting for a moving habitat. Instead of a single bounded suitable patch with abrupt edges, they assume a gradient-like environment, i.e., where habitat quality is a nondecreasing function of space. We discuss their approach in more detail in Chap. 15. Harsch et al. (2017) review IDE and reaction–diffusion models for moving habitats in the context of species management and give various examples of how to incorporate additional mechanisms, such as a shifting phenology, into IDE models for population dynamics in moving habitats.

12.4   Sessile Stages

Many organisms that possess clearly distinct growth and dispersal phases in their life cycle, and hence could be modeled appropriately by an IDE, exhibit overlapping generations, so that the simple model formulation in (2.1) may not be appropriate. We can still model the dynamics of such populations by IDEs if we write a system of equations, one for each stage in the life cycle; see Chap. 13. Here, we present a simpler scenario where the dynamics of a population can be captured by a relatively small extension of IDE (2.1), namely by including a nondispersing (sessile) component. We begin with a brief overview of several examples from the literature and then concentrate the analysis on a few select questions.

Annual Plants with Seed Bank

Even simple annual plants typically have a seed bank, where seeds may remain dormant and sessile in the soil for a number of years. Seeds disperse only in the year when they are produced by the plant. We denote by $N_t(x)$ the density of seeds in the soil at the beginning of the growing season in year t. Seeds remain dormant with probability $1 - g$ and germinate with probability g. Seeds that did not germinate survive until the next growing season with probability s. Then the seed density changes from year to year according to the equation

$$N_{t+1}(x) = s(1 - g)N_t(x) + \int K(x, y)F(gN_t(y))\mathrm{d}y\,, \qquad (12.51)$$

where $F(N)$ denotes the number of viable seeds produced by N seedlings. Similar and also much more elaborate IDEs for plants with seed banks have been formulated and studied from various perspectives. Allen et al. (1996) derive several models for a single and two competing species with nondifferentiable updating function. Latore et al. (1998) consider a structured seed bank by keeping track of when seeds were produced. Mistro et al. (2005b) consider seeds that can survive at most one year and formulate the corresponding model as a delayed IDE, where the density of plants in year $t + 1$ depends on their density in years t and $t - 1$. Li (2012) studies traveling waves in model (12.51), whereas Meyer (2012) and Meyer and Li (2013) generalize these results to an infinite system for age classes of seeds. Lutscher and Van Minh (2013) formulate a similar model and reduce it to two stages under several simplifying assumptions. Dewhirst and Lutscher (2009) study a model similar to (12.51) for trees in a heterogeneous landscape, where seeds that fall too close to the parent tree are considered nondispersing.

Aquatic Organisms

Many aquatic organisms, such as mussels and barnacles, have sessile adult stages that survive for many years and distribute their offspring in space. In addition, the dispersal distances of adults of many other species, such as crabs, are negligible compared to those of their offspring. Kanary et al. (2014) formulate a model for the density of European green crab along the coast of Nova Scotia as

$$N_{t+1}(x) = p(N_t(x))N_t(x) + s(N_t(x)) \int K(x - y)R(N_t(y))N_t(y)\mathrm{d}y, \quad (12.52)$$

where p, s, and R denote the (potentially density-dependent) probability of adult survival, probability of juvenile settling, and adult per capita reproduction, respectively. The dispersal term contains the local conditions at the locations of departure *and* arrival of the dispersal process. Kanary et al. (2014) consider a single and two competing species. They determine conditions for spread against the dominant current, as well as potential outcomes of competition. A corresponding stage-structured model for the same species by Marculis and Lui (2015) tracks mobile juveniles and sessile adults as well as a parasite infection. Gharouni et al. (2017) apply the same ideas but include stochastic variation in the IDE. Britton-Simmons and Abbott (2008) formulate and parameterize a model for the invasive brown algae *Sargassum muticum*. These organisms usually remain attached to the substratum, but their propagules detach and float to new locations. The authors couple the equation for the adult density with a stochastic equation for the availability of free space created by grazers in the system.

Birds and Mammals

In many bird and mammal species, juveniles stay with their parents until maturation, and spatial dispersal is often driven by crowding effects in the population. Veit and Lewis (1996) study how house finches spread westward after being released in the

New York area. Juveniles mature within the first year; both juveniles and adults may remain sessile or may disperse. The equation for the total density of birds in the spring of year t is

$$
\begin{aligned}
N_{t+1}(x) &= s_a(1 - p_a)N_t(x) + s_a p_a \int K_a(x, y)N_t(y)dy \\
&\quad + (1 - p_j)F(N_t(x)) + p_j \int K_j(x, y)F(N_t(y))dy .
\end{aligned}
\tag{12.53}
$$

The first line corresponds to adults that survive with probability s_a and disperse with probability p_a. The second line tracks juveniles that disperse with probability p_j. Both probabilities are density dependent. Adults produce juveniles according to growth function F. Veit and Lewis (1996) explore the consequences of an Allee effect on spread rates; see Sect. 12.6. Some analytical aspects of this model in the absence of an Allee effect are studied by Le et al. (2011), in particular the existence of monotone, continuous traveling waves.

Lutscher (2008) formulates a toy model, reminiscent of the models above, to study the effect of population density on the probability of dispersal as

$$
N_{t+1}(x) = g F(N_t(x)) + \int K(x, y)(1 - g)F(N_t(y))dy ,
\tag{12.54}
$$

where $g = g(aF(N))$ is the probability of remaining sedentary. It is assumed to be a decreasing function of population density. Parameter a measures how sensitive the dispersal probability is to population density. The case of constant g is analyzed in Volkov and Lui (2007), and further numerical simulations for the density-dependent case can be found in Leo (2007).

Critical Patch-Size

The analysis of a critical patch-size in the presence of nondispersing states has received only marginal attention (Lutscher 2008). Let us consider Eq. (12.54) with constant g on a bounded set Ω. Linearizing the equation at the trivial solution leads to the eigenvalue problem

$$
\lambda\phi(x) = Q_g\phi(x) = g R\phi(x) + (1 - g)R \int_\Omega K(x, y)\phi(y)dy ,
\tag{12.55}
$$

with $R = F'(0)$.

Operator Q_g is not compact for $g > 0$, so the existence of a dominant eigenvalue is not guaranteed by the previous theory. In the simple case, when R is constant in space, the problem is equivalent to

$$v\phi(x) = \int_{\Omega} K(x, y)\phi(y)dy\,, \qquad v = \frac{\lambda - gR}{R(1-g)}\,. \qquad (12.56)$$

This equation is of the form that we encountered in Chap. 3. Under the standard assumptions on K, there is a dominant eigenvalue $0 < v \le 1$ with positive eigenfunction. Accordingly, we obtain λ as

$$\lambda = vR(1-g) + Rg\,. \qquad (12.57)$$

Clearly, λ is an increasing function of g. As the propensity to remain sedentary increases, the likelihood of the population to persist also increases since dispersal carries a risk of loss but remaining sedentary does not. The situation can change when habitat quality varies temporally so that remaining sedentary can be risky (Lutscher 2008).

In a much more general setting, Jin et al. (2016) consider questions of population extinction and persistence in a discrete dynamical system on some Banach space if the next-generation operator can be written as a sum of a linear contraction and a compact operator. They show that persistence is governed by the dominant eigenvalue of an appropriately defined operator.

The Speed of Spatial Spread

The analysis of spreading speeds and traveling waves in the presence of sessile stages has received considerably more attention than the critical patch-size problem. The challenge in analyzing equations with immobile individuals is that the resulting next-generation operator fails to be compact. While the results of Weinberger (1982) on the existence of a spreading speed still hold, proving the existence of a traveling front requires additional arguments but can be done (Allen et al. 1996; Volkov and Lui 2007; Le et al. 2011; Lutscher and Van Minh 2013; Li 2012; Meyer 2012; Meyer and Li 2013; Le and van Nguyen 2017). In special cases, the profile of a traveling front can be computed explicitly (Allen et al. 1996; Lutscher 2008).

One surprising and biologically relevant aspect is the dependence of the spreading speed on parameter g, the probability of not dispersing. We study this dependence under the usual conditions that (i) there is no Allee effect and (ii) the dispersal kernel is exponentially bounded and of the form $K(x, y) = \widetilde{K}(x - y)$ with $\widetilde{K}(x) = \widetilde{K}(-x)$. Then the spreading speed of model (12.54) with constant g is given by formula (5.17) (Weinberger 1982). To emphasize the dependence on g, we write

$$\hat{c}(g) = \inf_{s>0} \frac{1}{s} \ln\left(R[(1-g)M(s) + g]\right), \qquad (12.58)$$

where M is the moment-generating function of $\widetilde{K}$. The following lemma is an extension of Lemma 3.1 from Lutscher (2008).

Lemma 12.1 *The spreading speed, $\hat{c}$, in (12.58) is monotonically decreasing in g, and $\hat{c}(1) = 0$. If $M(s)$ is defined on all of $\mathbb{R}$ (e.g., the Gaussian kernel), then $\hat{c}$ is continuous at $g = 1$. If $\lim_{s \to a} M(s) = \infty$ for some $a < \infty$ (e.g., the Laplace kernel), then $\lim_{g \to 1} \hat{c}(g) = \ln(R)/a > 0$. In the latter case, $\hat{c}$ is not continuous at $g = 1$.*

Proof Since $\widetilde{K}$ is even, we have $M(s) > 1$ for $s \neq 0$. Since the expression inside the logarithm in (12.58) is a decreasing function of g, so is $\hat{c}$. The statement $\hat{c}(1) = 0$ is obvious since $\lim_{s \to \infty} 1/s = 0$. Biologically, this statement simply means that without dispersal, the population does not spread.

Now suppose that the moment-generating function of some kernel is defined and continuous on all of $\mathbb{R}$. Pick any small $\varepsilon > 0$. Choose $\hat{s}$ such that $\ln(R)/\hat{s} < \varepsilon/2$. By continuity of the expression in square brackets, there is some $g < 1$ such that

$$\frac{1}{\hat{s}} \ln \left(R[(1 - g)M(\hat{s}) + g] \right) < \varepsilon. \tag{12.59}$$

In particular, then, $\hat{c}(g) < \varepsilon$. Hence, $\hat{c}$ approaches zero as $g \to 1$.

Next, suppose that the moment-generating function is defined only for $s \in (-a, a)$ and $M(s) \to \infty$ as $s \to a$. The infimum in (12.58) becomes a minimum over $0 < s < a$. Clearly, $\hat{c}(g) \geq \ln(R)/a$. The function $s \mapsto \ln(R)/s$ is continuous and decreasing for $s > 0$. For any $\varepsilon > 0$, there is some $\hat{s} < a$ such that $\ln(R)/\hat{s} < \ln(R)/a + \varepsilon$. Set $\tilde{s} = \frac{\hat{s}+a}{2}$. Then there is some $\hat{\varepsilon} > 0$ such that for $g \in (1 - \hat{\varepsilon}, 1]$ we have $(1 - g)M(\tilde{s}) + g < e^{\hat{s}\varepsilon}$. Now we calculate

$$\frac{\ln \left(R[(1 - g)M(\tilde{s}) + g] \right)}{\tilde{s}} \leq \frac{\ln \left(Re^{\hat{s}\varepsilon/2} \right)}{\hat{s}} < \frac{\ln(R)}{\hat{s}} + \varepsilon < \frac{\ln(R)}{a} + 2\varepsilon. \tag{12.60}$$

This proves the claim. In particular, $\hat{c}$ is not continuous at $g = 1$. $\square$

12.5   Multiple Dispersal Modes

Some empirical dispersal data exhibit patterns that are difficult to fit with the various dispersal kernels that we have seen so far. For example, seed trap data for heather plants reveal that a large fraction of seeds is concentrated near the release point, while at the same time, a relatively long tail exists (Bullock and Clarke 2000). Fitting an exponential or inverse power law kernel to the data leads to a concentration near zero and underestimates the tail. Instead, a linear combination of two exponential kernels captures both features. Their observation indicates that a (small) fraction of seeds dispersed differently from the rest. For example, these seeds may have been moved long distances by strong but rare winds. Since we know that the behavior of the tails of the dispersal kernel is crucial to determining the speed of spatial spread, the question is how big the influence of a small percentage of long-distance dispersers could be. We explore this question in the context of Reid's paradox of rapid tree migration (Clark et al. 1998a). Buckley et al. (2005) employ a mix of a Gaussian and a Laplace kernel to predict the spread of pine trees (*pinus nigra*) in New Zealand. Fort (2007) revisits Reid's paradox with mixed kernels and calculates front speeds in one- and two-dimensional models.

Reid's Paradox of Rapid Tree Migration

In 1899, Clement Reid asked how plants that "merely scatter their seeds" could have moved northward since the last glacial period to reach their current locations in northern Britain (Skellam 1951). Oak trees (*Quercus*), for example, covered a distance of about 1000 km. Oaks have a very high lifetime fecundity of about $R_0 = 10^7$ (Clark et al. 1998a), but they also have a generation time of at least 60 years. With the glacial period between 10,000 and 18,000 years ago, there would be no more than 300 generations for oaks to move this distance. At a constant speed, these data result in a speed of $c \approx 3.3$ km per generation. If we assume a Gaussian dispersal kernel with variance σ^2, the spread speed formula $c^* = \sqrt{2\sigma^2 \ln(R_0)}$ gives the required variance of

$$\sigma^2 = \frac{c^2}{2\ln(R_0)} \approx 0.34 \text{ km}^2 \tag{12.61}$$

or a mean dispersal distance of $\sqrt{2/\pi}\sigma \approx 0.47$ km. Clark et al. (1998b) empirically find a dispersal distance of 11.8 m for acorns. It is known that some birds and mammals can transport seeds and fleshy fruit over large distances (several kilometers) and thereby increase the spread rate, but this long transport is thought to be very rare. What difference would such rare events make?

To model this situation, we consider a model with two dispersal kernels, K_1 and K_2, that correspond to short- and long-distance dispersal, respectively. We denote by g and $1 - g$ the probabilities that an organism disperses according to dispersal kernels K_1 and K_2, respectively. Reproduction is unaffected by dispersal behavior. The linear IDE becomes

$$N_{t+1}(x) = R \int [gK_1(x - y) + (1 - g)K_2(x - y)]N_t(y)dy. \tag{12.62}$$

In the special case that $K_1(x) = \delta(x)$ is the delta distribution, i.e., individuals do not disperse, we recover the model from the previous section. For simplicity, we assume that both kernels are symmetric. The formula for the spreading speed, analogous to (12.58), is

$$\hat{c}(g) = \inf_{s>0} c(s, g) = \inf_{s>0} \frac{1}{s} \ln \left(R[g M_1(s) + (1-g) M_2(s)] \right). \tag{12.63}$$

The following lemma shows that, just as in Lemma 12.1, the spreading speed need not be continuous at $g = 1$.

Lemma 12.2 *Assume that K_i in (12.62) are even functions, and assume that their moment-generating functions $M_i(s)$ exist for at least one nonzero value of s. Denote by $\hat{c}(g)$ the spreading speed from (12.63) and by $s^* < \infty$ the value of s that minimizes the expression $c(s, 1) = \ln(R M_1(s))/s$. If $M_2(s^*)$ is finite, then $\hat{c}(g)$ is continuous at $g = 1$. If $M_2(s) \to \infty$ as $s \to a$ for $a < s^*$, then $\hat{c}(g)$ is not continuous at $g = 1$ and $\lim_{g \to 1^-} \hat{c}(g) = \ln(R M_1(a))/a$.*

Proof If $M_1(s) < M_2(s)$ for all s where M_2 exists, then the proof of Lemma 12.1 carries over with small modifications. The first part is even simpler. Suppose at first that $M_2(s^*)$ is finite. For every $\epsilon > 0$, there is some $\delta > 0$ such that for $g \in (1-\delta, 1]$, we have $|c(s^*, g) - \hat{c}(g)| < \epsilon$, since $c(s, g)$ is continuous in g for every fixed s where c is defined. Hence, $\lim_{g \to 1^-} |\hat{c}(g) - \hat{c}(1)| = 0$, so that the function is continuous at $g = 1$.

If $M_2(s) \leq M_1(s)$ for some s, we need a lower bound to prove discontinuity. Assume that $M_2(s) \to \infty$ as $s \to a < s^*$. The infimum in (12.63) becomes a minimum over $(0, a)$. Note that $\ln(R M_1(s))/s$ is decreasing on $(0, a)$. For every small enough $\epsilon > 0$, the set $[\epsilon, a-\epsilon]$ is compact. Hence, we can choose some $\delta > 0$ such that we have $|c(s, g) - c(s, 1)| < \epsilon$ for all $s \in [\epsilon, a-\epsilon]$ and $g \in (1-\delta, 1]$. Since $c(s, 1)$ is decreasing, we have $|\min_{\epsilon < s < a-\epsilon} c(s, g) - c(a-\epsilon, g)| < \epsilon$ for $g \in (1-\delta, 1]$. As $\epsilon \to 0$, we have $\delta \to 0$ and hence $\lim_{g \to 1^-} \hat{c}(g) = c(a, 1)$. Since $c(s, 1)$ is decreasing for $a < s < s^*$, we have $c(a, 1) > c(s^*, 1) = \hat{c}(1)$, so the function is not continuous at $g = 1$. $\qquad\square$

To apply this result to Reid's paradox, we assume that K_1 represents short-distance dispersal as a Gaussian kernel with (small) variance σ_1^2. If all acorns disperse a short distance ($g = 1$), we find a speed of $\hat{c}(1) = \sigma_1 \sqrt{2 \ln(R)}$ that occurs at $s^* = \sigma_1 \sqrt{2 \ln(R)}$. Next, we let K_2 represent long-distance dispersal as a Laplace kernel with (large) variance σ_2^2 or parameter $a = \sqrt{2}/\sigma_2$. If $a < s^*$, then even the tiniest fraction of long-distance dispersers will lead to a jump in the speed; see Fig. 12.7. For small reproduction ($R < e$), the variance of the Laplace kernel has to be larger than that of the Gaussian kernel for the jump in spreading speed at $g = 1$ to occur. However, for high reproduction ($R > e$), it is not even necessary that the variance be larger. Since the tails of the Laplace kernel naturally decay more slowly than those of the Gaussian kernel, high reproduction is sufficient to speed up the invasion considerably.

In a series of papers, Clark and coworkers (Clark 1998; Clark et al. 1998a,b) carefully demonstrate that data on tree dispersal are consistent with a sufficient fraction of seeds dispersing sufficiently long distances to explain the inferred spread

Fig. 12.7 Illustrating Lemma 12.2. The plot shows the minimal speed, $\hat{c}(g)$, of model (12.62) with $R = 1.5$ when K_1 is the Gaussian kernel with variance $\sigma^2 = 0.2$ and K_2 is either a Gaussian kernel with $\sigma^2 = 1$ (solid) or a Laplace kernel with $\sigma^2 = 1$ (dash-dot) or $\sigma^2 = 0.4$ (dashed).

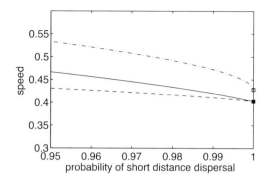

rate of trees after the last glacial period in Britain and North America. Their approach is built around the two-parameter family of dispersal kernels

$$K(x) = \frac{c}{2\alpha \Gamma(1/c)} e^{-|x/\alpha|^c}. \tag{12.64}$$

These kernels are heavy tailed for certain parameter values ($c < 1$); i.e., they do not possess a moment-generating function. In particular, their model does not have an asymptotic spreading speed but accelerating invasions when $c < 1$. In the next section, we present one mechanism for how such kernels may arise.

The spreading speed $\hat{c}(g)$ in (12.63) has another surprising property: it need not be a monotone function of g, as the following lemma shows.

Lemma 12.3 *Assume that K_i in (12.62) are even functions and that their moment-generating functions $M_i(s)$ exist for at least one nonzero value of s. Denote by $c_i^* = \min_{s>0} c_i(s) = \min_{s>0} \ln(RM_i(s))/s$, $i = 1, 2$, the spreading speeds for $g = 1$ and $g = 0$, respectively, and by s_i^* the argument at the minimum. If $M_1(s) \neq M_2(s)$ for all s between s_1^* and s_2^*, then $\hat{c}(g)$ is monotone in g. If $M_1(\bar{s}) = M_2(\bar{s})$ for some $\bar{s}$ between s_1^* and s_2^*, then $\hat{c}(g)$ is not monotone in g.*

Proof Without loss of generality, we assume $c_1^* \leq c_2^*$. For fixed s, the function $g \mapsto gM_1(s) + (1 - g)M_2(s)$ is nonincreasing if and only if $M_1(s) \leq M_2(s)$. If $M_1(s)$ and $M_2(s)$ do not intersect between s_1^* and s_2^*, then neither do $c_1(s)$ and $c_2(s)$. Since we assumed $c_1^* \leq c_2^*$, we have $M_1(s) \leq M_2(s)$ and $c_1(s) \leq c_2(s)$ for all s between s_1^* and s_2^*. Function $c(s, g)$ from (12.63) is convex in s for each g. Since $c_i(s)$ are convex functions, the minimum of $c(s, g)$ with respect to s occurs for some s_g^* between s_1^* and s_2^* for each g. Then we have, for $\tilde{g} > g$,

$$\hat{c}(g) = \frac{1}{s_g^*} \ln \left(R[gM_1(s_g^*) + (1 - g)M_2(s_g^*)] \right)$$

$$\geq \frac{1}{s_g^*} \ln \left(R[\tilde{g}M_1(s_g^*) + (1 - \tilde{g})M_2(s_g^*)] \right) \geq \hat{c}(\tilde{g}). \tag{12.65}$$

Hence, $\hat{c}$ is nonincreasing.

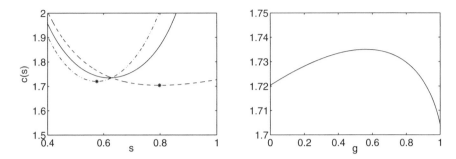

Fig. 12.8 Illustrating Lemma 12.3. The left plot shows the dispersion relations of the top-hat kernel (dashed) and the Laplace kernel (dash-dot), as well as $c(s)$ for the mixed kernel (solid). Stars on the curves indicate the minima. The right plot shows $\hat{c}(g)$ according to formula (12.63). Parameters are $\beta = 2.9$ for the top-hat kernel, $a = 1$ for the Laplace kernel, $g = 0.5$, and $R = 1.8$.

Next, we assume that there exists some $\bar{s}$ between s_1^* and s_2^* such that $M_1(\bar{s}) = M_2(\bar{s})$. We necessarily also have $c_1(\bar{s}) = c_2(\bar{s}) = c(\bar{s}, g)$ for all $g \in [0, 1]$. Since c_i are convex functions with a unique minimum at s_i^* and since $\bar{s}$ is between s_1^* and s_2^*, we must have $c(\bar{s}, g) > \min(c_1^*, c_2^*)$ and $c_1'(\bar{s})c_2'(\bar{s}) < 0$. Therefore, the derivative $\frac{\partial c}{\partial s}(\bar{s}, g)$ changes sign for $g \in [0, 1]$. By continuity, there is some $\bar{g}$ such that $\frac{\partial c}{\partial s}(\bar{s}, \bar{g}) = 0$. Since for every fixed g, $c(s, g)$ is a convex function with respect to s, the point $c(\bar{s}, \bar{g})$ must be the minimum. Since $c(\bar{s}, \bar{g}) > \min(\hat{c}(0), \hat{c}(1))$, we see that $\hat{c}(g)$ cannot be monotone. $\square$

In Fig. 12.8, we illustrate a nonmonotone curve $\hat{c}(g)$ that arises when we choose the top-hat and the Laplace kernel; see Table 10.1. The left plot shows how the dispersion relations intersect between their minima. The right plot illustrates the hump-shaped function $\hat{c}(g)$.

Continuously Distributed Dispersal Behavior

The previous approach of combining two dispersal behaviors generalizes to a continuous distribution of dispersal behaviors. For example, wind-dispersed seeds experience different wind speeds and turbulences, depending on when they are released. This situation can lead to heavy-tailed kernels (Clark et al. 1999).

We assume that the dispersal distance r (in two space dimensions) for each individual is distributed according to a Gaussian distribution with parameter α as

$$f(r|\alpha) = \frac{1}{\pi\alpha^2}e^{-r^2/\alpha^2} . \tag{12.66}$$

If we denote by $\tilde{f}(\alpha)$ the distribution of dispersal parameter α in the population, then the dispersal kernel for the population is given by

$$K(r) = \int f(r|\alpha)\tilde{f}(\alpha)\mathrm{d}\alpha . \tag{12.67}$$

We write $\tilde{\alpha} = L/\alpha$, where L is a length-scale parameter. If we assume that $\tilde{\alpha}$ is gamma-distributed as

$$\tilde{f}(\tilde{\alpha}) = \frac{\tilde{\alpha}^{p-1}\mathrm{e}^{-\tilde{\alpha}}}{\Gamma(p)} , \tag{12.68}$$

we can explicitly evaluate the integral above and obtain (Clark et al. 1999)

$$K(r) = \frac{p}{\pi L \left(1 + \frac{r^2}{L}\right)^{p+1}} . \tag{12.69}$$

This distribution tends to the Gaussian kernel as $p \to \infty$ and to the Cauchy kernel as $p \to 0$.

The procedure of integrating an individual dispersal kernel over the distribution of the dispersal parameter in the population is used by several other researchers. Yamamura (2002) considers Gaussian dispersal with gamma-distributed settling times in a two-dimensional environment and finds accelerating invasions. Skarpaas and Shea (2007) consider the spread of wind-dispersed invasive weeds. They use the WALD model (7.61) for dispersal under fixed conditions and integrate over distributions of the physical parameters in the model to obtain a seasonal dispersal kernel. Petrovskii and Morozov (2009) use a diffusion equation for individual movement, resulting in a Gaussian kernel for individuals, and consider normally distributed diffusion coefficients in the population, which leads to heavy-tailed dispersal kernels. Stover et al. (2014) consider similar questions from the point of view of dispersal heterogeneity: how do certain distributions of dispersal parameters in the population affect population spread rate and critical reproduction rate in the face of advection; see Sect. 12.2. They give expressions of the moments of the dispersal kernel in terms of moments with which the dispersal trait is distributed in the population. Among other things, they find that dispersal heterogeneity increases spread rates.

12.6 Allee Effects

There are relatively few application studies of IDEs with Allee effect. Those that exist are largely based on numerical simulations since analytical calculations are typically impossible. In this section, we highlight some modeling aspects, analytical results where possible, and a few numerical outcomes.

Pair-Formation Model for an Allee Effect

Veit and Lewis (1996) study the spread of house finches in eastern North America since 1940. In particular, they address one curious observation of this invasion: during the first 10–20 years, the range radius expanded extremely slowly. Over the next 10 or so years, the speed of expansion picked up, and since about 1975, the spread has been quite fast; see Fig. 2A in Veit and Lewis (1996). In Chap. 5, we saw that the speed of a spreading front approaches its asymptotic value fairly quickly; see, e.g., Fig. 5.3. Why would it take so long in the case of the house finches?

Veit and Lewis (1996) argue that two factors are crucial in this slow initial spread: an Allee effect and a density-dependent dispersal probability. They use a mechanistic pair-formation model, similar to a chemical reaction, where one male and one female form a pair according to the law of mass action. We use τ to denote the continuous time variable during the breeding season. The equations for the male (m) and female (f) unpaired birds and the pairs (p) for the reaction $m + f \xrightarrow{v} p$ are

$$\dot{m} = -vmf, \quad \dot{f} = -vmf, \quad \text{and} \quad \dot{p} = vmf, \tag{12.70}$$

where the dot denotes the derivative with respect to τ. We denote the density of birds at the beginning of the breeding season by N and assume a 1:1 sex ratio. Then $m(0) = f(0) = N/2$ and $p(0) = 0$. Hence, we have $m(\tau) = f(\tau)$ for all τ. We simplify and write $n = m + f$ for the total number of unpaired birds. Then the equations become

$$\dot{n} = -\frac{v}{2}n^2 \quad \text{and} \quad \dot{p} = -\frac{1}{2}\dot{n} . \tag{12.71}$$

Their solutions are

$$n(\tau) = \frac{N}{1 + \frac{v\tau}{2}N} \quad \text{and} \quad p(\tau) = \frac{N^2}{\frac{4}{v\tau} + 2N} . \tag{12.72}$$

Only pairs that find an appropriate nesting site can breed. Veit and Lewis (1996) choose $(1 + p/\bar{p})^{-1}$ as the probability of finding a nesting site, where $\bar{p}$ denotes the pair density that leads to a 50% probability. If each breeding pair produces on average c offspring, the total number of offspring produced from N individuals is

$$F(N) = \frac{cp}{1 + p/\bar{p}} = \frac{cN^2}{\frac{4}{v\tau^*} + 2N + N^2/\bar{p}} , \tag{12.73}$$

where τ^* is the length of the breeding season.

To complete their model, Veit and Lewis (1996) assume that newborn birds become mature at the end of their first year, and that adults and juveniles have

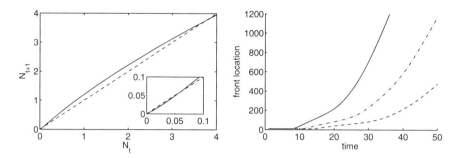

Fig. 12.9 Illustration of the house finch model (12.53) by Veit and Lewis (1996). **Left:** The nonspatial updating function $N_{t+1} = s N_t + F(N_t)$ with F as in (12.73). **Right:** The front location of an invasion for different lengths of the breeding season. Parameters are $c = 1.5$, $s = 0.67$, $\bar{p} = 1.5$, and $\nu = 1$ and $\tau^* = 16$ (dashed), $\tau^* = 26.6$ (dash-dot), and $\tau^* = 80$ (solid). The kernel parameters are $a = 0.47$ and $b = 41$ for juveniles and $a = 0.86$ and $b = 77$ for adults. We chose $N^* = 10$ and tracked the level set where $N_t(x_t) = 0.1$.

a density-dependent probability of dispersal, given by $p_A(N) = p_J(N) = \min(N/N^*, 1)$ for some threshold density N^*. Their final model is the one in (12.53) with growth function F in (12.73). They choose the modified Weibull kernel

$$K = \frac{a}{2 b \, \Gamma(1/a)} \exp\left(-\left(\frac{x}{b}\right)^a\right), \qquad (12.74)$$

with different parameters for juveniles and adults. They estimate model parameters from data and show that the rate of range expansion is close to zero because bird densities need to build up in the first years after introduction. Then the rate accelerates; see Fig. 12.9 here and Fig. 8A in Veit and Lewis (1996). If either of the two components, Allee effect and density-dependent dispersal, is dropped from the model, this accelerating effect is not present or at least not strong enough to provide a good fit to the data. We note that the modified Weibull kernel with the parameter estimates is not exponentially bounded, since parameters a for juveniles and adults are both less than one. Hence, we cannot expect a constant speed of spread.

The Timing of Mate Finding and Dispersal

Hurford et al. (2006) also consider mate finding as a mechanism that can generate a (component) Allee effect. They ask how the timing of mate finding relative to dispersal affects the spread rate of a population. Their work is inspired by the observation that wolves (*Canis lupus*) recolonized the greater Yellowstone area fairly slowly. Slower than expected spread rates can indicate an Allee effect. Since population density is low at the front of an invasion, mate finding could have a particularly strong effect there. Trying to find a mate after dispersing even farther ahead of the front could be harder than before dispersing.

Hurford et al. (2006) keep the population dynamics aspects of their model fairly simple and concentrate on mate finding. They write the equation for the total population as

$$N_{t+1}(x) = F(N_t(x)) + D_t(x), \qquad (12.75)$$

where F describes the growth of the local density of established, nondispersing individuals and D_t the density of new "breeding units" that establish after dispersal. They assume that individuals disperse only if the local density reaches some critical threshold, N_c, and that local population density will not decline below N_c once it has reached this threshold. We assume density-independent growth below this threshold and choose the local growth function as $F(N) = \min(RN, N_c)$.

The region where dispersers are produced is $\Omega_t = \{x \mid N_t(x) \geq N_c\}$. We denote the density of dispersers produced at $y \in \Omega_t$ by $G_t(y)$. Within a (relatively small) local "search area" ϕ, the total density of dispersers is approximately $\phi G_t(y)$. Under the assumption of a 1:1 sex ratio, the number of pairs formed locally before dispersal is $\psi \phi G_t^2 / 2$, where ψ is the probability of pair formation. Pairs disperse together according to some dispersal kernel K. The density of new breeding units is then

$$D_t(x) = \sigma \psi \frac{\phi}{2} \int_{\Omega_t} G_t^2(y) K(x - y) dy, \qquad (12.76)$$

where σ is the average number of wolves in a newly formed pack.

Alternatively, if wolves disperse as individuals and form breeding pairs after dispersal, analogous considerations lead to the term

$$D_t(x) = \sigma \psi \frac{\phi}{2} \left(\int_{\Omega_t} G_t(y) K(x - y) dy \right)^2. \qquad (12.77)$$

The model becomes analytically tractable if we assume that the number of dispersers produced is a constant (γ) whenever $N \geq N_c$ and if we choose the Laplace kernel. When mate finding occurs before dispersal, the complete model reads

$$N_{t+1}(x) = r N_t(x) + A \int_{\Omega_t} \frac{a}{2} e^{-a|x-y|} dy, \quad \text{with} \quad A = \sigma \psi \phi \gamma^2 / 2. \qquad (12.78)$$

This model has a particular class of solutions for which one can explicitly calculate the speed of spread. First, we assume that the region $x \leq 0$ is populated by wolves at threshold density levels and that the region $x > 0$ is empty, i.e., $N_0 = N_c \chi_{(-\infty, 0]}$, where χ is the characteristic function. Then we calculate $N_1 = A \exp(-ax)/2$ for $x > 0$. Next, we assume that at time t, we have

$$\begin{cases} N_t(x) \geq N_c, & x < x_t, \\ N_t(x) = N_c e^{-a(x-x_t)}, & x \geq x_t. \end{cases} \qquad (12.79)$$

Fig. 12.10 Illustration of the invasion front (12.79) in model (12.78). The front advances by the distance calculated in (12.81) per time step. Behind the front, we only plot the critical density, N_c.

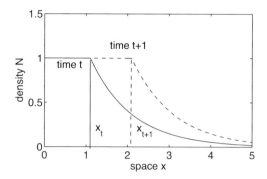

In particular, we have $\Omega_t = (-\infty, x_t]$. Then we calculate the next generation as

$$
\begin{cases}
N_{t+1}(x) \geq N_c, & x < x_t, \\
N_{t+1}(x) = (RN_c + A/2)\, e^{-a(x-x_t)}, & x \geq x_t.
\end{cases} \tag{12.80}
$$

In particular, we have $\Omega_{t+1} = (-\infty, x_{t+1}]$ for some x_{t+1}, and N_{t+1} decays exponentially for $x > x_{t+1}$ with the same rate as N_t. Hence, the invading front has just been shifted by $x_{t+1} - x_t$; see Fig. 12.10.

We can calculate the speed explicitly from the relation $N_{t+1}(x_{t+1}) = N_c$. The expression is

$$
x_{t+1} - x_t = \frac{1}{a} \ln \left(R + \frac{A}{2N_c} \right). \tag{12.81}
$$

Corresponding calculations for the model in which pair formation occurs after dispersal reveal that the invasion front does indeed spread more slowly under that scenario (Hurford et al. 2006).

Minimal Founding Population Size

Dispersal typically decreases the peaks of a given population density. When the growth function has a strong Allee effect, this decrease could push the population density below the Allee threshold so that the population will decline in subsequent generations, even when the initial population density was above the Allee threshold at some location; see, e.g., Fig. 1 in Lutscher and Petrovskii (2008) or Fig. 2 in Goodsman and Lewis (2016). This effect is somewhat similar to but also significantly different from the mechanism behind the critical patch-size in Chap. 3. In both cases, dispersal moves individuals from locations where the population grows to locations where it declines. In the case of the critical patch-size, population growth and decline are determined by the local habitat quality inside versus outside the suitable patch. In the case of the Allee effect, they are determined by whether the local population density is above or below the Allee threshold. Just like there is a minimal size of suitable habitat for a population without an Allee effect to persist, there is a minimal size of the initial population density for a population with a strong Allee effect to persist. Goodsman and Lewis (2016) present a simple model for a strong Allee effect and derive an explicit formula for this minimal founding population size. We present some of their ideas here.

We choose the power function $F(N) = RN^\gamma$ with $\gamma > 1$ to describe population growth. This function induces a strong Allee effect since $F(N) < N$ for $0 < N < N_a$ with Allee threshold

$$N_a = \left(\frac{1}{R}\right)^{\frac{1}{\gamma - 1}} . \tag{12.82}$$

We have $F(N_a) = N_a$ and $F(N) > N$ for $N > N_a$. The zero state is locally stable; the Allee threshold state is unstable. There is no stable positive state. As long as we are only interested in whether the population will grow and persist or decline and become extinct, we do not need there to be a positive stable state. (We need to assume, however, that there is no overcompensation; see Schreiber 2003.) The IDE

$$N_{t+1}(x) = R \int_{-\infty}^{\infty} K(x - y)[N_t(y)]^\gamma \, dy \tag{12.83}$$

with localized initial condition and Gaussian dispersal kernel can be solved explicitly (Goodsman and Lewis 2016). We assume that an initial population of n_0 individuals starts dispersing from $x = 0$ according to a Gaussian kernel, $G(x; \sigma^2)$, with variance σ^2 and mean zero. Then the first-generation density is

$$N_1(x) = n_0 G(x, \sigma^2) . \tag{12.84}$$

The expression for the second generation,

$$N_2(x) = RN_0^\gamma G(x, \sigma^2)^\gamma * G(x, \sigma^2) , \tag{12.85}$$

can be simplified since the power of a Gaussian distribution is again a Gaussian distribution and the convolution of two Gaussian distributions is again a Gaussian. We find

$$G(x, \sigma^2)^\gamma = \frac{1}{\sqrt{\gamma}(\sqrt{2\pi\sigma^2})^{\gamma-1}} G(x, \sigma^2/\gamma) , \tag{12.86}$$

and therefore

$$N_2(x) = \frac{Rn_0^\gamma}{\sqrt{\gamma}(\sqrt{2\pi\sigma^2})^{\gamma-1}} G\left(x, \frac{\gamma + 1}{\gamma}\sigma^2\right) . \tag{12.87}$$

This process can be iterated to find the general solution

$$N_t(x) = R^{\frac{\gamma^{t-1}-1}{\gamma-1}} n_0^{\gamma^{t-1}} G\left(x, \frac{\gamma^t - 1}{(\gamma - 1)\gamma^{t-1}}\sigma^2\right) h(t),$$

(12.88)

where

$$h(t) = \prod_{i=2}^{t}\left(\frac{1}{\sqrt{\gamma}\left(2\pi\sigma^2\frac{\gamma^{i-1}-1}{(\gamma-1)\gamma^{i-2}}\right)^{(\gamma-1)/2}}\right)^{\gamma^{t-i}}.$$

(12.89)

Goodsman and Lewis (2016) derive the corresponding formula in two spatial dimensions.

We want to find the minimal size of the founding population for the population to eventually remain above the Allee threshold, i.e., $N_t \geq N_a$ as $t \to \infty$. We set the expression in (12.88) equal to (12.82) and solve for n_0. We find

$$n_0 = N_a\sqrt{2\pi\sigma^2\frac{\gamma^t - 1}{(\gamma - 1)\gamma^{t-1}}}^{1/\gamma^{t-1}}\left(\frac{1}{h(t)}\right)^{1/\gamma^{t-1}}.$$

(12.90)

In the limit as $t \to \infty$, we find the minimal size of the founding population as

$$n_0^* = N_a\prod_{i=2}^{\infty}\sqrt{\gamma}^{\gamma^{1-i}}\sqrt{2\pi\sigma^2\frac{\gamma^{i-1}-1}{(\gamma-1)\gamma^{i-2}}}^{(\gamma-1)\gamma^{1-i}}.$$

(12.91)

This formula is somewhat unwieldy and because of the infinite sum also numerically tricky, but it can be simplified more and approximated as follows. First, we take logarithms on both sides of the expression to get

$$\ln(n_0^*) = \ln(N_a) + \frac{\ln(\gamma)}{2}\sum_{i=2}^{\infty}\gamma^{1-i} + \frac{\gamma - 1}{2}\sum_{i=2}^{\infty}\gamma^{1-i}\ln\left(2\pi\sigma^2\frac{\gamma^{i-1}-1}{(\gamma-1)\gamma^{i-2}}\right).$$

(12.92)

Then we use the geometric series to see that $(\gamma - 1)\sum_{i=2}^{\infty}\gamma^{1-i} = 1$. Finally, we split the infinite sum into the finite sum from $i = 2$ to $i = M$ and the infinite sum from $i = M + 1$ to ∞. Since $\gamma > 1$, the term $(\gamma^{i-1} - 1)/\gamma^{i-2}$ can be approximated by γ for large enough i. Another application of the geometric series and taking exponentials on both sides of the equation finally gives

$$n_0^* = N_a\sqrt{2\pi\sigma^2}\frac{\sqrt{\gamma}^{\frac{1}{\gamma-1}+\frac{1}{\gamma^{M-1}}}}{\sqrt{\gamma - 1}}\exp\left(\frac{\gamma - 1}{2}\sum_{i=2}^{M}\gamma^{1-i}\ln\left(\frac{\gamma^{i-1}-1}{\gamma^{i-2}}\right)\right);$$

(12.93)

see Goodsman and Lewis (2016) for the corresponding formula in two spatial dimensions. The minimum founding population depends linearly on the Allee threshold and the standard deviation of the Gaussian kernel. In two spatial dimensions, the relationship with the variance is linear.

Goodsman and Lewis (2016) apply this formula to estimate the minimum founding population size of mountain pine beetle. This forest insect experiences a strong Allee threshold because only a large number of beetles attacking simultaneously will overcome the defenses of the tree. A small number of attacking beetles will be washed out by tree sap (Heavilin and Powell 2008). Goodsman and Lewis (2016) use the previous model by Heavilin and Powell (2008), who implement the corresponding beetle growth function as

$$F(N) = \frac{N^2}{\alpha^2 + N^2} S, \tag{12.94}$$

where N denotes the number of attacking beetles and S is the density of healthy trees. For small enough N, this function is approximated by the power function $F(N) \approx SN^2/\alpha^2$, so the theory derived above applies. The model by Heavilin and Powell (2008) includes, in addition, a stage-structured description of the tree density; see Chap. 13.

The growth function in (12.94) is also used by Schofield (2002) (see also Barton and Turelli 2011) in a numerical comparison of continuous- and discrete-time models for a Wolbachia infection spreading in fruit flies. Using a modified Nicholson–Bailey IDE model, Goodsman et al. (2016) model how aggregation of mountain pine beetle can overcome an Allee effect induced by tree defenses. Sullivan et al. (2017) combine an Allee effect with overcompensation to obtain pulsating traveling waves; see Chap. 6. Otto (2017) also combines an Allee effect with overcompensation and finds nonspreading solutions.

A completely different representation of the Allee effect is used by Etienne et al. (2002). These authors define an upper and lower threshold, $\underline{N}$ and $\overline{N}$, for population growth and choose N to be a linear function in the interval $[\underline{N}, \overline{N}]$ and zero outside. They study numerically the stabilizing effect of dispersal on locally unstable dynamics.

12.7   Dispersal and Spread in Two Dimensions

Almost all of the theory and examples presented so far were formulated in idealized one-dimensional space. Real ecological systems are two- or three-dimensional. To what extent and how does the theory carry over to higher dimensions, and what needs to be considered in the application to data? The theoretical results on critical patch-size from Chap. 3 carry over to higher dimension (Van Kirk 1995), but explicit calculations are generally not feasible. An exception is the case where the dispersal kernel is the Green's function of a solvable partial differential equation (e.g., as in Sect. 3.2). The dispersal success approximation from Chap. 9 or other higher-order expansions provide an alternative way forward. Phillips and Kot (2015) apply various approximation and numerical techniques to evaluate persistence conditions on two-dimensional habitats. The theoretical results on asymptotic spreading speeds by Weinberger (1982) and most subsequent authors are formulated in higher space dimensions; see Chap. 5. Instead of traveling waves, we have to consider *planar traveling waves* in higher dimension, i.e., waves that travel in one direction in space and are constant in perpendicular directions. There are, however, some subtle issues to relate dispersal data to population density or distance moved. These are addressed by Lewis et al. (2006). We present some of their material here; see also Lewis et al. (2016).

We consider the linear equation

$$N_{t+1}(\mathbf{x}) = R \int_{-\infty}^{\infty} \int_{-\infty}^{\infty} K(\mathbf{x} - \mathbf{y}) N_t(\mathbf{y}) d\mathbf{y} , \qquad (12.95)$$

with $\mathbf{x}, \mathbf{y} \in \mathbb{R}^2$. The dispersal kernel $K(\mathbf{x})$ gives the probability *per unit area* of finding an individual that began dispersing at the origin. If K is isotropic, i.e., dispersal is equally likely in all directions, then the distribution of dispersal distances, i.e., the probability of moving a distance *per unit length*, is given by $\widetilde{K}(r) = 2\pi r K(r)$, where $r = \|\mathbf{x}\|$. While the IDE uses K, empirical data are typically recorded as $\widetilde{K}$, e.g., as histograms. We will see how $\widetilde{K}$ arises in calculations for the speed of planar traveling fronts.

The calculations for the point-release scenario from Sect. 5.2 with a Gaussian kernel carry over almost verbatim to the two-dimensional case. The only difference is that the scaling factor in the denominator does not have a square root; see (2.26). With the initial condition $N_0(\mathbf{x}) = \delta(\mathbf{x})$, we find the solution of (12.95) as

$$N_t(\mathbf{x}) = \frac{R^t}{2\pi\sigma^2 t} \exp\left(-\frac{\|\mathbf{x}\|^2}{2\sigma^2 t}\right) . \qquad (12.96)$$

Because of radial symmetry, we can define a threshold value $\widetilde{N}$ and locations $\mathbf{x}_t$ via $N_t(\mathbf{x}_t) = \widetilde{N}$ and calculate the asymptotic speed

$$c^* = \lim_{t \to \infty} \frac{\|\mathbf{x}_t\|}{t} = \sqrt{2\sigma^2 \ln(R)} . \qquad (12.97)$$

Figure 12.11 (left plot) shows how a level set expands over time. A planar front in direction $\mathbf{u}$ is a solution of the IDE of the form

$$N_t(\mathbf{x}) = W(\mathbf{x} \cdot \mathbf{u} - ct). \qquad (12.98)$$

Hence, in direction $\mathbf{u}$, the front looks like a one-dimensional profile, and in the perpendicular direction, the density is constant in space. Figure 12.11 (right plot)

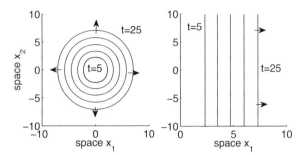

Fig. 12.11 Level sets of population spread in two-dimensional space with scaled Beverton–Holt growth function and Gaussian dispersal kernel. **Left:** A radially symmetric initial condition. **Right:** A planar traveling wave in the x_1-direction. Parameters are $R = 1.5$ and $\sigma^2 = 0.1$.

shows how a level set of a planar wave moves over time. When we substitute an exponential front profile, $W(z) = e^{-sz}$, into IDE (12.95), we find

$$W(\mathbf{x} \cdot \mathbf{u} - c(t + 1)) = R \int_{-\infty}^{\infty} \int_{-\infty}^{\infty} K(\mathbf{x} - \mathbf{y}) W(\mathbf{y} \cdot \mathbf{u} - ct) d\mathbf{y} \qquad (12.99)$$

or, after some manipulation,

$$e^{sc} = R \int_{-\infty}^{\infty} \int_{-\infty}^{\infty} K(\mathbf{z}) e^{s(\mathbf{z} \cdot \mathbf{u})} d\mathbf{z} =: R M_{\mathbf{u}}(s). \qquad (12.100)$$

We call $M_{\mathbf{u}}$ the directional moment-generating function. If K is isotropic, the speed of the front is independent of the direction, and we can choose $\mathbf{u} = (1, 0)$ to simplify calculations. We find the dispersion relation

$$e^{sc} = R \int_{-\infty}^{\infty} \int_{-\infty}^{\infty} K(\mathbf{z}) e^{sz_1} dz_1 dz_2 = R \int_{-\infty}^{\infty} K_{x_1}(z_1) e^{sz_1} dz_1, \qquad (12.101)$$

where K_{x_1} denotes the marginal distribution of K in the x_1-direction. The marginal distribution of the bivariate Gaussian in (2.26) is the univariate Gaussian in (2.25). Hence, the dispersion relations in the one- and two-dimensional cases are identical. Therefore, the minimal speeds of a traveling wave are also identical.

Marginal distributions are not always easy to calculate. It turns out that we can write the directional moment-generating function explicitly as

$$M_{(1,0)}(s) = \int_{-\infty}^{\infty} \int_{-\infty}^{\infty} K(r) e^{sz_1} dz_1 dz_2 = \int_{0}^{2\pi} \int_{0}^{\infty} K(r) e^{sr \cos \theta} r \, dr \, d\theta$$

$$= 2\pi \int_{0}^{\infty} K(r) r I_0(sr) dr = \int_{0}^{\infty} \tilde{K}(r) I_0(sr) dr , \qquad (12.102)$$

where I_0 is the zeroth-order Bessel function of the first kind. Hence, the distribution of dispersal distances arises naturally in the calculation of the dispersion relation.

Rather than choosing a particular parametric form for the dispersal kernel, Lewis et al. (2006) give (unbiased) estimators for the directional moment-generating function from dispersal distance data. They apply these formulas to dispersal data of the fruit fly *Drosophila pseudoobscura*; see Kot et al. (1996). They find that the spreading speed from the two-dimensional formula is higher than from the one-dimensional formula; i.e., if the *same* dataset is interpreted as two-dimensional dispersal distances, then the predicted speed is higher than when the dataset is interpreted as one-dimensional distances. Lewis et al. (2006) also apply this theory to teasel (*Dipsacus sylvestris*) via a stage-structured model; see Chap. 13.

Fort (2007) also compares spreading speeds in one and two dimensions, but in a different way. He uses the same kernel of dispersal distances in the two cases and proves that the speed is always slower in two dimensions. He gives an intuitive explanation for this phenomenon. We suppose that each individual moves exactly the same distance during one dispersal period. In one dimension, the dispersal kernel is the sum of two delta distributions. In two dimensions, the dispersal kernel is a "delta distribution" on a circle, so that the marginal distribution in any fixed direction is a uniform distribution up to the maximal distance. Hence, the marginal distribution sees most individuals make shorter moves in a given direction, and, as a result, the spread rate is lower. Fort (2007) also considers Reid's paradox (see above). He chooses the short-distance dispersal kernel as a uniform distribution with a radius of 15 m. For long-distance dispersal, he chooses a power law that he truncates at 10^4 m. Without long-distance dispersal, he obtains a spread rate of up to 10 m/year, whereas when a tiny proportion of 0.002 individuals use long-distance dispersal, the speed increases to almost 1000 m/year. This increase is on the same order of magnitude as required to explain Reid's paradox.

Fort et al. (2007) and Isern et al. (2008) consider an IDE for the Neolithic transition in Europe from hunter–gatherer groups to farmers. They consider and compare several models in two spatial dimensions with different dispersal kernels and mechanisms. Typical parameter values, obtained from archeological records, suggest a generation time of 32 years, a growth rate of $R = 2.2$ (with range [1.9, 2.6]), and a probability of moving of 0.38. They find speeds of 0.8–1.3 km/year or up to 50 km per generation, which compares quite well with data.

Coutinho et al. (2012) model the invasion of an African blowfly species (*Chrysomya albiceps*) in South America. They reduce a stage-structured model to a scalar model with a Ricker-type growth function. With the estimated parameter values, the steady state of the nonspatial model is unstable. They use a Gaussian dispersal kernel. In the one-dimensional model, starting from a point-release initial condition, they find a spreading front with a linearly determined speed of between 0.4 and 2 km/year. Since the steady state is unstable, they find an oscillating pattern behind the front, at least for small times. The same initial conditions in two dimensions give a radially symmetric spread pattern with the same speed. The instability leads to concentric rings that spread outward from the center of population introduction; see Fig. 12.12.

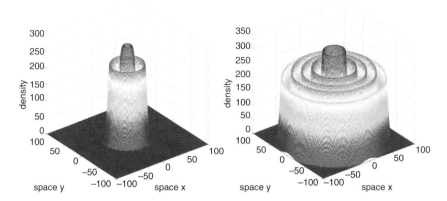

Fig. 12.12 Simulation of the blowfly model by Coutinho et al. (2012). The two-dimensional IDE has Ricker growth function $F(N) = 130N \exp(-0.1N)$ and a Gaussian dispersal kernel with variance $\sigma^2 = 8$. The initial condition is concentrated at zero. The plot on the left is after 5 generations; the plot on the right after 12.

12.8 Further Reading

Most examples in this chapter were either scalar equations, i.e., they represent the population as a single unstructured quantity, or they could be collapsed into a scalar equation. Many more interesting applications arise when we study systems of equations, either in the form of different stages of one population (see Chap. 13) or in the form of several interacting species (see Chap. 14).

The examples of models for partially sessile populations in Sect. 12.4 are by no means exhaustive. Many species have evolved quite particular life cycles with different strategies for dispersal and reproduction. IDEs present a very versatile modeling framework to adequately represent these life cycles. The potential downside is that a plethora of models needs to be studied and analyzed. Fortunately, most of the basic ideas of stability analysis and spreading speed analysis carry over to many model structures relatively easily.

Fedotov (2001) studies a model where individuals first reproduce locally and then disperse in space while their offspring remain at the location where they were produced until the next generation. In our notation, his model reads

$$N_{t+1}(x) = \int_{-\infty}^{\infty} K(x - y)N_t(y)\mathrm{d}y + F(N_t(x)) . \tag{12.103}$$

Fedotov (2001) uses a Hamilton–Jacobi framework to derive a formula for the speed of traveling waves for the linearized equation. He obtains an explicit expression when K is the sum of two delta distributions, i.e., $K(x) = (\delta(x - a) + \delta(x + a))/2$ for some fixed dispersal distance $a > 0$. For the same equation, Méndez et al. (2002) show how the spreading speed depends on the dispersal kernel. They also derive an approximate expression for the width of the transition front when the generation time is small.

Neubert and Parker (2004) model the local spread of scotch broom (*Cytisus scoparius*) in two stages. Initially, seeds follow ballistic dispersal; see (7.31). Subsequently, a proportion of seeds is carried by ants, with dispersal distances chosen from a Laplace kernel. Neubert and Parker (2004) show that the resulting two-stage kernel can be unimodal if the dispersal by ants is large compared to ballistic dispersal, or bimodal if dispersal by ants is relatively short. They then study a suite of increasingly complex models (including stochasticity and stage structure) to assess the invasion risk of scotch broom and its predicted spread rate.

Mistro et al. (2005a) carefully derive and study a model for the spread of the Africanized honey bee (*Apis mellifera adansonii*) in the Americas. Established colonies of these bees tend to remain sessile but can sometimes move as a colony, and they produce offspring that move and form new colonies elsewhere. Furthermore, the authors explicitly include the search for high-quality habitat in the dispersal kernel as follows. We denote by $q(x) \in [0, 1]$ the habitat quality at location x and by m some monotone bounded function on $[0, 1]$. For any given kernel $K(x)$ that depends on signed distance only, we then define the modified kernel

$$K^{[m]}(x, y) = \frac{K(x - y)m[q(x)]}{\int_{-\infty}^{\infty} K(z - y)m[q(z)]\mathrm{d}z} . \tag{12.104}$$

The numerator in this expression guarantees that individuals settle preferentially in places with higher habitat quality. The denominator ensures that the expression integrates to unity. The authors then study numerically how the different elements of the model interact to predict the spread rate of these bees.

Drury and Candelaria (2008) study the spread of the California sea otter (*Enhydra lutris nereis*) between 1914 and 1986. Otters had been almost extirpated from the California coast, but in 1914, one population was found in Point Sur. Under protection, this population has since spread north and south along the coast. Looking at the spread data, there seem to be differences in spread rates in direction (north vs. south) and time (before vs. after 1973). Drury and Candelaria (2008) formulate and fit integrodifference models with Laplace kernels for these different cases to data and compare them to reaction–diffusion models. They find that the IDE models generally fit considerably better.

Rodriguez (2010) studies fish dispersal in river networks. He models "sedentary" fish movement by a Laplace kernel with small variance and "mobile" fish dispersal by a Laplace kernel is large variance. He estimates parameters from experiments and indicates that such an estimate requires the release of about 4000 individuals. He then adapts his dispersal model to allow for dispersal barriers in the river network and discusses conditions under which passability can be reliably detected in empirical work.

Robinet et al. (2012) present a suite of models of increasing complexity for quantitative risk assessment for the spread of a pest species. Among others, they use a two-dimensional IDE model and parameterize it for the western corn rootworm (*Diabrotica virgifera virgifera*).

Part III
Extensions and Challenges

Chapter 13
Structured Populations

Abstract So far, we have treated populations as homogeneous: all individuals were assumed to be identical with respect to reproduction and dispersal. We only modeled the dynamics of a single density function. In reality, most populations are heterogeneous in many ways. Individuals differ with respect to age, size, gender, and other attributes, and their reproductive and dispersal behavior may depend on these attributes. The nonspatial dynamics of populations with complex life cycles have been successfully described by matrix models. In this chapter, we introduce and study spatially explicit matrix models to generalize the simple IDE to stage-structured populations. We present an in-depth analysis of the critical patch-size problem and the spreading speed for these equations, including several proofs that we omitted in the scalar case in earlier chapters. Throughout the chapter, we use a simple two-stage model for juveniles and adults to illustrate the theory. We close with an overview of the rich literature of applications of structured IDEs to real-world systems, in particular to species invasions.

13.1 Matrix Models

Most populations are heterogeneous in many ways. One of the simplest structures is to distinguish between nonreproductive juveniles and reproductive adults, but many more complex life cycles exist. For example, plant life cycles may include a seed bank, seedlings, and nonflowering and flowering individuals of different sizes and ages. Since reproductive output and dispersal behavior may differ between the different stages, we would like to include this stage-specific information into our models so that we can make accurate predictions. The nonspatial dynamics of such structured populations can be conveniently described by matrix models (Caswell 2001). We briefly review the most important aspects of this theory below. Then we generalize the scalar IDE (2.1) to stage-structured IDEs, or spatially explicit matrix models.

The different stages in the life cycle of an organism can be represented in a life-cycle graph, where vertices represent stages and directed edges indicate possible transitions between stages. For example, if we consider only the two stages of

© Springer Nature Switzerland AG 2019
F. Lutscher, *Integrodifference Equations in Spatial Ecology*, Interdisciplinary Applied Mathematics 49, https://doi.org/10.1007/978-3-030-29294-2_13

Fig. 13.1 Representation of a simple juvenile–adult two-stage life cycle as a graph.

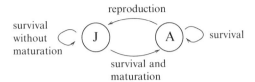

nonreproductive juveniles and reproducing adults, we need to specify up to four transitions: juveniles remain juveniles if they survive but do not mature, juveniles become adults if they survive and mature, adults remain adults if they survive, and adults produce juveniles; see Fig. 13.1. More generally, structuring populations by age or other relevant life-cycle stages can lead to highly complex life-cycle graphs.

Matrix models are the corresponding mathematical framework for capturing the dynamics of structured populations (Caswell 2001). For example, the juvenile–adult model representing the life-cycle graph in Fig. 13.1 can be written as

$$J_{t+1} = s_j(1 - g)J_t + RA_t , \qquad A_{t+1} = s_j g J_t + s_a A_t , \qquad (13.1)$$

where J_t and A_t stand for the number of juveniles and adults in generation t. Parameters s_j and s_a represent the probability of survival of juveniles and adults, respectively; g indicates the probability of maturation; and R is the per capita number of offspring for adults. In general, parameters could depend on population density. This model can be written in matrix notation as

$$\begin{bmatrix} J_{t+1} \\ A_{t+1} \end{bmatrix}_{t+1} = \begin{bmatrix} s_j(1-g) & R \\ s_j g & s_a \end{bmatrix} \begin{bmatrix} J_t \\ A_t \end{bmatrix} . \qquad (13.2)$$

More generally, for a population divided into ℓ stages, we denote the number or density of individuals at stage i and generation t by N_t^i, $i = 1, \ldots, \ell$, and the transition (including reproduction) from stage j to stage i by b_{ij}. Then the dynamics between generations can be written as

$$N_{t+1}^i = \sum_{j=1}^{\ell} b_{ij} N_t^j \qquad (13.3)$$

or, in matrix notation,

$$\mathbf{N}_{t+1} = B\mathbf{N}_t , \quad \text{where} \quad \mathbf{N} = [N^1, \ldots, N^\ell]^T , \quad B = [b_{ij}] . \qquad (13.4)$$

Superscript T stands for the transpose that turns the row vector into a column vector.

When the transition rates between stages are independent of population density, the entries of matrix B are constants and model (13.4) is linear. In that case, we can explicitly solve the recursion by taking powers of B, i.e., $\mathbf{N}_t = B^t \mathbf{N}_0$. To determine the long-term behavior of this solution, we make the important assumption that the nonnegative matrix B is primitive, i.e., that there is some power of B where all entries are positive (Caswell 2001). The biological meaning of this assumption is that individuals of every stage can be produced from any stage in finitely many time steps. This assumption can be satisfied in many applications by choosing stages appropriately, e.g., by excluding post-reproductive stages. Mathematically, this assumption is the crucial ingredient for proving the Perron–Frobenius theorem; see, e.g., Caswell (2001).

Theorem 13.1 (Perron–Frobenius) *Assume that B is primitive.*[1] *Then B has a simple, positive, strictly dominant eigenvalue with positive eigenvector. No other eigenvalue has a nonnegative eigenvector.*

The fate of the population described by the linear matrix model is completely determined by the dominant eigenvalue, μ, of matrix B. If $\mu < 1$, then the population will go extinct; if $\mu > 1$, then the population will grow indefinitely with asymptotic growth rate μ and the population structure given by the corresponding (right) eigenvector of B.

When the transition rates between stages depend on population density, the elements of $B = B(\mathbf{N})$ depend on the stage vector, and the model is nonlinear. In that case, we find steady states, i.e., solutions of the equation $\mathbf{N}^* = B(\mathbf{N}^*)$; we linearize matrix B at those states; and we determine the stability and bifurcation behavior from the spectrum of the linearization (Caswell 2001; Cushing 2014).

To include space in matrix models, we denote by x the spatial location in some domain of interest, Ω, and by $N_t^i(x)$ the spatial density of individuals at stage i of generation t. Again, we take a census of the population after the end of the dispersal phase. The life cycle begins with the growth phase, which is described by matrix B as above. To model dispersal, we denote by $K_{ij}(x, y)$ the dispersal kernel of individuals at stage i that were produced by individuals at stage j. Then the IDE reads

$$N_{t+1}^i(x) = \int_{\Omega} \sum_{j=1}^{\ell} K_{ij}(x, y) b_{ij} N_t^j(y) \mathrm{d}y. \tag{13.5}$$

With the notation $\bullet$ for the Hadamard product of entrywise matrix multiplication, these equations can be written more elegantly as

---

[1]The Perron–Frobenius theorem holds under the more general condition that B is *irreducible*. A matrix is irreducible if it is not conjugate to a block upper-triangular matrix (Caswell 2001). A typical example for a reducible matrix arises in populations with post-reproductive stages. Since these stages do not contribute to reproduction, their contribution to the pre-reproductive stages is zero. If we order the variables in the equations such that the post-reproductive stages are first, the resulting matrix will be block upper-triangular. In terms of the life-cycle graph, a matrix is reducible if there is a proper subset of vertices from which there are no edges (transitions) to nodes outside that subset.

$$\mathbf{N}_{t+1}(x) = \int_{\Omega} \mathbf{K}(x, y) \bullet B\mathbf{N}_t(y)\mathrm{d}y\,, \qquad (13.6)$$

where $\mathbf{K} = [K_{ij}]$ is the matrix of dispersal kernels.

We analyze these stage-structured IDEs in two steps. We begin with the question of extinction, persistence, and steady states on bounded domains. Then we turn to invasions on unbounded domains.

13.2   Persistence on a Bounded Domain

We begin the analysis of IDE (13.6) on bounded domains. We study steady states and their stability, returning to the question of finding the critical patch-size from Chap. 3 but now for a structured population. This analysis can be seen as generalizing both matrix models to include space and scalar spatial models to include population structure. We present here the more formal underpinnings of the theory that we omitted in Chap. 3. The exposition follows the work by Lutscher and Lewis (2004).

We explicitly denote the dependence of vital rates in matrix B on spatial location and density and write the IDE

$$\mathbf{N}_{t+1}(x) = \mathbf{Q}[\mathbf{N}_t](x) = \int_{\Omega} [\mathbf{K}(x, y) \bullet B(\mathbf{N}_t, y)]\mathbf{N}_t(y)\mathrm{d}y\,, \qquad (13.7)$$

where $\Omega \subset \mathbb{R}^n$ is bounded. We work in the product space $\mathscr{L}^2 = (\mathrm{L}^2(\Omega))^{\ell}$ of square-integrable functions and its positive cone. Accordingly, we make the following assumptions for the rest of this chapter.

(A1)   Elements of B are bounded, i.e., $0 \leq b_{ij}(\mathbf{N}, y) \leq b_{\max} < \infty$ for all i, j. Furthermore, each $b_{ij}(\mathbf{N}, y)$ is continuous with respect to y and continuously differentiable with respect to $\mathbf{N}$.

(A2)   If b_{ij} is nonzero, then k_{ij} satisfies $k_{ij} \in (\mathrm{L}^2(\Omega))^2$.

The first assumption is natural for biological populations; it results from even relatively weak self-limitation. Note that we do not assume that the per capita production functions converge to zero. The second assumption requires that all stages disperse. When we allow for sedentary stages, operator Q fails to be compact. We discuss several results on noncompact operators at the end of this chapter.

Lemma 13.1 *Under assumptions (A1)–(A2), operator $\mathbf{Q} : \mathscr{L}^2 \to \mathscr{L}^2$ as defined in (13.7) is positive and completely continuous. Furthermore, $\mathbf{Q}$ is strongly Fréchet differentiable at $\mathbf{N} = 0$ with respect to the positive cone. Its derivative is the positive, completely continuous linear operator given by*

$$\mathbf{Q}'[0]\phi(x) = \int_{\Omega} [\mathbf{K}(x, y) \bullet B(0, y)]\phi(y)\mathrm{d}y\,. \qquad (13.8)$$

Proof The proof of this lemma follows from Sects. 17.3, 17.5, and 17.8 as well as Theorem 17.1 in Krasnosel'skii and Zabreiko (1984); see Van Kirk and Lewis (1997) and Lutscher and Lewis (2004).   □

The next step is to prove an analogue of the Perron–Frobenius theorem for the linear operator $\mathbf{Q}'[0]$. A linear operator is called *superpositive* (Krasnosel'skii and Zabreiko 1984) if it has a simple positive dominant eigenvalue with positive eigenfunction, and no other eigenfunction is positive. If an operator is superpositive, then the stability of zero and the asymptotic behavior of the equation are determined by the dominant eigenvalue and the corresponding eigenfunction. To show that $\mathbf{Q}'[0]$ is superpositive, we require the following spatial version of primitivity.

(A3) Matrix $B(0, y)$ is primitive for each $y \in \Omega$. Furthermore, if $b_{ij}(0, y)$ is positive for some $y \in \Omega$, then it is positive for all $y \in \Omega$.

In biological terms, these assumptions mean that if individuals at stage j produce individuals at stage i somewhere in the domain, then they do so everywhere, possibly at different rates.

Theorem 13.2 *Assume that (A1)–(A3) hold. In addition, assume that there are constants $0 < \underline{\kappa} \le k_{ij} \le \overline{\kappa}$ on Ω for all pairs (i, j) for which b_{ij} is nonzero. Then $\mathbf{Q}'[0]$ is superpositive.*

Proof The proof of this theorem follows largely from applying the results from Chap. 2 in Krasnosel'skii (1964). At its core is an application of the Krein–Rutman theorem. The most important estimate is to show that there exists some constant m such that

$$(\underline{\kappa} b_{\min})^m \int \phi(y)\mathrm{d}y \le \mathbf{Q}'[0]^m \phi(x) \le m\ell(\overline{\kappa} b_{\max})^m \int \phi(y)\mathrm{d}y . \tag{13.9}$$

This estimate can be obtained in several steps. First, we use primitivity of B to show that if some power of B is positive, then so are all higher powers. Next, we use compactness of the domain to show that some finite power of $B(0, y)$ is positive simultaneously for all y. Finally, we use the lower bounds on the dispersal kernels to obtain the estimate. Then we use the results from Chap. 2 in Krasnosel'skii (1964) together with the properties of the space $\mathscr{L}^2$ to obtain the result; see Lutscher and Lewis (2004).   □

The positivity assumption in Theorem 13.2 implies that dispersers can reach any point in the domain from any other point within one dispersal period. This assumption seems unreasonable for some species and certain domains. We relax these assumptions in two steps. First, in the case where Ω is connected, we make the following assumption.

(A4) There is a nonnegative symmetric continuous function κ such that for all $k_{ij} \ne 0$ we have $\kappa(x, y) = \kappa(y, x) \le k_{ij}(x, y) \le \overline{\kappa}$, and there is a constant $\epsilon > 0$ such that for all $x \in \Omega$ the measure of the set $\{y \in \Omega \mid \kappa(x, y) \ge \underline{\kappa} > 0\}$ is at least ϵ.

This assumption covers two important cases. If dispersal distances are small compared to the patch-size, then $k_{ij}(x, y) = 0$ if $|x - y|$ is large. Then the lower bound has to be satisfied for y near x. On the other hand, some dispersal kernels are zero when $x = y$, e.g., the Weibull kernel (see Table 3.1). In that case, the symmetry condition implies that after two dispersal periods, individuals are back near where they started. Hence, the lower bound condition near $x = y$ holds after two dispersal periods.

Proposition 13.1 *Assume that Ω is connected and that (A1)–(A4) hold. Then $\mathbf{Q}'[0]$ is superpositive.*

In the second step, we deal with the case where Ω is a finite collection of disjoint connected components. We write $\Omega = \dot{\bigcup}_{\gamma=1,\ldots,\Gamma} \Omega_\gamma$ and assume that each Ω_γ is connected. The *connectivity matrix* $C = (c_{\alpha\beta})$ for continuous kernels k_{ij} is given by

$$
c_{\alpha\beta} = \begin{cases} 1, & \text{if for some} \quad x \in \Omega_\alpha, y \in \Omega_\beta, i, j : k_{ij}(x, y)b_{ij}(0, y) > 0, \\ 0, & \text{otherwise.} \end{cases}
$$
(13.10)

Then we need the following assumption.

(A5)   Matrix C is primitive.

Assumptions (A4) and (A5) together imply that an individual at stage i and point x can get to any other location $y \in \Omega$ and stage j through dispersal and production in finitely many generations. In mathematical terms, this means that operator $\mathbf{Q}'[0]$ is irreducible.

Proposition 13.2 *Let $\Omega = \dot{\bigcup}_{\gamma=1,\ldots,\Gamma} \Omega_\gamma$ be the disjoint union of connected components and assume that (A1)–(A5) hold. Then $\mathbf{Q}'[0]$ is superpositive.*

Proof The proof is tedious but not hard and has a nice biological interpretation. The goal is to prove inequality (13.9). The idea is to find a "connecting path": for any two points x, y in the domain and any two stages i, j in the life cycle, there is a sequence of points in the domain and life stages such that an individual at stage j and location y produces an individual at stage i and location x with positive probability in finitely many generations. Then we use compactness to find the smallest number of generations so that such a sequence exists for all points and life stages. Details are provided in Lutscher and Lewis (2004). □

The theory we have developed so far implies that if the assumptions are satisfied, then the stability of the zero solution is determined by the dominant eigenvalue of the linearized operator at zero; see (13.8). Calculating this dominant eigenvalue for particular examples is a different matter. The calculation can be carried out when all the dispersal kernels are identical Laplace kernels, i.e., when dispersal behavior is independent of stage, and when vital rates are independent of spatial location. In that case, the calculations from Sect. 3.2 can be extended to the stage-structured model. We present this calculation here and give a different example in the next section.

The eigenvalue equation with $B = B(0)$ on the domain $\Omega = [-L/2, L/2]$ is

$$\lambda\phi(x) = \mathbf{Q}'[0]\phi(x) = \int_{-L/2}^{L/2} \frac{a}{2} \exp(-a|x - y|)B\phi(y)dy. \tag{13.11}$$

Differentiating this equality twice and substituting as in Sect. 3.2 leads to the vector-valued boundary-value problem

$$\phi''(x) = -a^2 \left(\frac{1}{\lambda}B - I\right)\phi, \qquad \phi'(\pm L/2) \pm a\phi(\pm L/2) = 0, \tag{13.12}$$

where I denotes the identity matrix. The exponential ansatz $\phi(x) = e^{\xi x}\psi$ leads to the eigenvalue problem

$$\xi^2\psi = -a^2 \left(\frac{1}{\lambda}B - I\right)\psi. \tag{13.13}$$

Hence, we can express ξ via the eigenvalues μ_i of B and the eigenvalue λ of $\mathbf{Q}'[0]$ as

$$\xi^2 = -a^2 \left(\frac{\mu_i}{\lambda} - 1\right). \tag{13.14}$$

If the expression on the right-hand side is positive, then we get real solutions $\xi = \pm a\sqrt{1 - \mu_i/\lambda}$. Since dispersal is symmetric, the eigenfunctions of $\mathbf{Q}'[0]$ have to be symmetric on $[-L/2, L/2]$, which leads to the form

$$\phi(x) = \cosh\left(a\sqrt{1 - \mu_i/\lambda}\, x\right)\psi. \tag{13.15}$$

These solutions cannot satisfy the boundary conditions. We can rule out $\xi = 0$ equally easily. Therefore, the right-hand side in (13.14) must be negative, which implies $\lambda < \mu_i$. In particular, if the dominant eigenvalue of matrix B is less than unity, then λ is less than unity. In biological terms, if the population goes extinct in the nonspatial setting, then it will also go extinct in the spatial setting. Again, we see that boundary loss reduces the overall population growth rate.

When the right-hand side of (13.14) is negative, we obtain eigenfunctions of the form

$$\phi(x) = \cos\left(a\sqrt{\mu_i/\lambda - 1}\, x\right)\psi. \tag{13.16}$$

For the critical patch-size, we set $\lambda = 1$, take the largest eigenvalue of B, and solve as in Sect. 3.2 to get

$$L^* = \frac{2}{a\sqrt{\mu_1 - 1}} \arctan\left(\frac{1}{\sqrt{\mu_1 - 1}}\right). \tag{13.17}$$

From the explicit expression for L^*, we conclude that if matrix B depends on a parameter, say P, and if the dominant eigenvalue is an increasing function of that parameter, then the critical patch-size is a decreasing function of that parameter, i.e., $dL^*(P)/dP < 0$. This relationship actually holds in much more generality when the dispersal behavior of the different stages differs.

Lemma 13.2

1. *On a fixed domain, suppose that the matrix of production rates, $B(0, y; P)$, is nondecreasing in P. Denote $\lambda(P)$ as the dominant eigenvalue of $Q'[0]$. If at least one entry of $B(0, y; P)$ is strictly increasing in P, then so is $\lambda(P)$.*
2. *Fix P and let $\Omega = [0, L]$. Assume that the matrix of dispersal kernels is of the form $\mathbf{K}(x, y) = \mathbf{K}(x - y) > 0$, and denote $\lambda(L)$ as the dominant eigenvalue of $Q'[0]$. Then $\lambda(L)$ is a strictly increasing function of L.*
3. *If both previous conditions are satisfied, then $dL^*(P)/dP < 0$.*

The first two statements in the lemma are proved in Lutscher and Lewis (2004). The last statement follows from the implicit function theorem applied at the bifurcation point where $\lambda(L, P) = 1$. In biological terms, this result means that if at least one of the vital rates of the population increases, then the critical patch-size decreases.

13.3   Application

We apply some of the preceding ideas to a three-stage life-cycle model for the common lizard (*Lacerta vivipara*). This example was originally presented by Tricia Morris in her honor's thesis in 2017. The common lizard has three life stages (Galliard et al. 2010). The early juvenile stage (E) lasts for one year. After that year, lizards disperse to find their own territory and enter the late juvenile stage (L), which lasts for another year. At the beginning of their third year, they turn into adults (A) and can continue to live for several years. We denote by $\mathbf{N} = [E, L, A]^T$ the densities of the three stages. We consider linear, space-independent survival probabilities s_X for stage $X \in \{E, L, A\}$ and a linear, space-independent birth term with an average of b (female) offspring per (female) adult. We assume no mating limitation. Then the dynamics between years can be written as

$$\mathbf{N}_{t+1}(x) = \int_{\Omega} \mathbf{K}(x - y) \bullet B\mathbf{N}_t(y)dy$$

with

$$\mathbf{K}(x) = \begin{bmatrix} 0 & 0 & \delta(x) \\ K(x) & 0 & 0 \\ 0 & \delta(x) & \delta(x) \end{bmatrix} \quad \text{and} \quad B = \begin{bmatrix} 0 & 0 & b \\ s_E & 0 & 0 \\ 0 & s_L & s_A \end{bmatrix}.$$

Here, K denotes the dispersal kernel at the end of the early juvenile phase. Since only one of the three stages disperses, delta distributions appear at the other stage transitions. The resulting operator is not compact. Condition (A4) on the existence of a simultaneous positive subfunction is also violated. Nonetheless, we can continue with the analysis by a reduction argument.

The corresponding eigenvalue problem is

$$\lambda \phi_1(x) = b \phi_3(x),$$

$$\lambda \phi_2(x) = s_E \int_\Omega K(x - y) \phi_1(y) dy,$$

$$\lambda \phi_3(x) = s_L \phi_2(x) + s_A \phi_3(x).$$

Substituting the equations for ϕ_1 and ϕ_2 into the third equation, we find

$$\lambda^2 \phi_3(x) = s_L s_E b \int_\Omega K(x - y) \phi_3(y) dy + \lambda^2 s_A \phi_3(x)$$

and eventually

$$\tilde{\lambda} \phi_3(x) = \int_\Omega K(x - y) \phi_3(y) dy, \quad \tilde{\lambda} = \frac{\lambda^3 - \lambda^2 s_A}{s_L s_E b}.$$

This integral operator is compact and positive under the usual assumptions on K and Ω. Hence, a dominant eigenvalue $\tilde{\lambda} > 0$ exists. For every $\tilde{\lambda} > 0$, the above condition defines a unique $\lambda > s_A$. The largest value of λ corresponds to the largest value of $\tilde{\lambda}$. In particular, the persistence threshold is given by $\lambda = 1$ or $\tilde{\lambda} = \frac{1 - s_A}{s_L s_E b}$. Since the integral operator contains only dispersal, its dominant eigenvalue is bounded by $\tilde{\lambda} \leq 1$. Hence, we require $b > \frac{1 - s_A}{s_E s_L}$ for there to be a solution with $\lambda = 1$. This condition means that each adult has to produce at least one offspring that reaches adulthood for the population to persist. With the survival probabilities estimated at $s_E = 27.9\%$, $s_L = 65.4\%$, and $s_A = 50.6\%$ (Galliard et al. 2010), we require the average number of (female) offspring to be at least $b > 2.7$ for population persistence on a large enough domain. For a given b above this threshold, we can calculate the critical patch-size. If we choose the Laplace kernel with parameter a, we can use the results from Chap. 3 to find the critical patch-size as

$$L^* = \frac{2}{a \sqrt{\frac{s_L s_E b}{1 - s_A} - 1}} \arctan \left(\frac{1}{a \sqrt{\frac{s_L s_E b}{1 - s_A} - 1}} \right).$$

With a mean dispersal distance of $a^{-1} = 64.7$ m (Warner and Shine 2008) and $b = 3$, we find a critical patch-size of $L^* = 487.88$ m.

13.4 Nonlinear Analysis

In this section, we present some of the theory that allows us to prove the existence and uniqueness of a positive fixed point. It is based on bifurcation theory and monotone systems theory. We begin with an additional assumption on the vital rates for large population density.

(A6)   There is some matrix-valued function $B(\infty, y)$ such that

$$\| B(\mathbf{N}(y), y) - B(\infty, y) \| \leq \frac{\text{const.}}{\|\mathbf{N}\|} \qquad \text{for large} \quad \|\mathbf{N}\| . \qquad (13.18)$$

Lemma 13.3 *Assume that (A1), (A2), and (A6) are satisfied. Then, by Sect. 3.2.1 in Krasnosel'skii (1964) and Theorem 17.2 in Krasnosel'skii and Zabreiko (1984), operator* $\mathbf{Q}$ *has a strong asymptotic derivative at infinity. It is given by the completely continuous operator*

$$\mathbf{Q}'[\infty]\phi(x) = \int [\mathbf{K}(x, y) \bullet B(\infty, y)]\phi(y)\mathrm{d}y . \qquad (13.19)$$

Proposition 13.3 (Existence of Fixed Points) *Assume that (A1)–(A6) hold. Suppose that the spectral radius of* $\mathbf{Q}'[\infty]$ *is less than one and that the dominant eigenvalue of* $\mathbf{Q}'[0]$ *is greater than one. Then by Theorem 4.11 in Krasnosel'skii (1964), operator* $\mathbf{Q}$ *has a positive fixed point.*

Under additional assumptions about the vital rates, we can even get the uniqueness of the positive fixed point. We have already encountered these conditions in Chap. 4 for unstructured populations. When the population growth function was monotone and concave down, we obtained a unique positive fixed point. The following proposition generalizes this result to structured populations.

Proposition 13.4 (Uniqueness of Fixed Points) *Let the assumptions of Proposition 13.3 be satisfied. Assume in addition that the function*

$$\mathbf{N} \mapsto B(\mathbf{N})\mathbf{N} \qquad (13.20)$$

is increasing and that

$$t \mapsto B(t\mathbf{N}) \qquad (13.21)$$

is decreasing for $0 \leq t \leq 1$. *Then* $\mathbf{Q}$ *is concave and monotone. By Theorem 6.3 in Krasnosel'skii (1964), the positive fixed point is unique. By Theorem 6.6 in*

Krasnosel'skii (1964), every solution with $\mathbf{N}_0 \neq 0$ *converges to the positive fixed point.*

The concavity condition in the preceding proposition is fairly strong, as we will see in an example below. Weaker conditions are "strict sublinearity" (Zhao 1996) or "strict subhomogeneity" (Zhao 2003). When the concavity conditions are not satisfied, we can still obtain the existence of a (unique) fixed point by studying the behavior near the bifurcation point $\lambda = 1$. In order to apply the theory, we need to exclude the Allee effect; compare Chap. 4. The following assumption is the stage-structured formulation that accomplishes that goal.

(A7)   The production rates decrease with population density in any state; i.e., they satisfy $(\partial/\partial N_l)b_{ik} \leq 0$ for all i, k, l, and the inequality is strict for at least one set i, k, l.

We present two forms of the bifurcation results: one with respect to a parameter that affects the vital rates of the population and one with respect to domain length. We denote the parameter by P, the dominant eigenvalue of $\mathbf{Q}'[0]$ by $\lambda = \lambda(P)$, and the bifurcation point by $\lambda(P^*) = 1$. Proofs of both results can be found in Lutscher and Lewis (2004).

Lemma 13.4 (Bifurcation I)   *Assume that (A1)–(A7) are satisfied and that the spectral radius of $\mathbf{Q}'[\infty]$ is less than one, independent of P close to P^*. Assume that the production rates b_{ij} are nondecreasing in P and that at least one of the rates is increasing in P. Then there is a transcritical bifurcation at $P = P^*$; i.e., a continuous branch of solutions intersects the zero solution. The nonzero solution is positive for $P > P^*$.*

To present a bifurcation result with respect to domain length L, we denote $\lambda(L)$ as the dominant eigenvalue of $\mathbf{Q}'[0]$ and $\lambda(L^*) = 1$ as the bifurcation point. To formulate the result, it is convenient to introduce a generalization of the dispersal success function (see Chap. 9) to the case of structured populations. We define

$$s_{ij}(y) = \int_{\Omega} k_{ij}(x, y)\mathrm{d}x \tag{13.22}$$

as the dispersal success function of an individual at stage i produced from an individual at stage j.

Lemma 13.5 (Bifurcation II)   *Assume that (A1)–(A7) are satisfied and that the spectral radius of $\mathbf{Q}'[\infty]$ is less than one, independent of L close to L^*. Assume that the dispersal success functions s_{ij} are nondecreasing in L and that at least one of them is increasing in L. Then there is a transcritical bifurcation at $L = L^*$; i.e., a continuous branch of solutions intersects the zero solution. The nonzero solution is positive for $L > L^*$.*

13.5   Example: Juveniles and Adults

We return to the example of a simple population structure of juveniles and adults as in (13.2) and illustrate the conditions of the preceding lemmas and propositions as well as the resulting dynamics. We take advantage of the fact that the nonspatial model was studied in great detail by Neubert and Caswell (2000b) and that most of its dynamic behavior is well understood.

We begin with the scenario that only offspring production is density dependent and replace R with $R(1 + J + A)^{-1}$. Writing $\mathbf{N} = (J, A)^T$, the resulting matrix of vital rates and its limits at zero and infinity are

$$B(\mathbf{N}) = \begin{bmatrix} s_j(1-g) \, \frac{R}{1+J+A} \\ s_j g \qquad s_a \end{bmatrix}, \quad B(0) = \begin{bmatrix} s_j(1-g) \; R \\ s_j g \quad s_a \end{bmatrix}, \quad B(\infty) = \begin{bmatrix} s_j(1-g) \; 0 \\ s_j g \quad s_a \end{bmatrix}.$$
$$\tag{13.23}$$

The dominant eigenvalue, μ_1, of $B(0)$ is an increasing function of R. For $R = 0$, we have $\mu_1 = \max\{s_j(1-g), s_a\} < 1$ and $B(\infty) = B(0)$. When R is large enough, we have $\mu_1 > 1$. The critical value for $\mu_1(R^*) = 1$ is

$$R^* = \frac{(1 - s_a)(1 - s_j(1-g))}{s_j g}. \tag{13.24}$$

Furthermore, with this choice of density dependence, assumptions (A1), (A3), and (A6), as well as the assumption in Proposition 13.4, are satisfied.

We choose the domain $\Omega = [0, L]$ and let all dispersal kernels be equal to the Laplace kernel with identical variance. Then the remaining assumptions are also satisfied. Therefore, operator $\mathbf{Q}'[0]$ will have a dominant eigenvalue $\lambda = \lambda(R, L)$. If that eigenvalue is less than unity, then the population extinction state will be stable. If it is greater than unity, then there will be a unique, globally stable positive population persistence state. The condition $\lambda(R, L) > 1$ requires $R > R^*$ from (13.24) and $L > L^*$ from (13.17).

A more interesting dynamical scenario arises when density dependence follows the Ricker function; i.e., we replace R in (13.2) with $R \exp(-(J + A))$. This case is studied in detail by Lutscher and Lewis (2004). The resulting matrices, $B(0)$ and $B(\infty)$, are the same as in (13.23), but the monotonicity assumption in Proposition 13.4 is not satisfied. Lemmas 13.4 and 13.5 give the local existence and stability of a positive equilibrium near the bifurcation point. However, increasing parameter R can destabilize the positive steady state and lead to two-cycles and more complicated dynamic behavior. We illustrate some of these behaviors with the simplified model where all stages have the same dispersal behavior, described by the Laplace kernel. Hence, we study the equation

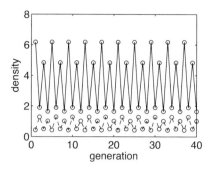 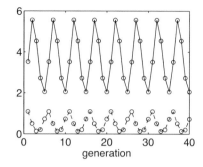

Fig. 13.2 Left: Stable four-cycle in the nonspatial matrix model with Ricker-type density-dependent reproduction ($R = 80$). **Right:** Stable invariant curve in the nonspatial matrix model with density-dependent maturation ($R = 70$). Solid lines are juveniles; dashed lines are adults. Circles indicate generations; connecting lines are for optical reasons only. In the right panel, the adult density is plotted at 20 times the actual density to make it visible on this scale. Parameters are $s_j = 0.5$, $s_a = 0.1$, and $g = 0.4$.

$$\mathbf{N}_{t+1}(x) = \int_{-L/2}^{L/2} \frac{a}{2} \exp(-a|x - y|) B(\mathbf{N}_t(y)) \mathbf{N}_t(y) dy, \qquad (13.25)$$

where $B(\mathbf{N})$ is the matrix in (13.23) but with the Ricker-type growth function $R \exp(-(J + A))$. We fix $L = 1$ and illustrate how the dynamics depend on the variance $\sigma^2 = 2/a^2$ of the dispersal kernel.

We fix parameters $s_j = 0.5$, $s_a = 0.1$, and $g = 0.4$. The critical value R^* for the nonspatial model to allow population persistence is $R^* \approx 3.15$. When $R^* < R < 14$, the nonspatial model has a unique stable persistence state. When $15 < R < 74$, we observe stable two-cycles, and for $74 < R < 100$ the system shows stable four-cycles; see left panel in Fig. 13.2.

As we turn to the spatial model, we fix $R = 80$. Then matrix B has the dominant eigenvalue $\mu_1 \approx 4.2012$. Since we fixed $L = 1$, we can solve relation (13.17) for dispersal parameter a and get a critical value of $a^* \approx 0.5697$, which translates into a critical variance of $(\sigma^2)^* \approx 6.1622$. When σ^2 is larger than this number, the zero state is stable and the population will not go extinct. As σ^2 decreases, the steady-state density increases; see top panels in Fig. 13.3. When $\sigma^2 < 4.5$, the steady state becomes unstable and a two-cycle emerges; see bottom left panel in Fig. 13.3. Finally, when $\sigma^2 = 0.05$, we even observe a four-cycle in the spatial model; see bottom right panel in Fig. 13.3.

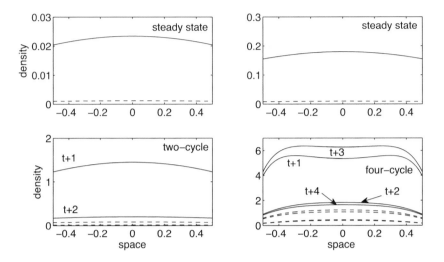

Fig. 13.3 Dynamics of the juvenile–adult model with density-dependent reproduction for different variances of the dispersal kernel. **Top left:** Stable persistence state for $\sigma^2 = 6$. **Top right:** Higher population densities at steady state with $\sigma^2 = 5$. **Bottom left:** Two-cycle for $\sigma^2 = 4$. **Bottom right:** Four-cycle for $\sigma^2 = 0.05$. For other parameters and details, see text. Solid lines represent juvenile densities; dashed lines represent adult densities.

Another interesting phenomenon arises when maturation instead of reproduction is density dependent (Neubert and Caswell 2000b). We keep R constant and replace g by $g \exp(-(J + A))$. Then the matrix and its limits at zero and at infinity are

$$B(\mathbf{N}) = \begin{bmatrix} s_j(1 - ge^{-(J+A)}) & R \\ s_j ge^{-(J+A)} & s_a \end{bmatrix}, \quad B(0) = \begin{bmatrix} s_j(1 - g) & R \\ s_j g & s_a \end{bmatrix}, \quad B(\infty) = \begin{bmatrix} s_j & R \\ 0 & s_a \end{bmatrix}.$$
$$(13.26)$$

The eigenvalues of $B(0)$ are the same as in the previous scenario, and the eigenvalues of $B(\infty)$ are both less than unity. When $R^* < R < 49$, there is a globally stable positive state. When $R > 50$, however, this state is unstable and there is a stable oscillating solution; see right panel in Fig. 13.2. This solution seems to be periodic with period five, but it is not since we have a discrete-time system. Instead, the solution converges to the discrete-time analogue of a periodic orbit (an invariant closed curve). We will discuss this behavior in more detail at the beginning of Chap. 14.

We consider the effects of space on these dynamics in exactly the same setting as for Eq. (13.25) above: a one-dimensional habitat of length one and all dispersal kernels identical Laplace kernels. Since the linearization of this model is the same as above, the persistence condition is exactly as above. For $0.3 < \sigma^2 < 6.16$, we observe a stable persistence state. As we decrease the variance further, the persistence state becomes unstable and an oscillating solution appears. This solution seems very close to a period-four solution but is qualitatively very different from

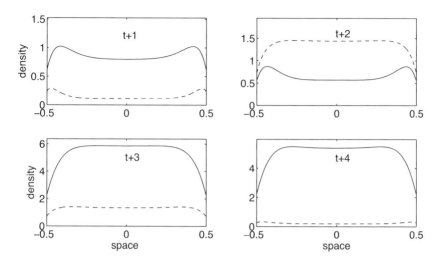

Fig. 13.4 Oscillatory dynamics of the juvenile–adult model with density-dependent maturation for $\sigma^2 = 0.01$. The four panels show the density of juveniles (solid) and the 20-fold density of adults (dashed) for four subsequent iterations after initial transients have disappeared. Subsequent iterations (modulo four) are indistinguishable from these four. Parameters are $s_j = 0.5$, $s_a = 0.1$, $g = 0.4$, $R = 80$, and $L = 1$.

the four-cycle that we observed in the previous scenario; see Fig. 13.4. Whereas the previous four-cycle had densities alternating between high and low from one generation to the next, this example shows two generations of high densities followed by two generations of low densities.

13.6 Traveling-Wave Speed in Unbounded Domains

When a population can persist locally, it is expected to spread spatially. We extend several ideas and results from the single-stage spread models in Chap. 5 to the stage-structured case. Not only is this theory well developed mathematically, it is also very widely and successfully applied to empirical data. We begin with a more heuristic approach to traveling-wave speeds and include a sensitivity analysis, which is useful and important in applications. We discuss the analysis behind the asymptotic spreading speed in Sect. 13.8.

Based on the approaches and results in Chap. 5, we assume that the dispersal kernels depend on distance only and are symmetric, i.e., $K(x, y) = K(|x - y|)$, and that vital rates are independent of spatial location. As before, we begin with the linear model, which we can consider as the linearization of (13.7) at the zero state. The equation reads

$$\mathbf{N}_{t+1}(x) = \int_{\mathbb{R}} [\mathbf{K}(x - y) \bullet B(0)] \mathbf{N}_t(y) \mathrm{d}y . \tag{13.27}$$

The traveling-wave ansatz $\mathbf{N}_t(x) = \mathbf{n}(\xi)$ with $\xi = x - ct$ and $c > 0$ leads to the relation

$$\mathbf{n}(\xi - c) = \int_{\mathbb{R}} [\mathbf{K}(\xi - \eta) \bullet B(0)] \mathbf{n}(\eta) \mathrm{d}\eta \,, \tag{13.28}$$

where $\eta = y - ct$. Since the equation is linear, we make the exponential ansatz

$$\mathbf{n}(\xi) = \mathrm{e}^{-s\xi} \phi \,, \tag{13.29}$$

where vector ϕ describes the relative abundances of the different stages in the traveling wave. Then we arrive at the eigenvalue equation

$$\mathrm{e}^{sc} \phi = \left[\int_{\mathbb{R}} \mathbf{K}(v) \mathrm{d}v \bullet B(0) \right] \phi = [\mathbf{M}(s) \bullet B(0)] \phi = \mathbf{H}(s) \phi \,, \tag{13.30}$$

where $\mathbf{M}$ denotes the matrix of moment-generating functions of the kernel matrix $\mathbf{K} = [k_{ij}]$ and $\mathbf{H}$ denotes the Hadamard product matrix. Alternatively, taking an exponential transform of (13.28) leads to the same result. Hence, we are looking for eigenvalues e^{sc} of matrix $\mathbf{H}(s)$.

Matrix $\mathbf{H}(s)$ has ℓ eigenvalues, denoted by $\lambda_i(s)$ with eigenvectors $\phi_i(s)$. By assumption, matrix $B(0)$ is nonnegative and primitive; by definition, $\mathbf{M}$ is positive. Therefore, $\mathbf{H}$ is nonnegative and primitive so that the Perron–Frobenius theorem applies. We denote by $\lambda_1(s)$ the dominant positive eigenvalue with positive eigenvector $\phi_1(s)$.

From any pair $\lambda_i(s)$, $\phi_i(s)$ we can form a solution of Eq. (13.27) as

$$\mathbf{N}_t(x) = \mathbf{n}(x - ct) = \mathrm{e}^{-s(x-ct)} \phi(s) = \lambda_i(s)^t \mathrm{e}^{-sx} \phi_i(s) \,. \tag{13.31}$$

Since the equation is linear, every linear combination of such solutions is again a solution. Dividing any linear combination

$$\mathbf{N}_t(x) = \sum_{i=1}^{\ell} \beta_i \lambda_i(s)^t \phi_i(s) \mathrm{e}^{-sx} \tag{13.32}$$

by the dominant eigenvalue, we find

$$\lim_{t \to \infty} \frac{\mathbf{N}_t(x)}{\lambda_1(s)^t} = \lim_{t \to \infty} \left[\beta_1 \phi_1(s) + \sum_{i=2}^{\ell} \left(\frac{\lambda_i(s)}{\lambda_1(s)} \right)^t \beta_i \phi_i(s) \right] \mathrm{e}^{-sx} = \beta_1 \phi_1(s) \mathrm{e}^{-sx} \,. \tag{13.33}$$

Since λ_1 is the dominant eigenvalue, all the fractions in the sum above converge to zero as $t \to \infty$.

We can then express the asymptotic speed of propagation via the dominant eigenvalue $\lambda_1(s)$ of $\mathbf{H}$ and the slope of the spatial decay as

$$c(s) = \frac{1}{s} \ln \lambda_1(s). \tag{13.34}$$

This expression is the generalization of (5.16) to stage-structured models. In fact, in the case of a single stage, i.e., $\ell = 1$, matrix $\mathbf{H}$ reduces to a single element, which is also the "dominant eigenvalue," and which is given by $b_{11}M(s)$. This is exactly the expression in (5.16).

We have assumed that the population can persist locally; i.e., the dominant eigenvalue of $B(0)$ is greater than unity. We have further assumed that all dispersal kernels are symmetric, so that each element of $\mathbf{M}$ is bounded below by unity. Clearly, both matrices are nonnegative. Therefore, the dominant eigenvalue of $\mathbf{H}$ is also greater than unity. Hence, $c(s) > 0$ and the population can spread.

Now we consider the case where the vital rates are nonincreasing functions of density, i.e.,

$$B(\mathbf{N}) \leq B(0). \tag{13.35}$$

Then any solution of the nonlinear IDE

$$\mathbf{N}_{t+1}(x) = \int_{\mathbb{R}} [\mathbf{K}(x - y) \bullet B(\mathbf{N}_t)] \mathbf{N}_t(y) dy \tag{13.36}$$

is bounded by the solution of the linear equation in (13.27) with the same initial condition. Furthermore, if $\mathbf{N}_0(x) \leq \beta_1 e^{-sx} \phi_1(s)$, then

$$\mathbf{N}_1(x) = \int_{\mathbb{R}} [\mathbf{K} \bullet B(0)] \mathbf{N}_0(y) dy \leq \beta_1 \mathbf{H}(s) \phi_1(s) e^{-sx} \tag{13.37}$$

$$\leq \beta_1 \lambda_1(s) e^{-sx} \phi_1(s) = \beta_1 e^{-s(x-c)} \phi_1(s). \tag{13.38}$$

Iteratively, we can show the inequality

$$\mathbf{N}_t(x) \leq \beta_1 e^{-s(x-ct)} \phi_1(s). \tag{13.39}$$

Finally, we note that any compactly supported initial condition can be bounded above by an exponential of the form $\beta_1 e^{-sx} \phi_1(s)$ by choosing β_1 large enough for every s for which the moment-generating functions $\mathbf{K}$ exist. This reasoning proves the following lemma.

Lemma 13.6 *Assume that condition (13.35) holds. Then the spreading speed of the nonlinear model is bounded above by the minimal speed of traveling waves of the linearized equation, i.e., by*

$$\hat{c} = \inf_{s>0} \frac{1}{s} \ln \lambda_1(s). \tag{13.40}$$

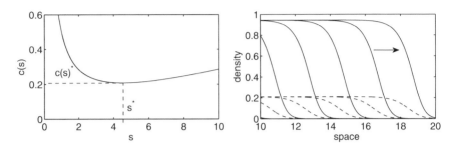

Fig. 13.5 **Left:** Dispersion relation $c(s)$ from (13.34) for the juvenile–adult model. Dashed lines indicate the minimum, $c(s^*)$, at s^*. **Right:** Advancing front of juveniles (solid) and adults (dashed) in the nonlinear model with density dependence in the growth rate. The density profile is plotted every 10 time steps. Parameters are $s_j = 0.5$, $s_a = 0.1$, $g = 0.4$, $R = 10$, and $\sigma^2 = 0.1$.

For an example, we return to the juvenile–adult model from Sect. 13.5. The matrix of vital rates is $B(0)$ as in (13.23). Let us consider a scenario where dispersal only happens at maturation and follows a Gaussian kernel with variance σ^2. None of the other stages disperses. Then the dispersal matrix is given by

$$\mathbf{K}(x) = \begin{bmatrix} \delta(x) & \delta(x) \\ \frac{1}{\sqrt{2\pi\sigma^2}} \exp(-\frac{x^2}{2\sigma^2}) \, \delta(x) \end{bmatrix}. \tag{13.41}$$

We calculate

$$\mathbf{H}(s) = \begin{bmatrix} 1 & 1 \\ \exp(-\sigma^2 s^2/2) & 1 \end{bmatrix} \bullet \begin{bmatrix} s_j(1-g) & R \\ s_j g & s_a \end{bmatrix} = \begin{bmatrix} s_j(1-g) & R \\ s_j g \exp(\sigma^2 s^2/2) & s_a \end{bmatrix}. \tag{13.42}$$

The dominant eigenvalue of $\mathbf{H}(s)$ is

$$\lambda_1(s) = \frac{1}{2}\left(s_j(1-g) + s_a + \sqrt{(s_j(1-g) - s_a)^2 + 4 s_j g R e^{\sigma^2 s^2/2}}\right). \tag{13.43}$$

We plot $c(s)$ in Fig. 13.5. With the chosen parameters, the minimum of $c(s)$ is $c^* = 0.208$ and occurs at $s^* = 4.53$. The dominant eigenvalue is $\lambda_1(s^*) = 2.56$, and a corresponding eigenvector is $\phi_1(s^*) = [1, 0.22]^T$.

To illustrate the advance of a traveling profile in the nonlinear equation, we let reproduction depend on density as in the previous section; i.e., we replace R by $R \exp(-J_t - A_t)$. We simulated the solutions of the IDE with initial conditions for juveniles and adults equal to 0.1 times the characteristic function of the negative half-line. Simulations develop the shape of the traveling profile after fewer than 10 iterations, and the distance per iteration is constant after 30 iterations. The resulting speed is $c = 0.198$, very close to the theoretically predicted value.

13.7 Sensitivity Analysis

As empirical measurements are always subject to various forms of error, we would like to know how the outcome of our calculations (here the minimal traveling-wave speed) depends on the input (here the parameters involved in our model). Answering this question is the goal of sensitivity analysis. *Sensitivity* measures the absolute change in the output quantity for a given change in the input quantity, whereas *elasticity* measures its relative change (Caswell 2001). In terms of the minimal traveling-wave speed $\hat{c} = c^*$ in (13.40) and some parameter p in the IDE, the two quantities are

$$\frac{dc^*}{dp} \quad \text{and} \quad \frac{p}{c^*}\frac{dc^*}{dp}, \tag{13.44}$$

respectively. The application of sensitivity analysis to spreading speeds in IDEs is developed by Neubert and Caswell (2000a).

Let p denote a parameter in IDE (13.27), either a population dynamic parameter in matrix $B(0)$ or a dispersal-related parameter in matrix $\mathbf{K}$. We consider the function of two variables

$$c(s, p) = \frac{1}{s}\ln(\lambda_1(s, p)). \tag{13.45}$$

We denote as $s^* = s^*(p)$ the value of s for which the minimum in (13.40) occurs, i.e., $c^*(p) = c(s^*(p), p)$. Then the sensitivity of c^* with respect to p is given by the chain rule as

$$\frac{dc^*}{dp} = \frac{\partial c}{\partial s}(s^*(p), p)\frac{ds^*}{dp}(p) + \frac{\partial c}{\partial p}(s^*(p), p). \tag{13.46}$$

Since c^* is the minimum with respect to s, the first term on the right-hand side vanishes. Hence, we get

$$\frac{dc^*}{dp} = \frac{1}{s\lambda_1}\frac{\partial \lambda_1}{\partial p}. \tag{13.47}$$

In other words, the sensitivity of c^* is a multiple of the sensitivity of λ_1, the dominant eigenvalue of $\mathbf{H}$. Formulas for the sensitivity of an eigenvalue with respect to matrix entries h_{ij} are well known. They can be expressed in terms of the right and left eigenvectors ϕ_1 and ψ_1, respectively, as (Caswell 2001)

$$\frac{\partial \lambda_1}{\partial h_{ij}} = \frac{\psi_{1,i}\phi_{1,j}}{\langle \psi, \phi \rangle}, \tag{13.48}$$

where $\phi_{1,j}$ indicates the jth entry of vector ϕ_1. To find the sensitivity of c^* with respect to a parameter p that may appear in one or more entries h_{ij} of matrix $\mathbf{H}$, we use the chain rule again. We arrive at the following sensitivity formula.

Lemma 13.7 *The sensitivity of c^* with respect to some parameter p is given by the expression*

$$\frac{dc^*}{dp} = \frac{1}{s^*\lambda_1} \sum_{i,j} \frac{\partial\lambda_1}{\partial h_{ij}} \frac{dh_{ij}}{dp} = \frac{1}{s^*\lambda_1} \sum_{i,j} \frac{\psi_{1,i}\phi_{1,j}}{\langle\psi,\phi\rangle} \frac{dh_{ij}}{dp}. \tag{13.49}$$

We return to the example in Sect. 13.5 to illustrate these ideas. With parameters as in Fig. 13.5, we obtain $c^* = 0.208$ with $s^* = 4.53$ and $\lambda_1 = 2.56$. Matrix $\mathbf{H}$ and its eigenvectors are given by

$$\mathbf{H}(s^*) = \begin{bmatrix} 0.3 & 10 \\ 0.558 & 0.1 \end{bmatrix}, \quad \phi_1 = \begin{bmatrix} 0.9753 \\ 0.2208 \end{bmatrix}, \quad \text{and} \quad \psi_1 = \begin{bmatrix} 0.2393 \\ 0.9710 \end{bmatrix}. \tag{13.50}$$

We calculate

$$\frac{dc^*}{dR} = \frac{1}{s^*\lambda_1} \frac{\psi_{1,1}\phi_{1,2}}{\langle\psi,\phi\rangle} = 0.0102 \tag{13.51}$$

and

$$\frac{dc^*}{d\sigma^2} = \frac{1}{s^*\lambda_1} \frac{\psi_{1,2}\phi_{1,1}}{\langle\psi,\phi\rangle} \frac{s_j g s^{*2}}{2} \exp\left(\frac{\sigma^2 s^{*2}}{2}\right) = 1.04. \tag{13.52}$$

Hence, an increase in R by a certain amount has a much smaller effect on c^* than an increase in σ^2 by the same amount. For example, increasing R from 10 to 10.1 and keeping all other parameters as in the previous example increases the speed from 0.208 to 0.209. Increasing σ^2 from 0.1 to 0.2, however, increases c^* to 0.29.

On the other hand, for the elasticities, we calculate

$$\frac{R}{c^*} \frac{dc^*}{dR} = 0.4887 \quad \text{and} \quad \frac{\sigma^2}{c^*} \frac{dc^*}{d\sigma^2} = 0.5014. \tag{13.53}$$

Therefore, an increase of either R or σ^2 by the same percentage has almost the same effect on the speed. For example, increasing either R or σ^2 from the previous baseline values by 10% increases c^* to 0.2174 and 0.218, respectively. We illustrate these results in Fig. 13.6, where we plot the profile of juveniles after 80, 90, and 100 time steps from the same initial condition with different parameter values. The front with increased R is steeper and reaches a higher steady-state value but spreads at (almost) the same speed as the front for increased σ^2.

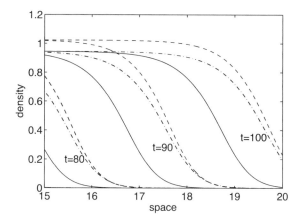

Fig. 13.6 Increasing parameter R or σ^2 in the juvenile–adult model with density-dependent reproduction increases the speed of spread. The density of juveniles is plotted after 80, 90, and 100 time steps, respectively, for baseline parameters from Fig. 13.5 (solid), increased reproductive rate (dashed), and increased variance of the Gaussian dispersal kernel (dash-dot). Increases are by 10%.

13.8   Spreading Speed for Structured Populations

Mathematically speaking, traveling waves are rather special solutions of IDEs. Their ecological interpretation is that the population is present in a very large region behind the invasion front. The asymptotic spreading speed (Definition 5.1) is a measure of the speed of propagation that is independent of the assumption of a traveling-wave profile. From an ecological point of view, it is more satisfying since it measures the spread of a population that is initially confined in space. From a mathematical point of view, this measure is more general than a traveling-wave speed but also more difficult to define and handle. In this section, we present the construction and basic concepts that lead to the definition of an asymptotic spreading speed as in Lui (1989a), but we refer to the original publication for the proofs. We consider the particular case of monostable dynamics on a one-dimensional domain; the general theory is given in Lui (1989a).

We study the recursion from (13.7) on the real line with spatially independent vital rates, so that the operator becomes

$$\mathbf{N}_{t+1} = \mathbf{Q}[\mathbf{N}_t] = \int_{\mathbb{R}} [\mathbf{K}(x - y) \bullet \mathbf{B}(\mathbf{N}_t)] \mathbf{N}_t(y) \mathrm{d}y , \qquad (13.54)$$

where $\mathbf{N} = [N^1, \ldots, N^\ell]^T$ as before. Throughout, we assume that B is nonnegative and primitive and that if $b_{ij} > 0$ for some $\mathbf{N} \geq 0$ then $b_{ij} > 0$ for all $\mathbf{N} \geq 0$. Whenever $b_{ij} > 0$, we also assume that the corresponding dispersal kernel $K_{ij} = K_{ij}(x - y)$ is a nonnegative symmetric bounded continuous function whose integral over the entire real line is unity and whose moment-generating function exists. To define the most important property of $\mathbf{Q}$, we need to introduce an order structure on $\mathbb{R}^\ell$ and an appropriate function space.

For $\mathbf{n}, \mathbf{m} \in \mathbb{R}^\ell$, we define $\mathbf{n} \geq \mathbf{m}$ ($\mathbf{n} \gg \mathbf{m}$) if $n^i \geq m^i$ ($n^i > m^i$) for all $1 \leq i \leq \ell$. Furthermore, $\mathbf{n} > \mathbf{m}$ is a short notation for $\mathbf{n} \geq \mathbf{m}$ and $n^i > m^i$ for at least one i. The origin in $\mathbb{R}^\ell$ is denoted by $\mathbf{0}$. For a given vector $\mathbf{m} > \mathbf{0}$, we define the function space

$$\mathscr{C}_{\mathbf{m}} = \{\mathbf{N} \mid N^i : \mathbb{R} \to [0, m^i] \text{ piecewise continuous}^2\}. \tag{13.55}$$

We extend the order relation to $\mathscr{C}_{\mathbf{m}}$ by saying $\mathbf{N} \geq \widetilde{\mathbf{N}}$ ($\mathbf{N} \gg \widetilde{\mathbf{N}}$) if $\mathbf{N}(x) \geq \widetilde{\mathbf{N}}(x)$ ($\mathbf{N}(x) \gg \widetilde{\mathbf{N}}(x)$) for all $x \in \mathbb{R}$. We can always identify $\mathbf{n} \in \mathbb{R}^\ell$ with the constant function $N^i(x) = n^i$. By the assumptions on the dispersal kernels, if $\mathbf{n}$ is a constant function, then $\mathbf{Q}[\mathbf{n}]$ is also constant.

We now make assumptions on the local population dynamics that are closely related to those that led to the existence and uniqueness of a fixed point on a bounded domain in Sect. 13.4.

(S1)   Function $\mathbf{N} \mapsto B(\mathbf{N})\mathbf{N}$ is nondecreasing for $\mathbf{N} \geq \mathbf{0}$.
(S2)   There is a steady state $\mathbf{N}^* > \mathbf{0}$; i.e., $\mathbf{N}^* = B(\mathbf{N}^*)\mathbf{N}^*$.
(S3)   For all initial conditions $\mathbf{n} \geq \mathbf{0}$ with $\mathbf{n} \neq \mathbf{0}$, the iteration $\mathbf{m}_{t+1} = B(\mathbf{m}_t)\mathbf{m}_t$ converges to $\mathbf{N}^*$.

With these assumptions, operator $\mathbf{Q}$ satisfies the following hypotheses from Lui (1989a):

1. $\mathbf{Q} : \mathscr{C}_{\mathbf{N}^*} \to \mathscr{C}_{\mathbf{N}^*}$, $\mathbf{Q}[\mathbf{0}] = \mathbf{0}$, $\mathbf{Q}[\mathbf{N}^*] = \mathbf{N}^*$.
2. $\mathbf{Q}$ commutes with translations; i.e., $\mathbf{Q}[\mathbf{N}(\cdot - y)](x) = \mathbf{Q}[\mathbf{N}](x - y)$.
3. $\mathbf{Q}$ is order preserving; i.e., if $\mathbf{N} \geq \widetilde{\mathbf{N}}$, then $\mathbf{Q}[\mathbf{N}] \geq \mathbf{Q}[\widetilde{\mathbf{N}}]$.
4. $\mathbf{Q}$ is continuous in the topology of uniform convergence on bounded subsets of $\mathbb{R}$.
5. For any $\mathbf{0} \ll \mathbf{n} \ll \mathbf{N}^*$, the iterates $\mathbf{Q}^{(k)}[\mathbf{n}]$ converge to $\mathbf{N}^*$ as $k \to \infty$.

Order-preserving operators have the following two important properties.

Lemma 13.8 (Proposition 2.1 in Lui 1989a)

1. Let $\mathbf{R}_1$ or $\mathbf{R}_2$ be an order-preserving operator. Assume that sequences $\{\mathbf{n}_k\}$ and $\{\mathbf{m}_k\}$ satisfy $\mathbf{n}_{k+1} \geq \mathbf{R}_1[\mathbf{n}_k]$ and $\mathbf{m}_{k+1} \leq \mathbf{R}_2[\mathbf{m}_k]$, respectively. Suppose further that $\mathbf{R}_1[\mathbf{u}] \geq \mathbf{R}_2[\mathbf{u}]$ for all $\mathbf{u}$. If $\mathbf{n}_0 \geq \mathbf{m}_0$, then $\mathbf{n}_k \geq \mathbf{m}_k$ for all k.

2. Let $\mathbf{R}$ be an order-preserving operator. Assume that $\mathbf{R}[\mathbf{n}_0] \geq \mathbf{n}_0$ and define $\mathbf{n}_{k+1} = \mathbf{R}[\mathbf{n}_k]$. Then $\mathbf{n}_k$ is nondecreasing in k.

---

[2]A function on some interval is *piecewise continuous* if the interval can be written as a finite number of subintervals such that the function is continuous on each open subinterval and has a finite limit at each endpoint of each subinterval.

We now present the construction that leads to the definition of the asymptotic spreading speed. We pick some vector $\mathbf{0} \ll \mathbf{m} \ll \mathbf{N}^*$ and continuous functions ϕ^i with the following properties:

1. $\phi^i : \mathbb{R} \to [0, m^i]$ is nonincreasing.
2. $\phi^i(s) = 0$ for $s \geq 0$.
3. $\phi^i(-\infty) = \lim_{s \to -\infty} \phi^i(s) = m^i$.

Then we define a sequence of functions $\mathbf{a}(c, \cdot)$ depending on a parameter c via

$$a_0^i(c, s) = \phi^i(s), \tag{13.56}$$

$$a_{k+1}^i(c, s) = \max\{\phi^i(s), \mathbf{Q}[\mathbf{a}_k(c, \cdot + s + c)](0)\}. \tag{13.57}$$

The operator defined by the right-hand side of (13.57) is order preserving since $\mathbf{Q}$ is. The following properties of the sequence $\mathbf{a}_k$ are fairly straightforward to see.

Lemma 13.9 (Lemma 2.2 in Lui 1989a)

1. $\mathbf{a}_k(c, s)$ is bounded between $\mathbf{0}$ and $\mathbf{N}^*$.
2. $\mathbf{a}_k(c, s)$ is nondecreasing in k and nonincreasing in c and s.
3. $\mathbf{a}_k(c, -\infty)$ exists, and $\mathbf{a}_k(c, -\infty) \geq \mathbf{Q}^{(k)}[\mathbf{m}]$.
4. $\mathbf{a}_k(c, \infty) = \mathbf{0}$ for all k.
5. $\lim_{k \to \infty} \mathbf{a}_k(c, s) = \mathbf{a}(c, s)$ exists and is nonincreasing in c and s. Furthermore, $\mathbf{a}(c, -\infty) = \mathbf{N}^*$.

Before we define the asymptotic spreading speed, we give a rough illustration of the operator defined in (13.57). Let us consider the point $s = 0$ and some given element $\mathbf{a}_k$. Then the operator will first shift $\mathbf{a}_k$ to the left by c units; then apply the growth and dispersal operator $\mathbf{Q}$ to it, which presumably will help the species spread back to the right again; and then evaluate the result at $s = 0$. If the spreading to the right is in some sense "larger" than the shift by c to the left, then the value of $\mathbf{a}_{k+1}$ at $s = 0$ will be larger than that of $\mathbf{a}_k$ at $s = 0$; if it is smaller, then the reverse will happen. Hence, we could guess that for small enough values of c, the value of $\mathbf{a}_k(c, 0)$ will grow with k, but for large values it might not. We could therefore try to define the spreading speed as the value of c between these two scenarios. This idea can be used but we have to apply it at $s = \infty$ rather than $s = 0$.

We define

$$c^* := \sup\{c \mid \mathbf{a}(c, \infty) = \mathbf{N}^*\}. \tag{13.58}$$

Then the following can be shown.

Lemma 13.10 (Lemmas 2.4, 2.5, 2.6, and 2.9 in Lui 1989a)

1. $\mathbf{a}(c, s) \equiv \mathbf{N}^*$ if and only if $c < c^*$.
2. $\mathbf{a}(c, \infty)$ is independent of the choice of $\mathbf{m}$ and ϕ^i.
3. $\mathbf{a}(c, \infty) = \mathbf{Q}[\mathbf{a}(c, \infty)]$.
4. $\mathbf{a}(c, \infty) = \mathbf{0}$ if $c > c^*$.

We can now formulate the theorem that demonstrates that c^* is indeed the spreading speed of operator $\mathbf{Q}$; compare Chap. 5, Definition 5.1.

Theorem 13.3 (Theorems 3.1 and 3.2 and Proposition 3.3 in Lui 1989a) *Let (S1)–(S3) as well as the assumptions on* $\mathbf{K}$ *hold. Assume that* $\mathbf{Q}$ *is bounded above by its linearization at zero, i.e.,* $\mathbf{Q}[\mathbf{N}] \leq \mathbf{Q}'[\mathbf{0}]\mathbf{N}$. *Assume furthermore that near zero,* $\mathbf{Q}$ *can be bounded below, e.g.,* $\mathbf{Q}[\mathbf{N}] \geq (1 - \epsilon)\mathbf{Q}'[\mathbf{0}]\mathbf{N}$ *for* $\|\mathbf{N}\|$ *small enough. Pick* $\mathbf{0} \neq \mathbf{N}_0 \in \mathscr{C}_{\mathbf{N}^*}$ *with compact support and define* $\{\mathbf{N}_t\}$ *through the recursion (13.54). Then the following hold for any small* $\epsilon > 0$:

$$\limsup_{t \to \infty} \max_{|x| > (c^* + \epsilon)t} \mathbf{N}_t(x) = \mathbf{0},$$

$$\liminf_{t \to \infty} \min_{|x| < (c^* - \epsilon)t} \mathbf{N}_t(x) = \mathbf{N}^*.$$

Finally, Theorems 3.4 and 3.5 in Lui (1989a) guarantee that the spreading speed defined in (13.58) can be calculated by the formula in (13.40).

The original results by Lui (1989a) require that the initial condition is sufficiently large on a sufficiently large domain. When there is no Allee effect in the IDE, the "hairtrigger effect" (Weinberger 1982) ensures that any nonnegative, nonzero initial condition will eventually meet these requirements. The original results also require the kernels to have compact support. This requirement is removed by Liang and Zhao (2007). The theory formulated by Lui (1989a) and earlier for the scalar case by Weinberger (1982) is substantially more general than the version presented here. We mention a few aspects. First, the formulation by Lui (1989a) is based on some abstract operator $\mathbf{Q}$ and its properties. The theory is well suited to applications to IDEs, as we saw, but not limited to those. For example, it also applies to a time-one map of (systems of) reaction–diffusion equations. The abstract theory has since been developed significantly further by several authors, e.g., Liang and Zhao (2010). Second, the work in Lui (1989a) does not address the question of the existence of traveling waves treated in the scalar case by Weinberger (1982), but many authors since then have worked on this problem. A survey on the theory of spreading speeds and traveling waves for monotone systems of equations can be found in Zhao (2009). Third, the formulation by Lui (1989a) includes the case of an Allee effect, which we excluded here in assumption (S3). As we have seen in Chap. 6, the elegant spread speed formula (13.40) cannot be expected to hold in the presence of an Allee effect. The asymptotic speed of spread still exists, but it may be negative and, most important, for a population to actually spread at this speed, its initial density has to be high enough over a large enough set. For details, see Theorems 3.1 and 3.2 in Lui (1989a). Finally, the general theory is formulated for spreading phenomena in several spatial dimensions and not only one-dimensional space as presented here. We then formulate a spreading speed in a given direction and consider planar traveling waves in that direction. Amor and Fort (2009) formulate an explicit two-dimensional model, derive the corresponding formula for the spreading speed based on the linear conjecture, and compare how one- and two-dimensional structured and unstructured spread rates differ.

13.9 Further Reading

The aspects of IDEs for stage-structured populations presented in this chapter are only the basics of the existing theory. Because these structured equations can describe such rich life histories, the number of modifications and interesting questions seems almost infinite. The following summaries are meant to point the interested reader to the existing body of literature.

Dynamics on Bounded Domains

Fagan and Lutscher (2006) apply the theory by Lutscher and Lewis (2004) to a two-stage model of swift fox (*Vulpes velox*) population dynamics to estimate the habitat size required for conservation of the species. This application includes the scenario that the bounded domain consists of two disjoint subdomains corresponding to two spatially separated patches of habitat. We will investigate these ideas further in Chap. 15 when we focus on ways to include spatial variation in IDE models.

A bifurcation theory for systems of IDEs, somewhat more abstract than the results presented here, is developed independently in Alzoubi (2007, 2010a,b). It is based on the general theory of global bifurcations by Rabinovitz. Another series of papers uses the same bifurcation-theoretic approach—in addition to numerical experiments—and expands the problems studied via IDEs in various directions (Robertson 2009; Robertson and Cushing 2011, 2012; Robertson et al. 2012). The authors study the question of spatial segregation between life stages in a flour beetle population in a homogeneous habitat. They propose a dispersal kernel very different from the kernels presented here so far. In their approach, the probability that an individual will move to a certain location depends only on that location and not on the individual's initial location. This approach implies that the domain is relatively small compared to the individual's dispersal ability, so that individuals can reach any location from any point within a single dispersal period. Furthermore, the authors assume that dispersal is density dependent so that the probability of moving to a certain location decreases with the density of individuals already there. For an example with juveniles and adults, the authors choose

$$K(x, y, J, A) = \frac{1}{2} \sin(x) \exp(-D(s_j J(x) + s_a A(x))) \tag{13.59}$$

on the domain $\Omega = [0, \pi]$. When $D = 0$, dispersal is density independent. Increasing parameter D leads to avoidance of areas where the density is high. It would be an interesting challenge to derive dispersal kernels for avoidance (or attraction) from mechanistic random-walk models in the spirit of Chap. 7.

When the life cycle of an organism consists of many stages, it is typical for dispersal to be limited to only a few stages or transitions between stages, as discussed in Sect. 12.4. Formulating a model with nondispersing stages is fairly simple: the corresponding dispersal kernel is a delta distribution; see, e.g., matrix **K** in (13.41). Analyzing the dynamic behavior when there are sedentary stages is much harder. The mathematical reason for this difficulty is that the operators involved fail to be compact, whereas many of the existing theorems about stability and bifurcation require compactness. Corresponding results about critical patch-size and existence of steady states can still be achieved in several cases. For example, operator **Q** in (13.7) may not be compact but some iterate of it is compact if individuals have to go through a dispersal stage after finitely many generations. In that case, we can apply the theory to the corresponding power of **Q**. Alternatively, we can sometimes decompose the operator into the sum of a compact and a nilpotent operator and obtain the desired results. Details of these two approaches are laid out in Lutscher and Lewis (2004). Jin et al. (2016) develop a general theory for population persistence and extinction when the next-generation operator can be written as the sum of a contraction and a compact operator. This theory applies to the structured population models studied here, provided that certain conditions hold. In particular, if we write **Q** as a sum of those transitions relating to reproduction and those that relate to survival, then the part that relates to survival only is a contraction. If all components of the part with reproduction disperse, then that part of the operator is compact.

Spread on Unbounded Domains

We need to formulate stage-structured models because for many species, individuals of different life-history stages exist simultaneously. The stage-structured model projects the densities of all stages per year or some other suitable time unit. It takes several of these time units to complete a generation. Bateman et al. (2015) introduce the concept of a "generational spreading speed," the distance that a population front covers from one generation to the next. The methods are similar to the ones introduced above, but the authors use graph reduction techniques to simplify the full annual stage-structure dynamics to generational dynamics.

One of the biggest challenges for the mathematical theory of spread on unbounded domains is the loss of compactness of the operator **Q** in model (13.54) when some of the dispersal kernels are delta distributions. While the asymptotic spreading speed can still be defined and can be shown to exist, including the formula from the linearization, the proof of existence of a traveling wave in Weinberger (1982) relies on a compactness argument. One way to prove the existence of traveling waves is to decompose the operator into a compact operator and a contraction if the model structure allows it. This idea is developed by Le et al. (2011) and Lutscher and Van Minh (2013).

An abstract, very general theory of spreading speeds and traveling waves in Banach lattices is developed by Liang and Zhao (2010). The authors replace the requirement of compactness by the requirement that operator **Q** decrease the Kuratowski measure of noncompactness of a set. This theory applies to a wide range of types of equations, not only IDEs. Fang and Zhao (2014) use an even weaker compactness assumption but apply their theory to continuous-time equations only. A more direct approach for IDEs is carried out by Meyer (2012) and Meyer and Li (2013). These authors derive a structured IDE with infinitely many stages. Their model is motivated by a plant species with a seed bank where seeds can lie dormant for arbitrarily long times. They prove the existence of a spreading speed and of traveling waves for their model. They also prove the usefulness of the linearization formula and they consider nonmonotone growth functions. An earlier model with infinitely many stages in a completely different context was derived by Powell et al. (2005); see below.

In some cases, a scalar IDE may contain a time delay; i.e., the density at time $t + 1$ is determined by the density not only at time t but also at previous times. Accordingly, operator Q may depend on $N_t, N_{t-1}, \ldots$. Such equations may be reformulated as a structured model with nondispersing stages, and the existence of spreading speeds and traveling waves may be proved in that setting (Lin and Li 2010). These authors also show the stability of the traveling-wave front.

Another dimension for structuring a population is sex. If males and females have significantly different dispersal behavior, then distinguishing between them could be important for correctly predicting population spread rates. Miller et al. (2011) formulate and analyze the first two-sex model for invasions. They derive a heuristic formula for the spreading speed and use numerical simulations to demonstrate its validity, but a mathematical proof of its correctness is still missing.

Disease stage may constitute yet another factor of the structure of a population. For example, Marculis and Lui (2015) formulate a model for green crab with a juvenile, a susceptible-adult, and an infected-adult stage. The most interesting aspect of this model in the present context is that there can be a semi-trivial boundary state where some stages are positive (e.g., juveniles and susceptible adults), whereas others are zero (e.g., infected adults). The existence of a semi-trivial boundary stage violates the assumption made by Lui (1989a) that there be a unique positive equilibrium. Numerical simulations by Marculis and Lui (2015) reveal that the equations may support *stacked waves*: a traveling wave that connects the zero state with the semi-trivial boundary state and a secondary wave that connects that boundary state with the positive stage. The phenomenon is closely related to the observation of traveling two-cycles in Sect. 11.4. The analysis of such stacked waves is still wide open. Bateman et al. (2017) provide more simulation results on stacked waves in a closely related model for green crab with a castrating parasite.

Applications to Spread and Biological Invasions

We need to formulate stage-structured models because for many species, individuals of different life history stages exist simultaneously. There are, however, other scenarios to which this theory applies. For example, Lui (1989b) calculates spread rates for a population genetics model with equations for allele frequencies in the female and male subpopulation. He also applies the theory to an epidemic model that tracks the progression of a disease between a host and a vector.

A stage-structured model projects the densities of all stages per year or some other suitable time unit. It takes several of these time units to complete a generation. Bateman et al. (2015) introduce the concept of a "generational spreading speed," the distance that a population front covers from one generation to the next. The methods are similar to the ones introduced above, but the authors use graph reduction techniques to simplify the full annual stage-structured dynamics to generational dynamics.

Neubert and Caswell (2000a) develop the theory and apply it to two herbaceous flowering plants with six (*Dipsacus sylvestris*) and eight (*Calathea ovandensis*) stages. Among other things, the authors find that the speed of spread is highly sensitive to long-distance dispersal, a phenomenon that we discussed in Sect. 12.5. Caswell et al. (2003) expand the theory to use order statistics to calculate sensitivities and apply the techniques to three bird species with three stages each, namely Pied Flycatcher (*Ficedula hypoleuca*), Starling (*Sturnus vulgaris*), and Sparrowhawk (*Accipiter nisus*). Neubert and Parker (2004) review the calculation of spread rates and sensitivity analysis of IDEs for risk analysis of invasive species and apply it to scotch broom.

Le Corff and Horvitz (2005) use a stage-structured IDE to study the effects on population spread of the mixed reproductive strategy of a tropical herb (*Calathea micans*) with obligate selfed or potentially outcrossed seeds and dispersal by ants.

Jacquemyn et al. (2005) study the spread of perennial tussock grass *Molinia caerulea* in a heathland, in particular the effect of fire on the success of this invasive heathland plant.

Buckley et al. (2005) develop a stage-structured model for a pine invasion in New Zealand. They represent long- and short-distance dispersal by a mixed dispersal kernel consisting of a Laplace and a Gaussian kernel. Based on sensitivity analysis, they give some management recommendations on how to slow the invasion.

Vellend et al. (2006) consider the dual effect of herbivores on the spread of the forest herb *Trillium grandiflorum*: deer consume the plant and thereby inhibit its growth, but they also transport its seed and thereby facilitate its spread. Their seven-stage IDE model illustrates how the additional dispersal mechanisms outweigh the negative effects on population growth at low levels of herbivory and can explain range expansion of the herb at the northern edge of its range.

Garnier and Lecomte (2006) use a stage-structured model for transgenic oilseed rape with a combination of dispersal kernels accounting for long- and short-distance dispersal to estimate the spread risk of feral plants. Garnier et al. (2008) measure probabilities and distances of roadside spread and include them into their previous model.

Jongejans et al. (2008) use a four-stage model of the thistle (*Carduus nutans*) and develop a variance-decomposition method to study the relative impact of population dynamics and dispersal mechanisms. They employ the WALD model to generate dispersal kernels that include a distribution of seed release height and hourly wind-speed data. In a later investigation of the same species, Shea et al. (2010) study several management options by comparing sensitivities of the spread rate with respect to parameters in three different environments, one native and two invaded. Zhang et al. (2011) follow up on this study by exploring how global warming can speed up invasions of this thistle. Caplat et al. (2012) also use the WALD model to generate a dispersal kernel, and they apply it to the invasion of Corsican pine in New Zealand. Caplat and Buckley (2012), based on Buckley et al. (2005), present a more general article on management applications of structured IDEs.

Soons and Bullock (2008) generate dispersal kernels from wind tunnel experiments and wind speeds measured in the field. They study in particular the effect of nonrandom seed release in two heathland plants, *Calluna vulgaris* and *Erica cinerea*. These plants release their seeds only when wind speeds exceed a certain threshold, and in particular during wind gusts. This mechanism leads to predicted wave speeds that are twice as high as with temporally random seed release.

Bullock et al. (2008) consider a stage-structured IDE for the rare annual herb *Rhinanthus minor* under four different management options (grazing versus cutting) and calculate corresponding spread rates.

Miller and Tenhumberg (2010) study the spread of the Diaprepes root weevil (*Diaprepes abbreviatus*) in Florida with a six-stage IDE. They find, among other things, that transient speeds of range expansion at the onset of an invasion can be higher than the asymptotic speed, and that measures to reduce the asymptotic speed may have little or no effect on transient speeds.

Gruess et al. (2011) formulate a two-stage IDE to evaluate the relative impacts of adult movement and juvenile dispersal, as well as harvesting, on the effectiveness of marine reserve networks.

Travis et al. (2011) compare analytical predictions of a stage-structured IDE with simulation results from an individual-based model for the invasive shrub *Rhododendron ponticum*. They argue in particular that the two modeling approaches should be seen as complementary and used in conjunction since each can inform the other.

Bullock et al. (2012) use climate modeling and the WALD model to predict future wind speeds and their variation. They then include this information in mechanistic dispersal kernels and predict the ability of wind-dispersed plants to keep up with climate change; see Sect. 12.3.

Matlaga and Davis (2013) use a stage-structured IDE model to estimate the invasive potential of the engineered bioenergy crop *Miscanthus* × *giganteus*.

Lamoureaux et al. (2015) consider a seven-stage IDE for the weedy grass *Nassella trichotoma* to evaluate the need for and cost effectiveness of various control strategies of this plant.

Structured IDEs are also used to model species spread and invasions in advective environments with asymmetric dispersal. Pringle et al. (2009) generalize some of the ideas from Sect. 12.2 to structured populations; in particular, they characterize the critical population growth rate for persistence under biased dispersal. Gharouni et al. (2015) derive a three-stage model for a green crab invasion with biased dispersal. Krkošek et al. (2007) fit several dispersal kernels to the northern and southern spread of sea otters in California. They find that heavy-tailed dispersal kernels that predict accelerating invasions fit the observations very well. Smith et al. (2009) reanalyze the sea otter data but introduce spatial heterogeneity by assuming that population vital rates differ between the northern and southern directions.

A series of papers develops models and fits data of the spread of the late blight disease in potatoes and tomatoes and its pathogen *Phytophthora infestans*. The quantity of interest is the density of lesions caused by the pathogen. These lesions are structured by age, and their area increases with age. Lesions "disperse" through spores that are moved by wind or splatter with rain drops. The model development with infinitely many stages is presented by Powell et al. (2005). Several other papers are parameterized by field experiments and study the effect of host plant diversity, spatial scale of heterogeneity, and weather (Skelsey et al. 2005, 2009a,b, 2010).

On a much larger scale, Heavilin and Powell (2008) model the spread of pine tree (*Pinus contorta* Douglas) death due to attacks from mountain pine beetle. They structure trees as juvenile (which are not susceptible), susceptible, and infected. Infections spread through beetle reproduction and flight. This model exhibits an Allee effect, as a susceptible tree can fend off a small enough number of beetles. The authors then obtain parameter estimates through a maximum likelihood procedure.

Chapter 14
Two Interacting Populations

Abstract Most biological populations do not exist in isolation but interact with other species in many ways that may increase or decrease their reproductive ability, affect their survival, or alter their dispersal behavior. Species interactions can lead to phenomena such as sustained population oscillations or competitive exclusion. In this chapter, we present some of the spatial aspects of population interaction in the context of IDEs. We begin with a brief background on nonspatial models before we move to study critical patch-sizes for predator and prey systems. Some of the most surprising and beautiful results in this section relate to dispersal-induced pattern formation in these systems. Spatial invasion dynamics of predator and prey show rich and complex behavior. We then present the phenomenon of anomalous spreading speeds in mutualism systems. Finally, we consider several aspects of persistence and invasion of competing species.

14.1 Species Interactions

Our treatment of population dynamics so far has considered only a single species. In reality, species often interact with others, and the interactions can be beneficial or detrimental for any of the species involved. The two types of interaction that are typically thought to be most influential in shaping biological communities are predation, where one species benefits from the presence of the other but the other suffers, and competition, where both species suffer from the presence of the other.

Predators benefit from their prey, whose population growth rate they reduce. Arguably the most salient feature of predator–prey interactions is that they provide a mechanism for sustained population oscillations in the absence of external forcing or stochasticity; see, e.g., Murray (2001) or Kot (2001). A special class of predator–prey (or consumer–resource) relationships are insect host–parasitoid systems (Hassell 1978). Whereas a predator usually kills prey for its own consumption, a parasitoid uses its host for food and shelter for its offspring but does not directly consume the host. It is the offspring that consumes the host (Kot 2001). Insect host–parasitoid systems often have discrete, nonoverlapping generations, e.g., many wasp species and their hosts or the caterpillars of many moth and butterfly species.

© Springer Nature Switzerland AG 2019

F. Lutscher, *Integrodifference Equations in Spatial Ecology*, Interdisciplinary Applied Mathematics 49, https://doi.org/10.1007/978-3-030-29294-2_14

Nonspatial difference equations have a distinguished history in modeling these population dynamics, starting from the pioneering work by Nicholson and Bailey (1935). More recently, IDE models have been used to understand the importance of space and dispersal on outbreak dynamics of forest insects (Cobbold et al. 2005).

Similar species often compete for resources, mutually inhibiting each other's growth. A famous early result in competition theory is the "competitive exclusion principle," which states that no two species can stably coexist on a single limiting essential resource; see, e.g., Murray (2001). In reality, we observe a great variety of similar species stably coexisting while competing for relatively few limiting resources, e.g., plants that compete for water and nutrients (Tilman 1982). Several proposed mechanisms that could resolve this paradox are based on spatial considerations (Tilman 1982). Nonspatial discrete-time models for competition have been studied by many, including early works by Nicholson (1954) and May (1973). IDEs for competing species were originally formulated and analyzed by Allen et al. (1996) and Hart and Gardner (1997) and have since received major theoretical attention; see, e.g., Lewis et al. (2002).

Mutualism is a form of interaction where each species is beneficial to the growth of the other(s). Compared to competition and predation, the systematic study of (mutually) facilitative interactions in ecological theory arose much later and is still much less developed (Boucher 1982; Kot 2001). Mathematically, however, the structure of models with mutual facilitation is closely related to that of the stage-structured models that we saw in Chap. 13 and—in the special case of only two species—also to competition.

In this chapter, we study some aspects of the spatial dynamics of two interacting populations in IDEs. We review some background theory about nonspatial models. This material can be found in many textbooks on mathematical ecology, e.g., Allen (2006), Edelstein-Keshet (2005), or Kot (2001). Then we select various topics from spatial dynamics and consider how spatial aspects affect these types of interactions.

14.2   Nonspatial Models for Two Species

We denote by N_t and P_t the densities of two interacting species in generation t. We model their respective growth functions between generations by nonnegative, smooth functions F and G. Then we obtain the general planar discrete-time dynamical system

$$N_{t+1} = F(N_t, P_t), \qquad P_{t+1} = G(N_t, P_t). \qquad (14.1)$$

We recover the respective single-species models as $N_{t+1} = F(N_t, 0)$, and similarly for P_t. Since we consider the two species to be different, we have the additional property that, in the absence of immigration, $F(0, P) = G(N, 0) = 0$. This condition reflects the fact that one species cannot create another. While the juvenile–adult model for a single species from Chap. 13 can be written in the general form (14.1), it does not possess this latter property.

We classify the kind of interaction according to how one species affects the growth function of the other. Specifically, we speak of

- *predation* of N by P if $\frac{\partial F}{\partial P} < 0$ and $\frac{\partial G}{\partial N} > 0$,
- *(inter-specific) competition* between N and P if $\frac{\partial F}{\partial P} < 0$ and $\frac{\partial G}{\partial N} < 0$, or
- *mutual facilitation* between N and P if $\frac{\partial F}{\partial P} > 0$ and $\frac{\partial G}{\partial N} > 0$.

We begin analyzing the qualitative behavior of (14.1) with the standard steps of determining fixed points by solving the equations

$$N^* = F(N^*, P^*), \qquad P^* = G(N^*, P^*) \tag{14.2}$$

and investigating their local stability via the eigenvalues of the Jacobian matrix,

$$J = \begin{bmatrix} \frac{\partial F}{\partial N} & \frac{\partial F}{\partial P} \\ \frac{\partial G}{\partial N} & \frac{\partial G}{\partial P} \end{bmatrix}, \tag{14.3}$$

evaluated at the fixed point. A fixed point is locally asymptotically stable if both eigenvalues of J are inside the unit circle, which, according to the Jury conditions, is equivalent to the following three conditions for the trace and determinant of J (Kot 2001):

$$1 - \operatorname{tr} J + \det J > 0,$$

$$1 + \operatorname{tr} J + \det J > 0, \tag{14.4}$$

$$1 - \det J > 0.$$

The Jury conditions are also helpful in determining the kind of bifurcation that may arise if model parameters are varied and cause a change in the stability properties of a fixed point (Kot 2001). If the first of these conditions is violated at a bifurcation point, we have an eigenvalue of $+1$. In the context of population dynamics, the resulting bifurcation is often a transcritical bifurcation. If the second condition is violated, we have an eigenvalue of -1. We may expect a flip bifurcation in this case. Violating the third condition indicates eigenvalues with nonzero imaginary part. This is a necessary condition for a Naimark–Sacker bifurcation, the discrete version of a Hopf bifurcation. This latter bifurcation can produce oscillating solutions that are frequently found in the dynamics of a predator and its prey. We begin our exploration of coupled IDEs with such a system.

14.3 Critical Patch-Sizes for Predator–Prey Systems

Insect host–parasitoid systems constitute the earliest and most important discrete-time predator–prey models, dating back to the work of Nicholson and Bailey (1935); see Kot (2001) for a comprehensive overview. The Nicholson–Bailey model is highly unstable. In fact, every initial population will eventually collapse. Several stabilizing mechanisms have been proposed and studied, including spatial dispersal; see Adler (1993) for a discussion of this topic and Allen et al. (2001) for a discrete-space simulation model. Cobbold et al. (2005) present the first IDE model for host–parasitoid dynamics in continuous space to determine the effects of patch size on these populations. Their model is parameterized for the forest tent caterpillar (*Malacosoma disstria*) and includes a detailed discussion of the timing of parasitism in the life cycle of the caterpillar. Bramburger and Lutscher (2019) study a predator–prey model with the cosine kernel from (3.24) via reduction to a finite-dimensional system. The two models make qualitatively similar predictions about the influence of patch size and spatial dispersal on the population dynamics. Hence, these insights are quite robust. We illustrate the theory and results with a model from Kot (1989), which is simpler than the one by Cobbold et al. (2005). Related models appear in Neubert and Kot (1992) and Wei and Lutscher (2013).

A standard model for prey (host, N_t) and predator (parasitoid, P_t) reads

$$N_{t+1} = N_t e^{r(1-N_t)} e^{-\rho P_t} , \qquad P_{t+1} = N_t \left(1 - e^{-\rho P_t} \right) , \qquad (14.5)$$

with parameters $r, \rho \geq 0$ (Beddington et al. 1975); but see May et al. (1981) for its criticism. In the absence of the predator, the prey grows according to the (scaled) Ricker function in (2.19). A fraction, $\exp(-\rho P)$, of prey escapes predation, whereas the remaining fraction supports predator growth. This formulation represents a random search pattern by the predator (May 1973). The occurrence of the exponential functions makes explicit calculations difficult. However, the linear approximation $(1 - e^{-\rho P}) \approx \rho P$ gives a qualitatively similar system where explicit calculations are possible. With this approximation and some scaling, we arrive at the model

$$N_{t+1} = F(N_t, P_t) = N_t e^{r(1-N_t-P_t)} , \qquad P_{t+1} = G(N_t, P_t) = \rho N_t P_t , \qquad (14.6)$$

which is one of a suite of models studied by Neubert and Kot (1992).

System (14.6) has up to three biologically relevant steady states: the trivial extinction state $(0, 0)$, the semi-trivial prey-only state $(1, 0)$, and the coexistence state

$$(N^*, P^*) = \left(\frac{1}{\rho}, 1 - \frac{1}{\rho} \right) . \qquad (14.7)$$

The trivial state is unstable. The prey-only state is stable against perturbations in N as long as $0 < r < 2$ and loses stability in a flip bifurcation with a two-cycle

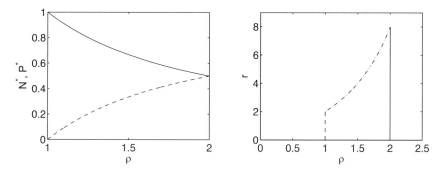

Fig. 14.1 Illustration of the coexistence state and its stability region of the nonspatial predator–prey model (14.6). **Left:** The solid (dashed) curve shows the prey (predator) densities from (14.7). **Right:** The steady state loses stability in a transcritical bifurcation along the dashed part of the boundary, in a flip bifurcation at the dash-dot curve, and in a Naimark–Sacker bifurcation along the solid line.

emerging at $r = 2$. It is stable against perturbations in P as long as $\rho < 1$. When $\rho > 1$, the prey-only state becomes unstable, the predator can invade, and the coexistence state is biologically meaningful.

The Jacobian matrix at the coexistence state is given by

$$J = \begin{bmatrix} 1 - r/\rho & -r/\rho \\ \rho - 1 & 1 \end{bmatrix}. \tag{14.8}$$

If $\rho > 1$, then the first Jury condition in (14.4) is satisfied. The last Jury condition is satisfied when $0 < \rho < 2$. At $\rho = 2$, we expect a bifurcation to sinusoidally oscillating solutions. These indicate the onset of predator–prey cycles. The remaining Jury condition holds if $0 < r < 4\rho/(3 - \rho)$. When $r = 4\rho/(3 - \rho)$, we expect a flip bifurcation. This scenario reflects the overcompensatory dynamics of the Ricker function for the prey alone and can lead to extinction of the predator (Neubert and Kot 1992). This stability analysis is summarized in Fig. 14.1.

We now introduce dispersal into the predator–prey model in (14.6) and study the effects of patch-size and dispersal ability on the existence and stability of steady states. We denote by Ω the spatial domain of interest and by K_N and K_P the dispersal kernels of prey and predator, respectively. As in Sect. 3.3, we choose the fixed one-dimensional domain $\Omega = [-1/2, 1/2]$ and scale the dispersal kernels accordingly. The resulting equations are

$$N_{t+1}(x) = \int_{\Omega} K_N(x, y) F(N_t(y), P_t(y)) \mathrm{d}y,$$
$$P_{t+1}(x) = \int_{\Omega} K_P(x, y) G(N_t(y), P_t(y)) \mathrm{d}y. \tag{14.9}$$

We consider the trivial extinction state, the semi-trivial prey-only state, and the coexistence state separately.

The Extinction State

The extinction state $(0, 0)$ is a steady state of the model. The linearization at this state yields the eigenvalue problem

$$\lambda\phi(x) = R \int_{\Omega} K_N(x, y)\phi(y)dy ,$$

$$\lambda\psi(x) = 0 ,$$

(14.10)

with $R = \partial F/\partial N(0, 0)$. Since the solution for the second equation is trivial, the stability of the extinction state reduces to the stability problem for a single species that we studied in Chap. 3. In particular, there is a critical patch-size, above which the prey population can grow when rare. If the prey disperses according to the Laplace kernel with parameter $a_N > 0$, then the critical patch-size for persistence of the prey is given by (3.14), i.e.,

$$L_N^* = \frac{2}{a_N \sqrt{R - 1}} \arctan\left(\frac{1}{\sqrt{R - 1}}\right) .$$

(14.11)

The Prey-Only State

When the extinction state is unstable, the system possesses a prey-only steady state $(N^*(x), 0)$, at least under some suitable conditions on the growth function $F(N, 0)$; see Chap. 4. The linearization at this state leads to the eigenvalue problem

$$\lambda\phi(x) = \int_{\Omega} K_N(x, y)[F_N(N^*(y), 0)\phi(y) + F_P(N^*(y), 0)\psi(y)]dy ,$$ (14.12)

$$\lambda\psi(x) = \int_{\Omega} K_P(x, y)G_P(N^*(y), 0)\psi(y)dy ,$$ (14.13)

where $F_N = \partial F/\partial N$, and similarly for F_P and G_P. We note that for a predator–prey model, we typically have $G_N(\cdot, 0) = 0$. This property reflects the fact that the predator cannot grow in the absence of the prey.

We could study stability conditions with respect to perturbations in the prey only, i.e., perturbations of the form $(\phi, 0)$. This approach would lead to the single-species stability considerations from Chap. 4. Instead, we are interested in perturbations with respect to the predator so that we can answer the question of when the predator can persist in the system. We will therefore assume that the prey-only state is stable with respect to perturbations in the prey only. In particular, we assume that all the eigenvalues of (14.12) with $\psi = 0$ are strictly inside the unit circle.

Since the system decouples, it suffices to study (14.13) separately. If λ is an eigenvalue of (14.13), the Fredholm alternative (Keener 2000) guarantees the existence of a solution to (14.12) provided λ is not also an eigenvalue of Eq. (14.12) with $\psi = 0$. Near the stability boundary $|\lambda| = 1$, this condition is satisfied since we assumed that the prey-only state is stable with respect to prey-only perturbations.

The eigenvalue problem

$$\lambda \psi(x) = \int_\Omega K_P(x, y) \rho N^*(y) \psi(y) dy \qquad (14.14)$$

cannot be solved explicitly because of the unknown function N^*. However, the integral operator is positive, so that a dominant eigenvalue exists; it determines the stability of the state $(N^*, 0)$. Numerical results by Cobbold et al. (2005) indicate that there is a critical patch-size, $L_P^* > L_N^*$, above which the predator can persist in the system. Similar results can be found in Bramburger and Lutscher (2019). Since N^* is positive, one can use the change of variables in (3.41) and arrive at the equivalent, symmetric eigenvalue problem

$$\lambda \tilde{\psi}(x) = \rho \int_\Omega \sqrt{N^*(x)} K_P(x, y) \sqrt{N^*(y)} \tilde{\psi}(y) dy, \qquad (14.15)$$

provided K is symmetric. This symmetrized version could be more convenient for numerical procedures (Kot and Phillips 2015).

For explicit but approximate calculations, we obtain a tractable, spatially implicit model by applying the averaging ideas from Chap. 9 (Cobbold et al. 2005). The integral operator in (14.14) can be bounded below and above by replacing $N^*(y)$ with $\min(N^*(y))$ and $\max(N^*(y))$, respectively. Accordingly, the eigenvalue in (14.14) is bounded below and above by the eigenvalues of the respective resulting operators. Hence, by continuity, there is a constant $\bar{N} \in [\min(N^*(y)), \max(N^*(y))]$ so that the eigenvalue in (14.14) is exactly the eigenvalue of the operator with $N^*(y)$ replaced by $\bar{N}$. We use the average dispersal success approximation, $\bar{N}_{ap}$, from (9.11) as an approximation for $\bar{N}$.

The explicit expression is

$$\bar{N}_{ap} = \bar{S}_N \bar{N}_{ap} e^{r(1-\bar{N}_{ap})} \quad \text{or} \quad \bar{N}_{ap} = 1 + \frac{\ln \bar{S}_N}{r}, \qquad (14.16)$$

where $\bar{S}_N$ is the average dispersal success (9.3) for the prey species. Since $\bar{S}_N < 1$, we have $\bar{N}_{ap} < 1$. The eigenvalue problem approximating (14.14) is then

$$\lambda \psi(x) = \rho \bar{N}_{ap} \int_\Omega K_P(x, y) \psi(y) dy. \qquad (14.17)$$

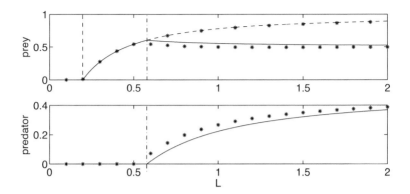

Fig. 14.2 Comparison of the approximate (curves) and exact (stars) steady-state average densities of prey (above) and predator (below). In the absence of the predator (dashed curve in upper plot), the prey density increases with domain length. The approximation by the average dispersal success is very close. When the domain is long enough for the predator to persist, the prey density decreases as the predator density increases and the upper branch of the prey density is unstable. The approximation is less accurate. The vertical lines denote the critical patch-size for prey persistence (dashed) and for predator persistence (dash-dot) according to the dispersal success approximation. Parameters are $r = 1$, $\rho = 2$, $a_N = 5$, and $a_P = 10$.

If K_P is a Laplace kernel, we obtain an explicit (approximate) critical patch-size for the predator as

$$L_P^* = \frac{2}{a_P\sqrt{\rho\bar{N}_{\text{ap}} - 1}} \arctan\left(\frac{1}{\sqrt{\rho\bar{N}_{\text{ap}} - 1}}\right). \tag{14.18}$$

We compare the exact and approximate densities and critical patch-sizes for prey and predator in Fig. 14.2; see also Fig. 6 in Cobbold et al. (2005).

The Coexistence State

When the domain is large enough to support predator and prey, we ask whether the coexistence state, $(N^*(x), P^*(x))$, is stable or unstable and how patch-size affects the transition to predator–prey oscillations. The eigenvalue problem at the coexistence state is

$$\lambda\phi(x) = \int_\Omega K_N(x, y)[F_N^*\phi(y) + F_P^*\psi(y)]\mathrm{d}y,$$

$$\lambda\psi(x) = \int_\Omega K_P(x, y)[G_N^*\phi(y) + G_P^*\psi(y)]\mathrm{d}y, \tag{14.19}$$

where F_N^* is the usual partial derivative evaluated at the coexistence state, and the other expressions are similar; see (14.13). This eigenvalue problem is much harder to study than the ones before because the two equations are fully coupled. We expect nonreal eigenvalues to appear. More important, a new phenomenon arises in that the eigenfunction of an eigenvalue at a bifurcation point need not be of one sign when the domain is large relative to dispersal distances. In that case, an instability can lead to spatial pattern formation. We briefly treat the simpler case of a relatively small domain here and devote the next section to studying pattern formation.

Aside from using numerical methods, we can calculate approximate stability conditions for the coexistence state from the average dispersal success approximation (Cobbold et al. 2005) or exact conditions by using the separable cosine kernel (Bramburger and Lutscher 2019). We outline some approximation ideas here.

The average dispersal success approximation of the IDE system in (14.9) with predator–prey dynamics (14.6) is (Cobbold et al. 2005)

$$N_{t+1} = \bar{S}_N N_t e^{r(1-N_t)} e^{-\rho P_t} , \qquad P_{t+1} = \bar{S}_P N_t \left(1 - e^{-\rho P_t}\right), \qquad (14.20)$$

where $\bar{S}_N$ and $\bar{S}_P$ are the average dispersal successes of the two species, respectively. The approximation of the positive steady state is given by

$$\bar{N}_{ap} = \frac{1}{\rho \bar{S}_P} , \qquad \bar{P}_{ap} = 1 - \frac{1}{\rho \bar{S}_P} + \frac{\ln(\bar{S}_N)}{r} . \qquad (14.21)$$

The stability conditions of the coexistence state for this spatially implicit model according to (14.4) are

$$1 - \frac{1}{\rho \bar{S}_p} + \frac{\ln(\bar{S}_N)}{R} > 0, \quad 4 + \ln(\bar{S}_N) + r - \frac{3r}{\rho \bar{S}_P} > 0, \quad \ln(\bar{S}_N) + r - \frac{2r}{\rho \bar{S}_P} > 0.$$

These conditions are the same as if we had substituted the steady-state values in (14.19) by their average dispersal success approximations and then used the eigenvalue approximation ideas from Sect. 9.3.

In Fig. 14.3, we illustrate how the spatially implicit stability conditions with dispersal loss from the domain differ from the nonspatial condition in Fig. 14.1. We also illustrate how the spatial system can have sustained oscillations of the (average) population densities. We see that dispersal loss requires a higher predation rate for the predator to persist in the system and allows for stable coexistence at a higher predation rate as well. When prey growth is very small ($r \approx 0$), dispersal loss from the bounded domain makes it hard for prey and predator to persist.

14.4   Pattern Formation in Predator–Prey Systems: Theory

The phenomenon of pattern formation or, more precisely, diffusion-driven instability, was discovered by Turing in a reaction–diffusion equation for a chemical

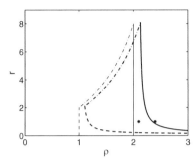 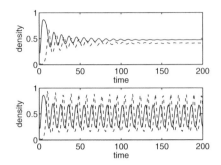

Fig. 14.3 **Left:** Comparison of the stability region for the spatially implicit predator–prey model according to the average dispersal success approximation (thick) and the nonspatial model (thin) as in Fig. 14.1. Dashed lines correspond to the transcritical bifurcation, dash-dot lines to the flip bifurcation, and solid lines to the Naimark–Sacker bifurcation. Parameter values are $S_N = 0.95$, $S_P = 0.9$. Stars correspond to parameter values used in the plots in the right. **Right:** Time series of the averaged densities of prey (solid) and predator (dashed) of the spatially explicit IDE system on a domain of length one. Each species disperses according to a Laplace kernel with parameters $a_N = 5$ and $a_P = 10$, so that the average dispersal successes are the same as those in the plot on the left. Population dynamics parameters are $r = 1$, $\rho = 2.1$ (top), and $\rho = 2.4$ (bottom).

reaction (Turing 1952). It has since been studied in many applications, from embryo development to animal coat patterns; see Murray (2002) for an excellent introduction to the mathematical theory, biological background, and implications. The basic setup requires two interacting species: one is an "activator," which enhances growth, and the other is an "inhibitor," which represses growth.[1]

In a nonspatial (well-mixed) scenario, these two species may coexist at a (globally) stable steady state. In a spatial setting, this stable state may become unstable if the activator acts (disperses) on a much shorter range than the inhibitor. If a small perturbation from the steady state arises, the growth of the activator occurs locally (since it does not spread far) before the inhibitor (which disperses far) can catch up and keep the activator in check. As a result, the system reaches a new stable steady state with spatially varying densities. Besides the necessary activator–inhibitor structure,[2] dispersal-driven instability requires that the dispersal ranges of the two species differ by at least an order of magnitude.

Spatial patterns of various extent and several characteristics are frequently observed in spatial ecology. Since prey activate predator growth and predators

---

[1] Strictly speaking, there are four possibilities: a species can activate itself and the other, it can inhibit itself and the other, it can activate itself and inhibit the other, or it can inhibit itself and activate the other. Classically, we speak of an activator (inhibitor) if the species activates (inhibits) itself and the other. An activator–inhibitor system is one that consists of an activator and an inhibitor in the classical sense.

[2] Diffusion-driven instability can also arise when one species activates itself and inhibits the other, and the other species inhibits itself and activates the other. This combination is sometimes referred to as positive feedback.

inhibit prey growth, prey and predator are candidates for ecological analogues to activator and inhibitor, respectively. The difficulty is that while their effect on the other species is clear from the biological interaction, their effect on themselves varies, depending on the population dynamics assumptions made. In particular, the classical MacArthur–Rosenzweig model does not support dispersal-driven instabilities (Okubo and Levin 2001). Several fairly strong assumptions, such as an Allee effect in the prey and self-limitation in the predator, are required for predator–prey equations to show activator–inhibitor structure (Murray 2002). It is also not easy to confirm that the dispersal abilities of a prey and a predator differ by an order of magnitude. For those reasons, the theory of dispersal-driven instabilities was rarely applied in ecology. More recently, empirical observations (Rietkerk and van de Koppel 2008) and theoretical results (Fasani and Rinaldi 2011; Alonso et al. 2002) have brought this mechanism of pattern formation back into ecological theory.

Much of the theory of dispersal-driven instabilities carries over from reaction–diffusion equations to IDEs (Kot and Schaffer 1986; Kot 1989; Neubert et al. 1995). It turns out that the conditions for pattern formation in IDEs are less strict than in reaction–diffusion equations if the dispersal behavior (represented by the dispersal kernel) has particular characteristics. In addition, novel bifurcation phenomena arise in IDEs that cannot be observed in reaction–diffusion equations. We present some of the main results and observations from Neubert et al. (1995).

Since we want the domain to be large with respect to the dispersal scale of at least one of the species, we consider the IDE on the entire real line, thereby also simplifying the analysis considerably. Hence, we begin with the system

$$
\begin{aligned}
N_{t+1}(x) &= \int_{-\infty}^{\infty} K_N(x, y) F(N_t(y), P_t(y)) \mathrm{d}y, \\
P_{t+1}(x) &= \int_{-\infty}^{\infty} K_P(x, y) G(N_t(y), P_t(y)) \mathrm{d}y.
\end{aligned}
\tag{14.22}
$$

We will show in the process that pattern formation requires N and P to have a predator–prey relation.

We assume that the nonspatial model has a coexistence state (positive steady state)

$$
N^* = F(N^*, P^*), \quad P^* = G(N^*, P^*)
\tag{14.23}
$$

that is locally asymptotically stable; i.e., the Jacobian matrix,

$$
J = \begin{bmatrix} \frac{\partial F}{\partial N} & \frac{\partial F}{\partial P} \\ \frac{\partial G}{\partial N} & \frac{\partial G}{\partial P} \end{bmatrix}_{|(N^*, P^*)} = \begin{bmatrix} a_{11} & a_{12} \\ a_{21} & a_{22} \end{bmatrix},
\tag{14.24}
$$

satisfies the Jury conditions in (14.4):

$$1 - (a_{11} + a_{22}) + (a_{11}a_{22} - a_{12}a_{21}) > 0\,,$$
$$1 + a_{11} + a_{22} + (a_{11}a_{22} - a_{12}a_{21}) > 0\,, \qquad (14.25)$$
$$1 - (a_{11}a_{22} - a_{12}a_{21}) > 0\,.$$

Since we are working on the entire real line, the spatially constant functions $N^*(x) = N^*$, $P^*(x) = P^*$ constitute a steady state of the spatial equations (14.22). The linearization at this state gives the equations

$$n_{t+1}(x) = \int_{-\infty}^{\infty} K_N(x - y)[a_{11}n_t(y) + a_{12}p_t(y)]dy\,,$$
$$\qquad (14.26)$$
$$p_{t+1}(x) = \int_{-\infty}^{\infty} K_P(x - y)[a_{21}n_t(y) + a_{22}p_t(y)]dy\,.$$

Taking Fourier transforms as in (5.9) results in a two-dimensional system for each mode ω separately, namely

$$\begin{bmatrix} \widehat{n}(\omega) \\ \widehat{p}(\omega) \end{bmatrix}_{t+1} = \begin{bmatrix} \widehat{K}_N(\omega) & 0 \\ 0 & \widehat{K}_P(\omega) \end{bmatrix} \begin{bmatrix} a_{11} & a_{12} \\ a_{21} & a_{22} \end{bmatrix} \begin{bmatrix} \widehat{n}(\omega) \\ \widehat{p}(\omega) \end{bmatrix}_t = \widehat{\mathbf{K}} J \begin{bmatrix} \widehat{n}(\omega) \\ \widehat{p}(\omega) \end{bmatrix}_t\,, \qquad (14.27)$$

where J is the Jacobian matrix from (14.24) and $\widehat{\mathbf{K}}$ is the diagonal matrix of the Fourier transforms of the dispersal kernels. The Fourier transform of a dispersal kernel is simply the moment-generating function from (5.15) evaluated at a purely imaginary argument, i.e., $\widehat{K}(\omega) = M(i\omega)$. The stability conditions analogous to (14.25) for system (14.27) are

$$1 - (\widehat{K}_N a_{11} + \widehat{K}_P a_{22}) + \widehat{K}_N \widehat{K}_P (a_{11}a_{22} - a_{12}a_{21}) > 0\,,$$
$$1 + (\widehat{K}_N a_{11} + \widehat{K}_P a_{22}) + \widehat{K}_N \widehat{K}_P (a_{11}a_{22} - a_{12}a_{21}) > 0\,, \qquad (14.28)$$
$$1 - \widehat{K}_N \widehat{K}_P (a_{11}a_{22} - a_{12}a_{21}) > 0\,,$$

where $\widehat{K}_N, \widehat{K}_P$ depend on ω.

Before we continue, we list a few observations and make a few simplifications with regard to the dispersal kernels. For $\omega = 0$, matrix $\widehat{\mathbf{K}}$ is the identity matrix, so that conditions (14.28) reduce to (14.25). By assumptions and continuity, the coexistence state is stable with respect to perturbations of small $|\omega|$. The Fourier transform of a symmetric dispersal kernel is real valued. We shall only consider symmetric kernels here. Since the dispersal kernel is a probability density function, the absolute value of its Fourier transform cannot exceed unity. Hence, we have $-1 \leq \widehat{K}(\omega) \leq 1$. It turns out that the signs of the Fourier transforms play a crucial role in the conditions for pattern formation.

We say that pattern formation arises if least one of the conditions in (14.28) for the spatial system is violated, while the conditions in (14.25) for the nonspatial system are satisfied. We derive these necessary conditions here, independent of the particular dispersal kernels. In the next section, we illustrate sufficient conditions that will depend on the details of the dispersal behavior. As before, we expect different patterns to emerge from an instability depending on which of the three conditions is violated. If the first condition is violated, we speak of a "plus-one" bifurcation, and we expect a stable pattern to emerge. In the case of the second condition, we speak of a "minus-one" bifurcation, and we expect a spatial pattern with a temporal two-cycle. In the last case, we speak of a Naimark–Sacker or Hopf bifurcation.

As a general condition, we find the requirement

$$a_{11} + a_{22} < 1 + \det J < 2 \tag{14.29}$$

by combining the first and third conditions in (14.25). For a potential instability, we consider the three conditions in (14.28) separately.

Proposition 14.1 *If* $0 < \widehat{K}_N, \widehat{K}_P \leq 1$, *then the necessary conditions for a plus-one bifurcation are*

$$a_{12}\, a_{21} < 0 \quad and \quad (a_{11} - 1)(a_{22} - 1) < 0. \tag{14.30}$$

Proof Condition (14.29) can be satisfied if $a_{11}, a_{22} \leq 1$ and at least one of these two inequalities is strict. If, however, $a_{11} > 1$, then necessarily $a_{22} < 1$ and vice versa. We suppose now that inequalities (14.25) are satisfied but the first inequality of (14.28) is reversed, so that

$$1 - (\widehat{K}_N a_{11} + \widehat{K}_P a_{22}) + \widehat{K}_N \widehat{K}_P (a_{11}a_{22} - a_{12}a_{21}) < 0. \tag{14.31}$$

Since $\widehat{K}_N \widehat{K}_P \geq 0$, we can substitute $a_{11}a_{22} - a_{12}a_{21} > a_{11} + a_{22} - 1$ from (14.25) and rearrange terms to find

$$\widehat{K}_N (1 - \widehat{K}_P)a_{11} + \widehat{K}_P (1 - \widehat{K}_N)a_{22} > 1 - \widehat{K}_N \widehat{K}_P. \tag{14.32}$$

If this inequality is satisfied for some $a_{11}, a_{22} \leq 1$, then it is satisfied for $a_{ii} = 1$. However, substituting $a_{ii} = 1$ gives the condition

$$\widehat{K}_P (1 - \widehat{K}_N) > 1 - \widehat{K}_N, \tag{14.33}$$

which is impossible by the bounds on $\widehat{K}_P$. Therefore, we cannot have $a_{11}, a_{22} \leq 1$ but must instead have $(a_{11} - 1)(a_{22} - 1) < 0$, which is the second condition in (14.30). If we expand this condition, we find

$$1 - (a_{11} + a_{22}) + (a_{11}a_{22} - a_{12}a_{21}) + a_{12}a_{21} < 0. \tag{14.34}$$

Comparing with the first inequality in (14.25), we see that necessarily $a_{12}a_{21} < 0$.

$\square$

   The first condition in (14.30) indicates that the two species have to act as predator and prey for pattern formation to occur. The second condition states that one species has to self-activate and the other self-inhibit. Hence, the conditions closely follow the conditions in reaction–diffusion systems (Murray 2002). A different set of conditions arises when the Fourier transforms of both kernels are negative.

Proposition 14.2 *If* $-1 \le \widehat{K}_N, \widehat{K}_P < 0$, *then the necessary conditions for a plus-one bifurcation are*

$$a_{12}\,a_{21} < 0 \quad and \quad (a_{11} + 1)(a_{22} + 1) < 0. \tag{14.35}$$

As before, the two species have to act as predator and prey. The second condition is different from the classical condition above, but again, one of the two species has to be self-destabilizing with $a_{ii} < -1$. The proof of Proposition 14.2 is analogous to the previous proof; for details, see Kot (1989). Instead, we look at the mixed case.

Proposition 14.3 *If* $0 < \widehat{K}_N \le 1$ *and* $-1 \le \widehat{K}_P < 0$, *then the necessary conditions for a plus-one bifurcation are*

$$a_{12}\,a_{21} < 0 \quad and \quad (a_{11} > 1 \ or \ a_{22} < 1). \tag{14.36}$$

If the signs of the Fourier transforms are exchanged, i.e., $0 < \widehat{K}_P \le 1$ *and* $-1 \le \widehat{K}_N < 0$, *then the indices in condition (14.36) are also exchanged.*

These conditions are much less stringent than in the previous two cases. We provide the detailed proof following Neubert et al. (1995).

Proof We suppose that the first inequality in (14.28) is violated. In the first step, we prove that we necessarily have

$$\widehat{K}_N a_{11} + \widehat{K}_P a_{22} > \widehat{K}_N + \widehat{K}_p. \tag{14.37}$$

We suppose that the opposite holds. Then

$$\begin{aligned}
(1 - \widehat{K}_N)(1 - \widehat{K}_P) &= 1 - (\widehat{K}_N + \widehat{K}_P) + \widehat{K}_N \widehat{K}_P \\
&\le 1 - (\widehat{K}_N a_{11} + \widehat{K}_P a_{22}) + \widehat{K}_N \widehat{K}_P \\
&< 1 - (\widehat{K}_N a_{11} + \widehat{K}_P a_{22}) + \widehat{K}_N \widehat{K}_P \det J < 0.
\end{aligned} \tag{14.38}$$

The second-last inequality follows from the stability condition $\det J < 1$. However, by assumption on the Fourier transforms, we have $(1 - \widehat{K}_N)(1 - \widehat{K}_P) > 0$. Hence, (14.37) must hold. Rearranging this condition gives

$$\widehat{K}_N(a_{11} - 1) + \widehat{K}_P(a_{22} - 1) > 0. \tag{14.39}$$

This condition cannot be satisfied if $a_{11} < 1$ and $a_{22} > 1$. Hence, one of these two inequalities must be reversed, which gives the second condition in (14.36).

Finally, to show that only a predator–prey relationship (i.e., $a_{12} a_{21} < 0$) will allow for the first inequality in (14.28) to be reversed, we define the function

$$I(a_{11}, a_{22}) = 1 - (\widehat{K}_N a_{11} + \widehat{K}_P a_{22}) + \widehat{K}_N \widehat{K}_P (a_{11} a_{22} - a_{12} a_{21}). \qquad (14.40)$$

Setting the partial derivatives to zero, we find that an extremum can occur only at $(a_{11}^*, a_{22}^*) = (1/\widehat{K}_N, 1/\widehat{K}_P)$. To investigate the nature of the critical point, we compute the Hessian matrix

$$\begin{bmatrix} 0 & \widehat{K}_N \widehat{K}_P \\ \widehat{K}_N \widehat{K}_P & 0 \end{bmatrix}. \qquad (14.41)$$

At the critical point, this matrix is positive definite, so that the point is a minimum. The function value at this minimum is $I(a_{11}^*, a_{22}^*) = -\widehat{K}_N \widehat{K}_P \, a_{11} a_{22}$. Hence, the function assumes negative values only if $a_{11} a_{22} < 0$. □

The case of a minus-one bifurcation is similar to the case above. We summarize the results from Neubert et al. (1995).

Proposition 14.4

1. *If $0 < \widehat{K}_N, \widehat{K}_P \leq 1$, then the necessary conditions for a minus-one bifurcation are*

$$a_{12} a_{21} < 0 \quad and \quad (a_{11} + 1)(a_{22} + 1) < 0. \qquad (14.42)$$

2. *If $-1 \leq \widehat{K}_N, \widehat{K}_P < 0$, then the necessary conditions for a minus-one bifurcation are*

$$a_{12} a_{21} < 0 \quad and \quad (a_{11} - 1)(a_{22} - 1) < 0. \qquad (14.43)$$

3. *If $0 < \widehat{K}_N \leq 1$ and $-1 \leq \widehat{K}_P < 0$, then the necessary conditions for a minus-one bifurcation are*

$$a_{12} a_{21} < 0 \quad and \quad (a_{11} < 1 \; or \; a_{22} > 1). \qquad (14.44)$$

Finally, we show that no Naimark–Sacker bifurcation can occur.

Proposition 14.5 *If inequalities (14.25) are satisfied, then the third inequality of (14.28) automatically holds.*

Proof Adding the first two inequalities in (14.25) results in $\det J > -1$. Together with the third inequality there, we find $|\det J| < 1$. Since $|\widehat{K}_N|, |\widehat{K}_P| \leq 1$, we have

$$|\widehat{K}_N \widehat{K}_P \det J| \leq |\det J| < 1. \qquad (14.45)$$

Table 14.1 Summary of necessary conditions for pattern formation.

	$0 < \widehat{K}_N, \widehat{K}_P \le 1$	$-1 \le \widehat{K}_N, \widehat{K}_N < 0$	$-1 \le \widehat{K}_P < 0 < \widehat{K}_N \le 1$
Plus-one	$(a_{11} - 1)(a_{22} - 1) < 0$	$(a_{11} + 1)(a_{22} + 1) < 0$	$a_{11} > 1$ or $a_{22} < 1$
Minus-one	$(a_{11} + 1)(a_{22} + 1) < 0$	$(a_{11} - 1)(a_{22} - 1) < 0$	$a_{11} < 1$ or $a_{22} > 1$

The predator–prey relationship $a_{12}a_{21} < 0$ is necessary in all cases. In the case $-1 \le \widehat{K}_N < 0 < \widehat{K}_P \le 1$, the conditions for plus-one and minus-one are simply reversed from the ones stated in the last column

Hence, the third inequality of (14.28) holds. $\square$

In summary, pattern formation requires a predator–prey relationship, i.e., $a_{12}a_{21} < 0$. If the Fourier transforms of the kernels have the same sign, then the pattern formation conditions are similar to those for reaction–diffusion equations; if they are of opposite sign, then the conditions are much less stringent. We list these conditions in Table 14.1. We illustrate the conditions and resulting patterns in the next section.

14.5   Pattern Formation in Predator–Prey Systems: Illustration

In this section, we illustrate the theoretical results from the previous section and explore how dispersal behavior and population dynamics interact to generate spatial patterns of various forms. We present three examples: the first two are based on the predator–prey model in (14.6). They are a plus-one bifurcation in the absence of a self-activator and a minus-one bifurcation (see definitions in the preceding section). Both of these have no equivalent in reaction–diffusion equations. The third example is a discrete-time analogue to the known reaction–diffusion predator–prey model with Allee effect of the prey and self-limitation in the predator (Murray 2002).

To find sufficient conditions for patterns to emerge, we need to choose specific dispersal kernels for the two species. Following Neubert et al. (1995), we shall use various combinations of the Laplace kernel

$$K(x) = \frac{a}{2}e^{-a|x|} \qquad \text{with} \qquad \widehat{K}(\omega) = \frac{a^2}{a^2 + \omega^2} \tag{14.46}$$

and the (special case of the) double gamma kernel (see also Table 3.1)

$$K(x) = \frac{b^2}{2}|x|e^{-b|x|} \qquad \text{with} \qquad \widehat{K}(\omega) = \frac{b^2(b^2 - \omega^2)}{(b^2 + \omega^2)^2} \tag{14.47}$$

for our illustrations. Whereas the Fourier transform of the Laplace kernel is positive, the Fourier transform of the double gamma kernel is negative for $\omega > b$. According to the results from the preceding section, we consider several cases.

The Plus-One Bifurcation

We return to the example from (14.6) with the nonspatial population dynamics

$$N_{t+1} = N_t e^{r(1-N_t-P_t)} \quad \text{and} \quad P_{t+1} = \rho N_t P_t . \tag{14.48}$$

The Jacobian matrix at the steady state $(N^*, P^*) = (1/\rho, 1 - 1/\rho)$ has diagonal entries $a_{11} = 1 - r/\rho$ and $a_{22} = 1$; see (14.8). Hence, there is no plus-one bifurcation when both kernels have positive Fourier transforms (Proposition 14.1). There can be a plus-one bifurcation when both kernels have negative Fourier transforms and $r/\rho > 2$ (Proposition 14.2) or when the two Fourier transforms have opposite signs with $\widehat{K}_N < 0 < \widehat{K}_P$ (Proposition 14.3). We take a closer look at the latter case.

We choose the Laplace kernel (with parameter a) for the predator and the double gamma kernel (with parameter b) for the prey. Since the Fourier transforms have opposite signs if $\omega > b$, we expect that by choosing b small enough, the instability condition can be satisfied. In general, for a plus-one bifurcation, we want the first inequality in (14.28) reversed; hence, we are looking for values of ω such that the dispersion relation

$$\mathscr{D}(\omega): = 1 - \widehat{K}_N \left(1 - \frac{r}{\rho}\right) - \widehat{K}_P + \widehat{K}_N \widehat{K}_P \left(1 + r - \frac{2r}{\rho}\right) \tag{14.49}$$

is negative. The explicit expression for $\mathscr{D}$ with the chosen kernels becomes

$$1 - \frac{b^2(b^2 - \omega^2)}{(b^2 + \omega^2)^2} \left(1 - \frac{r}{\rho}\right) - \frac{a^2}{a^2 + \omega^2} + \frac{b^2(b^2 - \omega^2)}{(b^2 + \omega^2)^2} \frac{a^2}{a^2 + \omega^2} \left(1 + r - \frac{2r}{\rho}\right) < 0 . \tag{14.50}$$

The plots in Fig. 14.4 show the shape of the double gamma kernel for three different values of b and the corresponding dispersion relation $\mathscr{D}$. For the smallest

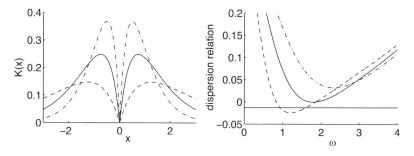

Fig. 14.4 **Left:** Double gamma dispersal kernel for prey with parameters $b = 0.8$ (dashed), $b = 1.35$ (solid), and $b = 2$ (dash-dot). **Right:** The corresponding dispersion relation from (14.49) for a plus-one bifurcation and the same line styles as in the left plot. The predator disperses according to a Laplace kernel with parameter $a = 10$. The population dynamics parameters are $r = 1$ and $\rho = 1.5$.

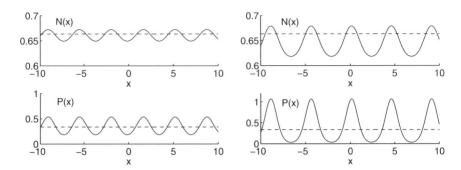

Fig. 14.5 Example of patterns arising from a plus-one bifurcation in the predator–prey model. The spatially homogeneous state (dashed) is unstable; the patterned state (solid) is stable. Parameters are $b = 1.3$ (left column) and $b = 1.0$ (right column). The other parameters are as in Fig. 14.4. The plots were obtained by simulating the IDE for 6000 time steps on a domain of length 200 from initial conditions chosen as small random perturbations of the homogeneous steady-state values.

and largest values of b, we have min $\mathscr{D} < 0$ and min $\mathscr{D} > 0$, respectively. At some intermediate value of b, the dispersion relation just touches the axis, i.e., min $\mathscr{D} = 0$. Patterns emerge when the pattern formation condition is satisfied; see Fig. 14.5. The plots show the spatially homogeneous state (dashed) as well as the stable patterns for two different values of prey dispersal.

The peaks of the predator population coincide with the peaks of the prey population. The amplitude of the pattern increases as b decreases, and the period increases slightly. At (and near) the bifurcation point, where the dispersion relation touches the axis, we expect the period to be determined by the frequency ω^* that satisfies $\mathscr{D}(\omega^*) = 0$ and $\mathscr{D}'(\omega^*) = 0$. Numerically, we find this critical value to be $b \approx 1.35$ with $\omega^* \approx 1.8$. Over an interval of length 20, we calculate $20\omega^*/(2\pi) \approx 5.7$. In the figure, we count more than 5.5 but less than 6 periods. This agreement seems to be reasonably good.

We can push the calculations one step further to obtain relations for the bifurcation point. For an explicit expression, we multiply (14.50) by the common denominator and divide by b^6. Then we obtain for $\mathscr{D}(\omega) = 0$ a cubic equation in $\tilde{\omega} = (\omega/b)^2$, namely

$$\tilde{\omega}^3 + \tilde{\omega}^2 \left(3 - \frac{r}{\rho}\right) + \tilde{\omega}\frac{r}{\rho}\left(1 - \frac{a^2}{b^2}(\rho - 1)\right) + \frac{a^2}{b^2}\left(\frac{r}{\rho}(\rho - 1)\right) = 0. \quad (14.51)$$

We note that the pattern formation condition depends only on the ratio of the dispersal parameters a/b. The tangency condition at the bifurcation point gives us a second equation via differentiation, namely

$$3\tilde{\omega}^2 + 2\tilde{\omega}\left(3 - \frac{r}{\rho}\right) + \frac{r}{\rho}\left(1 - \frac{a^2}{b^2}(\rho - 1)\right) = 0. \quad (14.52)$$

Numerically, we could solve the two conditions simultaneously, but there is another way to proceed. Equations (14.51) and (14.52) have a very simple structure with respect to the two parameter expressions r/ρ and $(\rho - 1)$. We can solve for these and obtain a parametric curve in the r-ρ parameter plane. More precisely, we solve (14.52) for $\frac{a^2}{b^2}(\rho - 1)\frac{r}{\rho}$ and substitute into (14.51), which can then be solved for

$$\frac{r}{\rho} = \frac{2\tilde{\omega}(\tilde{\omega}^2 - 3)}{(1 - \tilde{\omega})^2}. \tag{14.53}$$

Substituting this expression back into (14.52) gives

$$\frac{a^2}{b^2}(\rho - 1) = \frac{\tilde{\omega}(\tilde{\omega} + 1)(3 - \tilde{\omega})}{2(\tilde{\omega}^2 - 3)}. \tag{14.54}$$

Hence, we obtain the parametric curve (Neubert et al. 1995)

$$r = \frac{\tilde{\omega}}{(1 - \tilde{\omega})^2}\left[2(\tilde{\omega}^2 - 3) + \frac{b^2}{a^2}\tilde{\omega}(\tilde{\omega} + 1)(3 - \tilde{\omega})\right], \quad \rho = 1 + \frac{b^2}{a^2}\frac{\tilde{\omega}(\tilde{\omega} + 1)(3 - \tilde{\omega})}{2(\tilde{\omega}^2 - 3)}. \tag{14.55}$$

Plotting this curve in the stability region for the nonspatial system, we see that the region of possible pattern formation increases as the ratio of dispersal scales a/b increases (Fig. 14.6). Likewise, for a fixed dispersal ratio, pattern formation is more likely when the prey growth rate is higher.

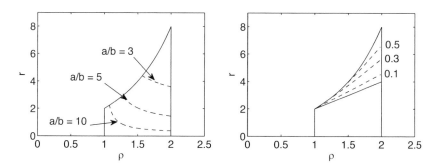

Fig. 14.6 Illustration of the conditions for dispersal-driven pattern formation in the predator–prey system (14.6). Patterns emerge above the dashed lines. **Left:** Level curves (14.55) of critical dispersal ratios a/b for the onset of pattern formation through a plus-one bifurcation. Higher values of the prey/predator dispersal ratio a/b enlarge the region of pattern formation. **Right:** Level curves (14.65) of critical dispersal ratios a_P/a_N for the onset of pattern formation through a minus-one bifurcation. Lower values of the prey/predator dispersal ratio a_P/a_N enlarge the region of pattern formation. There is no pattern formation below the line $r = 2\rho$.

The Minus-One Bifurcation

For a minus-one bifurcation with population dynamics as in (14.6), we choose parameters $r > 2\rho$ such that $a_{11} < -1$. According to Table 14.1, the necessary conditions for a minus-one bifurcation are satisfied when the Fourier transforms of both kernels are positive. Hence, we choose Laplace kernels (14.46) for predator and prey with parameters a_P and a_N, respectively. In that case, we can calculate all the critical values in detail. We use the general form of the Jacobian matrix as in (14.24), since the formulas tend to be clearer and shorter.

For a minus-one bifurcation, we reverse the second inequality in (14.28), which leads to the condition that the dispersion relation

$$\mathscr{D}(\omega): \ = 1 + \widehat{K}_N a_{11} + \widehat{K}_P a_{22} + \widehat{K}_N \widehat{K}_P \det J \qquad (14.56)$$

be negative. Substituting the expressions for the Fourier transforms gives the condition

$$1 + \frac{a_{11} a_N^2}{a_N^2 + \omega^2} + \frac{a_{22} a_P^2}{a_P^2 + \omega^2} + \frac{a_N^2}{(a_N^2 + \omega^2)} \frac{a_P^2}{(a_P^2 + \omega^2)} \det J < 0. \qquad (14.57)$$

Multiplying by the common denominator and dividing by a_N^2, we find the equivalent quadratic inequality in $\tilde{\omega} = (\omega/a_N)^2$ as

$$\tilde{\omega}^2 + \tilde{\omega} \left((1 + a_{11}) + (1 + a_{22}) \frac{a_P^2}{a_N^2} \right) + \frac{a_P^2}{a_N^2} (1 + \operatorname{tr} J + \det J) < 0. \qquad (14.58)$$

The minimum of the quadratic occurs at

$$\tilde{\omega}_m = -\frac{1}{2} \left((1 + a_{11}) + (1 + a_{22}) \frac{a_P^2}{a_N^2} \right). \qquad (14.59)$$

Since $\tilde{\omega} = (\omega/a_N)^2$ has to be positive, and since $a_{11} + 1$ is negative whereas $a_{22} + 1$ is positive, we obtain a necessary condition for the relative dispersal distances as

$$\frac{a_P^2}{a_N^2} < -\frac{a_{11} + 1}{a_{22} + 1} = \left| \frac{a_{11} + 1}{a_{22} + 1} \right|. \qquad (14.60)$$

The condition indicates that the dispersal distance of the prey $(1/a_N)$ must be smaller than that of the predator $(1/a_P)$.

A sufficient condition for pattern formation is that the quadratic in (14.58) be negative at $\tilde{\omega}_m$, i.e.,

$$\left((a_{11}+1)\frac{a_N}{a_P} + (a_{22}+1)\frac{a_P}{a_N} \right)^2 > 4(1 + \operatorname{tr}J + \det J). \tag{14.61}$$

This condition is satisfied when a_P/a_N is large *and* when it is small. However, because of the previous condition, we are only interested in the case where it is small. The right-hand side is positive by the Jury conditions for the nonspatial system. When taking roots, we need to respect the fact that the term inside the square on the left is negative when a_P/a_N is small enough. Therefore, we obtain the equivalent inequality

$$(1+a_{22})\frac{a_P^2}{a_N^2} + 2\frac{a_P}{a_N}\sqrt{1 + \operatorname{tr}J + \det J} + (1+a_{11}) < 0. \tag{14.62}$$

This condition is satisfied in the limit as $\frac{a_P}{a_N} \to 0$. There is a unique positive value of this fraction so that the inequality turns into an equality, and this is the critical ratio of the dispersal distances needed for pattern formation. After more tedious algebra, we find this upper bound to be

$$\frac{a_P}{a_N} < \frac{\sqrt{-a_{12}a_{21}} - \sqrt{1 + \operatorname{tr}J + \det J}}{1 + a_{22}}. \tag{14.63}$$

We now return to the particular example of system (14.6). The critical value where the dispersion relation and its derivative are zero is given by the pair of equations

$$\tilde{\omega}^2 + \tilde{\omega}\left(2 - \frac{r}{\rho} + 2\frac{a_P^2}{a_N^2} \right) + \frac{a_P^2}{a_N^2}\left(4 + r - \frac{3r}{\rho} \right) = 0,$$

$$2\tilde{\omega} + \left(2 - \frac{r}{\rho} + 2\frac{a_P^2}{a_N^2} \right) = 0. \tag{14.64}$$

We derive parametric expressions for r and ρ as we did in the previous section, namely

$$r = \frac{a_N^2}{a_P^2}\left(\tilde{\omega}^2 + (6\tilde{\omega}+2)\frac{a_P^2}{a_N^2} + 4\frac{a_P^4}{a_N^4} \right), \qquad \rho = \frac{r}{2(\tilde{\omega}+1+a_P^2/a_N^2)}. \tag{14.65}$$

These curves are plotted in the stability region in r-ρ space in Fig. 14.6.

The explicit expression for the critical value in terms of model parameters is

$$\frac{a_P}{a_N} < \frac{1}{2}\left(\sqrt{r - \frac{r}{\rho}} - \sqrt{4 + r - \frac{3r}{\rho}} \right). \tag{14.66}$$

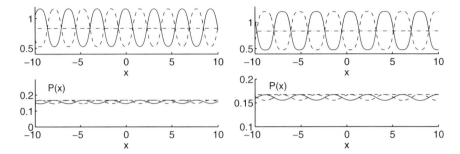

Fig. 14.7 Example of a pattern arising from a minus-one bifurcation in the predator–prey model. The spatially homogeneous state (dash-dot) is unstable; the patterned state is a stable two-cycle (solid and dashed for odd and even time steps, respectively). Parameters are $a_P = 1$ (left column) and $a_P = 0.5$ (right column). Other parameters are $r = 2.6$, $\rho = 1.2$, and $a_N = 10$. The figure was obtained by simulating the IDE for 6000 time steps on a domain of length 200 from initial conditions chosen as small random perturbations of the homogeneous steady-state values.

The plots in Fig. 14.7 illustrate the spatially heterogeneous two-cycle that emerges from the minus-one bifurcation as the spatially homogeneous state (dash-dot) becomes unstable. The peaks of the prey population correspond to the troughs of the predator population in the same time step. The amplitude and the wavelength of the pattern increase as the ratio of the dispersal rates decreases.

The Classical Case

We close this section by presenting a discrete-time analogue of the classical ecological predator–prey model for pattern formation, where the prey has an Allee effect and the predator dynamics are self-limited (Murray 2002). Our model is purely phenomenological, driven by the attempt to mimic these properties and at the same time keep the model complexity reasonably low.

We choose to model the nonspatial dynamics of prey (N_t) and predator (P_t) in generation t as

$$N_{t+1} = F(N_t, P_t) = \frac{R N_t^2}{1 + N_t^2} \frac{1}{1 + P_t}, \quad P_{t+1} = G(N_t, P_t) = \frac{\rho N_t P_t}{(1 + P_t)^2}.$$
$$(14.67)$$

The denominator in the first fraction is a slight variant (rescaling) of the model in (2.22) for the Allee effect. The probability of the predator consuming its prey is given by $(1 + P_t)^{-1}$, which has the same qualitative shape as the negative exponential that we used in the previous section. May (1973) proposed this probability as a model for a "clumped searching pattern" by the predator. Function G is the product of the available prey at the beginning of the season, the probability of catching prey,

$P_t/(1 + P_t)$, and the self-limitation term $(1 + P_t)^{-1}$, which we know from the classical Beverton–Holt updating function.

Our model has up to three steady states with $P = 0$, namely the trivial state and, provided $R > 2$, two semi-trivial states with $N^\pm = (R \pm \sqrt{R^2 - 4})/2$. The trivial state is stable, while the smaller of the two semi-trivial states is unstable. The state $(N^+, 0)$ is stable with respect to perturbations in N but unstable with respect to perturbations in P if $\rho N^+ > 1$. In the latter case, the predator can invade from low density.

A positive coexistence state is given by the equations

$$1 + N^2 = \frac{R}{\sqrt{\rho}} \sqrt{N}, \qquad (1 + P)^2 = \rho N. \tag{14.68}$$

The first of these equations has the unique solution

$$N = \left(\frac{R}{4\sqrt{\rho}} \right)^{2/3} = \sqrt{1/3} \quad \text{if} \quad R = \frac{4\sqrt{\rho}}{3^{3/4}}. \tag{14.69}$$

For R larger than this threshold, there are two solutions of the equation for N above, and consequently also for P. Hence, the relation between R and ρ defines a saddle-node bifurcation curve in parameter space. We are interested only in the larger of the two positive solutions for N since we expect the smaller one to be unstable.

The Jacobian matrix at a positive steady state (N, P) is given by

$$J = \begin{bmatrix} \frac{2}{1+N^2} & -\sqrt{\frac{N}{\rho}} \\ \frac{P}{N} & \frac{1-P}{1+P} \end{bmatrix}. \tag{14.70}$$

We observe that for N close enough to the bifurcation point $N = \sqrt{1/3} < 1$, the top left entry in the Jacobian matrix is greater than unity. This means that the prey is indeed an activator in the classical sense. The predator is an inhibitor because the bottom right entry is always less than unity.

We have to verify the stability conditions. Unfortunately, this is hard or impossible to do analytically. Instead, we plot the various threshold curves in Fig. 14.8. In

Fig. 14.8 Illustration of the stability region of model (14.67). See text for details.

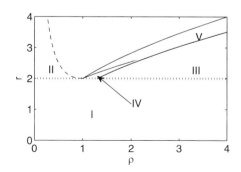

the region labeled I, below the dotted line, the trivial state is the only steady state. It is globally asymptotically stable. Above the dotted line, the two semi-trivial steady states exist. In region II, the predator cannot invade the state $(N^+, 0)$. Above the dashed curve, as well as above the dotted line to the right of where it meets the dashed curve, invasion is possible. In region III, however, there is no coexistence state. In this region, the predator can invade but will then drive the prey to extinction, and consequently itself as well. A positive coexistence state exists in regions IV and V. In region IV, coexistence is unstable because the third Jury condition is violated. The two populations will cycle and eventually both go extinct. Finally, region V denotes the part of parameter space where the positive state is stable and conditions for dispersal-driven instability are satisfied.

In region V, we have $a_{11} > 1$ and $a_{22} < 1$, so that the conditions for a plus-one bifurcation from Table 14.1 are satisfied for kernels with positive Fourier transform. As in the previous section, we choose Laplace kernels for both species with parameters a_N and a_P for prey and predator, respectively. Reversing the first inequality in (14.28) leads to the condition that the dispersion relation

$$\mathscr{D}(\omega): \ = 1 - \widehat{K}_N a_{11} - \widehat{K}_P a_{22} + \widehat{K}_N \widehat{K}_P \det J \tag{14.71}$$

be negative. Substituting the expressions for the Fourier transforms gives the condition

$$1 - \frac{a_{11} a_N^2}{a_N^2 + \omega^2} - \frac{a_{22} a_P^2}{a_P^2 + \omega^2} + \frac{a_N^2}{(a_N^2 + \omega^2)} \frac{a_P^2}{(a_P^2 + \omega^2)} \det J < 0. \tag{14.72}$$

The calculations from here on are exactly as in the case of the minus-one bifurcation in the previous example. The quadratic inequality in $\tilde{\omega} = (\omega/a_N)^2$ is

$$\tilde{\omega}^2 + \tilde{\omega} \left((1 - a_{11}) + (1 - a_{22}) \frac{a_P^2}{a_N^2} \right) + \frac{a_P^2}{a_N^2} (1 - \operatorname{tr} J + \det J) < 0. \tag{14.73}$$

The minimum of the quadratic occurs at

$$\tilde{\omega}_m = -\frac{1}{2} \left((1 - a_{11}) + (1 - a_{22}) \frac{a_P^2}{a_N^2} \right), \tag{14.74}$$

which can only be positive if

$$\frac{a_P^2}{a_N^2} < \frac{a_{11} - 1}{1 - a_{22}}. \tag{14.75}$$

As expected, the dispersal distance of the prey $(1/a_N)$ must be much smaller than that of the predator $(1/a_P)$. The same steps as in the previous example eventually lead to the sufficient condition

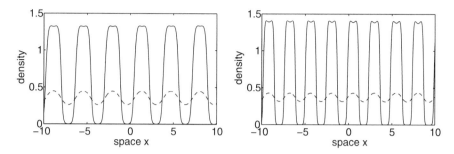

Fig. 14.9 Stable patterns for prey (solid) and predator (dashed) in model (14.67) with Laplace dispersal kernels for both species. Parameter values are $r = 3$ and $\rho = 2.5$; see region V in Fig. 14.8. The necessary condition for pattern formation from (14.75) is $a_P/a_N < 0.42$. The sufficient condition from (14.76) is $a_P/a_N < 0.11$. We choose $a_P = 1$ and $a_N = 10$ (left plot) and $a_N = 20$ (right plot). The plots were obtained by iterating the spatial equation from a small perturbation of the steady-state value for 5000 (left) or 10,000 (right) generations.

$$\frac{a_P}{a_N} < \frac{\sqrt{-a_{12}a_{21}} - \sqrt{1 - \text{tr}J + \det J}}{1 - a_{22}}. \tag{14.76}$$

Figure 14.9 shows the results of two numerical simulations with different ratios of a_P/a_N. As the dispersal distance of the prey decreases (i.e., a_N increases), more and narrower peaks develop. We also observe that the peaks in the prey population are not concave but rather double peaks with a small trough in between.

14.6 Spreading Phenomena in Predator–Prey Systems

We now turn to the question of how predator–prey interactions affect the spread of one or both of these interacting species. This question relates to the spread of invasive species, since most species consume existing resources (which results in a predator–prey relationship) and compete with existing species (an aspect that we will study in Sect. 14.8). There is a large body of literature for predator–prey invasions in reaction–diffusion systems, reviewed and summarized by Lewis et al. (2016), that reveals—mostly through numerical simulation—a treasure trove of complex dynamical behavior. The first investigation of this question in the IDE framework dates back to Kot (1992), who studied "waves of pursuit and evasion" via numerical simulations. A handful of authors have studied various aspects of predator–prey spread (Fagan et al. 2005; Dwyer and Morris 2006), but the literature is not nearly as vast as for reaction–diffusion equations.

We saw in Chaps. 5 and 11 that, in the simplest scenario, we can think of an invasion front as a spatial progression of a transition from an unstable to a stable state of the system. We also saw that proofs of the existence of a spreading speed and a traveling wave were based on some aspect of monotonicity (see also Chap. 13).

Then it becomes clear, even from an abstract viewpoint, why the study of spreading phenomena in predator–prey systems is so much more complicated than in single-species models: there are more stable and unstable states and even periodic orbits to connect, and the interactions between predator and prey are by definition not monotone because the off-diagonal entries in the community matrix (14.3) are of opposite sign. Hence, there have been very few analytical results in this area to date.

We explore and illustrate some of the spreading phenomena that arise in predator–prey systems, following the work of Kot (1992). We use the (nondimensional) interaction model from (14.6) and incorporate spatial dispersal as in the previous section; i.e., we study

$$N_{t+1}(x) = \int_{-\infty}^{\infty} K_N(x - y) N_t(y) e^{r(1-N_t(y)-P_t(y))} dy,$$
$$\tag{14.77}$$
$$P_{t+1}(x) = \int_{-\infty}^{\infty} K_P(x - y) \rho N_t(y) P_t(y) dy.$$

When $\rho < 1$, the predator cannot persist in the system. Asymptotically, then, we have a single equation for N, and we know that there exists a spreading speed, c_N^*, determined by the linearization at zero. There exist monotone traveling waves for speeds $c \geq c_N^*$ if $0 < r < 1$ and potentially nonmonotone waves for $1 < r < 2$; see Chap. 5. If K_N is the Gaussian kernel with variance σ_N^2, then $c_N^* = \sigma_N \sqrt{2r}$. In the following, we are only interested in the case $\rho > 1$.

Predator Invasion with Stable Coexistence

We choose parameter values r and ρ such that the coexistence state, (N^*, P^*), is stable for the nonspatial model; see Fig. 14.1. We expect to see a transition from zero to this state. Indeed, Fig. 14.10 shows two traveling waves. First, the prey density increases from zero to one, its single-species steady state. Then the predator density increases from zero to P^* (with decaying oscillations), while the prey decreases to N^*. Each wave moves at a constant speed. Since the initial prey-only wave moves faster than the second wave with the predator, the spatial extent of the region where we observe the unstable $(1, 0)$ state increases over time. Eventually, the predator seems to invade this prey-only state. We could hope to determine the speed of this invasion by the linearization at $(1, 0)$. When we linearize at this state, the equations decouple, and the predator equation becomes

$$P_{t+1}(x) = \rho \int_{-\infty}^{\infty} K_P(x - y) P_t(y) dy.$$
$$\tag{14.78}$$

Taken independently, this equation has a spreading speed c_P^*, which, for the Gaussian kernel, is given by $c_P^* = \sigma_P \sqrt{2 \ln(\rho)}$. For the simulations in Fig. 14.10,

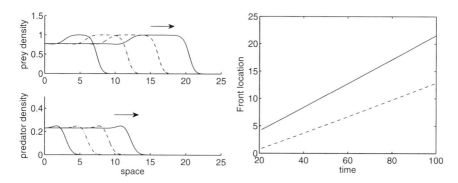

Fig. 14.10 **Left:** Simulation of model (14.77) for prey (top) and predator (bottom). The densities are plotted every 20 generations. **Right:** Front location of the prey (solid) and predator (dashed), obtained from tracking a level set. Parameters are $r = 0.5$, $\rho = 1.3$. Both dispersal kernels are Gaussian kernels with variance $\sigma_N^2 = \sigma_P^2 = 0.05$.

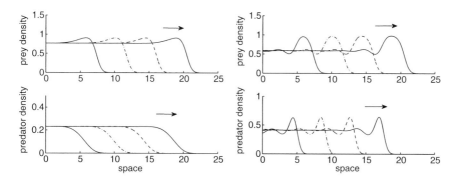

Fig. 14.11 Simulation of model (14.77) when the speed of the predator is limited by that of the prey. **Left:** Predator dispersal is increased to $\sigma_P^2 = 0.5$. **Right:** Predation is increased to $\rho = 1.7$. All other parameters are as in Fig. 14.10.

the theoretical values $c_N^* = 0.223$ and $c_P^* = 0.162$ are in quite good agreement with the numerical results $c_N^{\text{num}} = 0.221$ and $c_P^{\text{num}} = 0.159$.

There are obvious limits to this heuristic. If we choose σ_P^2 or ρ large enough, we can make $c_P^* > c_N^*$, so that the predator would spread faster than the prey, which is biologically impossible. Figure 14.11 shows that the spread rate of the predator is indeed limited to that of the prey in isolation. If predator dispersal has a large variance, so that $c_P^* = 0.512 > c_N^* = 0.22$, we observe a monotone predator front spreading at the speed of the prey, namely $c_P^{\text{num}} = c_N^{\text{num}} = 0.220$. If the growth rate of the predator is increased, the nonspatial model exhibits decaying oscillations to the stable coexistence state. The speed predicted from the linearization is $c_P^* = 0.230$. In the spatial model, we see a nonmonotone front that echoes the decaying oscillations and that moves at (almost) the speed of the prey ($c_P^{\text{num}} = 0.216$).

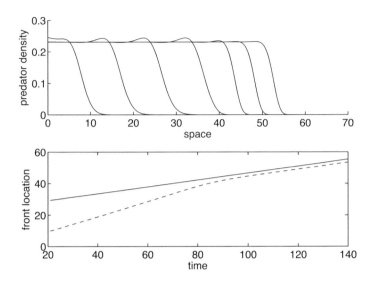

Fig. 14.12 Simulation of model (14.77) when the prey has a head start and $c_P^* > c_N^*$. **Top:** The predator spreads quickly until it catches up with the front of the prey and then slows down to the speed of the prey. **Bottom:** The location of the fronts (solid for prey and dashed for predator) shows the two phases of predator spread. Parameters are as in the left plot in Fig. 14.11. The prey is initially present for $x < 25$, the predator for $x < 0$.

In the previous simulations, we initialized predator and prey at the same spatial location. Let us suppose that the prey has a head start and the predator spread rate from (14.78) predicts $c_P^* > c_N^*$. Will the predator catch up with a spreading prey? The result in Fig. 14.12 not only demonstrates that the predator will catch up with the prey and then slow down to the speed of the prey, it also shows that the actual speed of the predator during the catch-up phase, $c_P^{\mathrm{num}} = 0.490$, (almost) reaches the speed predicted by $c_P^* = 0.512$.

Biological Control of Invasions

While the presence of a predator reduces the growth of its prey, the numerical simulations in the previous section indicated that it does not slow the spatial spread of the prey. In a reaction–diffusion system, a predator may slow or reverse the spread of its prey only if the prey exhibits an Allee effect (Owen and Lewis 2001). We illustrate these ideas in the IDE framework.

We replace the Ricker-type growth function for the prey in system (14.77) by the Allee growth function from (2.22). Our system becomes

$$N_{t+1}(x) = \int_{-\infty}^{\infty} K_N(x-y)\frac{RN_t^2(y)}{1+(R-1)N_t^2(y)}e^{-P_t(y)}dy\,,$$

$$P_{t+1}(x) = \int_{-\infty}^{\infty} K_P(x-y)\rho N_t(y) P_t(y)dy\,,$$ (14.79)

with growth parameter $R > 2$. The nonspatial system has three prey-only states $(0,0)$, $((R-1)^{-1},0)$, and $(1,0)$; see Sect. 2.2. The predator can invade the latter state if $\rho > 1$. The coexistence state is given by

$$N^* = \frac{1}{\rho}\,, \quad e^{P^*} = \frac{R/\rho}{1+(R-1)\rho^{-2}}\,.$$ (14.80)

This state is stable when it emerges but loses stability to a Naimark–Sacker bifurcation as ρ increases. Further increases of ρ lead to a global bifurcation, after which all solutions approach the trivial state. The behavior of the nonspatial model reflects that of its continuous-time analogue in Chap. 9 of Kot (2001). For our spatial explorations, we fix parameters R and ρ such that the coexistence state is locally stable.

We initialize the numerical simulations by setting the prey to its positive stable state $N = 1$ on the interval $[-10, 10]$ and the predator to $P = 0.1$ on the interval $[-1, 1]$. We choose $R = 4$, which is larger than the critical value for the prey to spread in the absence of its predator; see Chap. 6. Figure 14.13 shows the location of the fronts of the two species in the positive direction. When σ_P^2 is very small, the predator cannot catch up with the prey. We see two separate fronts. When σ_P^2 is intermediate, the predator catches up with the prey and both slow down, but continue

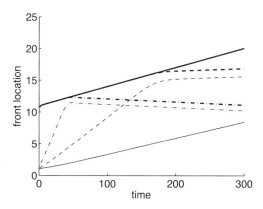

Fig. 14.13 Front location of the prey (thick) and predator (thin) for model (14.79). When $\sigma_P^2 = 0.001$ (solid), the predator spreads more slowly than the prey. When $\sigma_P^2 = 0.01$ (dashed), the predator catches up and slows the prey down. When $\sigma_P^2 = 0.1$ (dash-dot), the predator catches up quickly and both species retreat. Both species have a Gaussian dispersal kernel. Parameters are $\sigma_N^2 = 0.1$, $R = 4$, and $\rho = 1.45$.

spreading. When σ_P^2 is larger, both species start to retreat after the predator has caught up with the prey. Similar and additional results in the context of reaction–diffusion equations are synthesized by Fagan et al. (2002) and Lewis et al. (2016), who discuss invasion theory and biological control more broadly.

Fagan et al. (2005) derive and analyze a complex IDE model for a consumer–resource model with several stages for the recolonization of the pumice plain after the 1980 eruption of Mount St. Helens. The prairie lupin (*Lupinus lepidus*) started to recolonize the devastated areas and spread in a patchy way with high-density core areas and low-density edge patches. Caterpillars of two moth species (*Filatima* sp. and *Staudingeria albipenella*) are specialist herbivores on the lupins. Since lupins at high density have a decreased nutrient content and can develop chemical defenses against the caterpillars, the net growth rate of the moths exhibits an inverse density dependence with lupins. Depending on parameter values, the IDE model shows three different dynamical outputs: (1) a successful co-invasion of lupins and moths, (2) a collapse of both species, and (3) a stationary situation, where lupins and moths neither expand in space nor contract; see Fig. 6 in Fagan et al. (2005).

14.7   Spreading Speeds in Cooperative Systems

Mutual facilitation, the reciprocally supportive interaction between species, has long been largely ignored in ecological theory but is increasingly recognized as ubiquitous; see Chap. 13 in Kot (2001) and the references therein. Examples include seed dispersal by birds, pollination of plants (Lutscher and Iljon 2013), and intraguild mutualism between consumers (Assaneo et al. 2013).

Models for mutualism give rise to "cooperative" or "order preserving" systems: an increase in any one of the state variables leads to an increase in the other state variables. This property was a crucial ingredient in proving the existence of a (joint) spreading speed of a structured population (Chap. 13) and was missing from predator–prey interactions (Sect. 14.6) where no corresponding result holds. The study of spreading speeds in general cooperative systems is surprisingly more subtle than what we saw in structured populations (Weinberger et al. 2002; Li et al. 2005; Weinberger et al. 2007). There is typically more than one nontrivial steady state to which a traveling front can connect. Furthermore, different components of the system may spread at different speeds. Finally, one may observe an "anomalous" spreading speed (Weinberger et al. 2007). An anomalous spreading speed occurs if one of the species spreads faster in the system than it would in isolation and faster than any of the other species involved. Since its discovery, this phenomenon has been studied in reaction–diffusion equations (Holzer 2014) and has been applied to biological invasions of a species with two morphs (Elliott and Cornell 2012). We present some of these ideas and phenomena with simple examples but refer to the original literature for complete statements of theorems and proofs.

We model nonspatial population dynamics as a Beverton–Holt function (2.11) with positive steady state $C > 0$, i.e.,

$$N_{t+1} = F(N_t) = \frac{RN_t}{1 + (R - 1)N_t/C}, \tag{14.81}$$

with $R > 1$. Function F is monotone in N, and the interval $[0, C]$ is invariant for the dynamics. Next, we assume that the parameters are nondecreasing functions of some other population density, $\tilde{N}$, i.e., $R = R(\tilde{N})$ and $C = C(\tilde{N})$, and write $F(N, \tilde{N})$, accordingly. The derivative of F with respect to $\tilde{N}$ is

$$\frac{\partial F}{\partial \tilde{N}} = \frac{1}{(1 + (R - 1)N/C)^2} \left[R'N \left(1 - \frac{N}{C}\right) + \frac{R^2 N^2 C'}{C^2} \right]. \tag{14.82}$$

Under the monotonicity assumptions for R and C, this expression is positive for $0 \leq N \leq C$.

We use this observation to model a mutualistic interaction between two species. We denote their densities in generation t by $N_{i,t}$ with $i = 1, 2$. The equations are

$$\begin{aligned} N_{1,t+1} &= F_1(N_{1,t}, N_{2,t}), \\ N_{2,t+1} &= F_2(N_{1,t}, N_{2,t}), \end{aligned} \tag{14.83}$$

where $F_1(N_1, N_2)$ is as above, with N_1 and N_2 taking the roles of N and $\tilde{N}$, respectively, and F_2 has the same functional form with the roles of N_1 and N_2 exchanged. As an example, we choose simple linear functions for R and C. For F_i, we write $R(\tilde{N}) = R_i + a_i\tilde{N}$ and $C(\tilde{N}) = 1 + b_i\tilde{N}$. All parameters are nonnegative, and $R_i > 1$.

This system has the four steady states $(0, 0)$, $(1, 0)$, $(0, 1)$, and (N_1^*, N_2^*), where

$$N_1^* = \frac{1 + b_1}{1 - b_1 b_2} = C(N_2^*) > 1, \qquad N_2^* = \frac{1 + b_2}{1 - b_1 b_2} = C(N_1^*) > 1. \tag{14.84}$$

Since F_i are monotone in N_i, the set $(0, 0) \leq (N_1, N_2) \leq (N_1^*, N_2^*)$ is invariant for the dynamics. By the calculation in (14.82), F_i are nondecreasing with respect to N_j $(j \neq i)$ in this set when R' is small enough. Then the system is cooperative.

We incorporate space and dispersal in the usual way with dispersal kernel K_i for species N_i and write

$$\begin{aligned} N_{1,t+1}(x) &= \int_{-\infty}^{\infty} K_1(x - y)F_1(N_{1,t}(y), N_{2,t}(y))\,dy, \\ N_{2,t+1}(x) &= \int_{-\infty}^{\infty} K_2(x - y)F_2(N_{1,t}(y), N_{2,t}(y))\,dy. \end{aligned} \tag{14.85}$$

We explore the spreading behavior of the two species in this system. We begin by outlining why the theory presented in Chap. 13 does not necessarily apply.

The steady states of the nonspatial system are spatially constant steady states of system (14.85). In the sense of the order of continuous functions (Sect. 13.8), we have

$$(0, 0) \leq (0, 1) \ll (N_1^*, N_2^*) \text{ and } (0, 0) \leq (1, 0) \ll (N_1^*, N_2^*). \tag{14.86}$$

The existence of the semi-trivial, "in-between" steady states violates one important assumption from Chap. 13. The linearization of the system at $(0, 0)$ is

$$\begin{aligned}
n_{1,t+1}(x) &= \int_{-\infty}^{\infty} K_1(x - y) R_1 n_{1,t}(y) dy, \\
n_{2,t+1}(x) &= \int_{-\infty}^{\infty} K_2(x - y) R_2 n_{2,t}(y) dy.
\end{aligned} \tag{14.87}$$

The two equations decouple, so the operator is reducible. In addition, the nonlinear system is not bounded by its linearization at $(0, 0)$. In that sense, an Allee-like effect is induced by cooperation. Hence, two more assumptions from Chap. 13 are violated. What then can we say about spread rates in the cooperative system?

The cooperative nature of the system gives the bounds $F_1(N_1, 0) \leq F_1(N_1, N_2) \leq F_1(N_1, N_2^*)$. These translate into bounds on the rate at which species N_1 can spread in the system: it will spread at least (at most) at the speed of the single population model with growth function $F_1(N_1, 0)$ ($F_1(N_1, N_2^*)$). The analogous reasoning holds for species N_2. The spreading speeds of the single-species models that bound the solutions of the system are linearly determined according to the theory from Chap. 5. We denote the lower bounds as $c_i^{\min}$ and may assume $c_1^{\min} \leq c_2^{\min}$. If K_i are Gaussian kernels with variance σ_i^2, we have $c_i^{\min} = \sigma_i \sqrt{2 \ln(R_i)}$. The corresponding upper bounds are $c_i^{\max} = \sigma_i \sqrt{2 \ln(R_i + a_i N_j^*)}$. We obtain two cases: (1) When $c_1^{\max} < c_2^{\min}$, the first species will be slower than the second even if the second species supports the first. Consequently, there cannot be a joint spreading speed. (2) When $c_1^{\max} > c_2^{\min}$, the first species in the presence of the second could be faster than the second species on its own. Consequently, there could be a joint spreading speed if the support of the second species is strong enough.

The two plots in Fig. 14.14 illustrate that both cases occur. With parameters as in the figure caption, we have $c_1^{\min} = 0.285 < 0.333 = c_2^{\min}$. In the left plot, we also have $c_1^{\max} = 0.316 < c_2^{\min}$, whereas in the right plot, we find $c_1^{\max} = 0.384 > c_2^{\min}$. On the left, species 2 spreads ahead at speed $c_2^{\min}$, and a plateau forms, into which species 1 invades. On the right, both species spread at the same rate, namely $c_2^{\min}$.

The formation of the plateau prompts us to study the spread of one species into an environment occupied by another, cooperating species. The semi-trivial state $(0, 1)$ is unstable and the coexistence state (N_1^*, N_2^*) is stable for system (14.85), and there

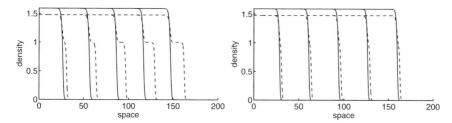

Fig. 14.14 Two scenarios of spread in the cooperating species model. **Left:** When $a_1 = 0.1$ is small, the upper bound of the speed of species 1 (solid) is below the lower bound of the speed of species 2 (dashed). Species 1 is slower. **Right:** When $a_2 = 0.4$ is large enough, the upper bound for species 1 exceeds the lower bound for species 2. Both species spread at the same rate. Parameters are $R_1 = 1.5$, $R_2 = 2$, $a_2 = 0.2$, $b_1 = 0.4$, $b_2 = 0.3$, $\sigma_1^2 = 0.1$, and $\sigma_2^2 = 0.08$. Both kernels are Gaussian kernels.

is no other homogeneous steady state between the two, in the sense of (14.86). We can shift the point $(0, 1)$ to $(0, 0)$ by the transformation $(N_1, N_2) \mapsto (N_1, N_2 - 1)$, so that the unstable state becomes the origin and the stable point remains positive in all components. The linearization at the unstable state is given by

$$n_{1,t+1}(x) = \int_{-\infty}^{\infty} K_1(x - y)(R_1 + a_1)n_{1,t}(y)dy \,,$$

$$n_{2,t+1}(x) = \int_{-\infty}^{\infty} K_2(x - y)[\hat{b}n_{1,t}(y) + n_{2,t}(y)/R_2]dy \,,$$

(14.88)

with $\hat{b} = b_1(R_1 - 1)/R_1$. This system is cooperative and reducible.

Cooperative and reducible systems with any finite number of species, truncated so that densities remain bounded, are studied in a series of papers by Weinberger et al. (2002), Li et al. (2005), and Weinberger et al. (2007), which also corrects a mistake from Weinberger et al. (2002). The authors establish that there are, in general, two spreading speeds: (1) a slowest speed, c^*, so that all components spread at least at speed c^* and at least one component does not spread faster, and (2) a fastest speed, c_f^*, such that no component spreads faster and at least one component spreads no slower. When the system is irreducible, Lui (1989a) shows that $c^* = c_f^*$; see Chap. 13. Weinberger et al. (2007) beautifully explain which aspect of Lui's proof breaks down when the system is reducible. Li et al. (2005) prove that c^* is characterized as the slowest speed of a family of traveling waves. The calculation of c^* and c_f^* from formulas such as (5.17) or (13.40) is more subtle, and, in fact, such formulas do not generally exist. To understand this last statement, we introduce some notation.

The linearization of a cooperative, reducible system can be brought into Frobenius form (for notation, see Chap. 13)

$$\mathbf{N}_{t+1} = \int_{-\infty}^{\infty} \mathbf{K} \bullet B\mathbf{N}_t(y)dy \,,$$

(14.89)

where B is a block lower-triangular matrix with nonnegative entries. As in (13.30), we can form matrix $\mathbf{H}(s)$, which is also block lower triangular. We denote the diagonal blocks by $H_{ii}(s)$, with their dominant eigenvalues $\lambda_i(s)$. We can define numbers

$$\tilde{c}_i = \inf_{s>0} \frac{1}{s} \ln(\lambda_i(s)). \tag{14.90}$$

These are the "component spreading speeds" at which component i would spread if all other components were absent (Lui 1989a; Weinberger et al. 2007). Since the system is cooperative, the maximum of these numbers is a lower bound for c_f^*. However, it is possible that c_f^* is strictly greater than this maximum, which has been termed an anomalous spreading speed (Weinberger et al. 2007). Before we explain this phenomenon in detail, we mention one affirmative answer that follows from Remark 2 after Theorem 4.1 in Weinberger et al. (2007).

Proposition 14.6 *If* $\lambda_1(0) > 1$ *and* $\lambda_i(0) < 1$ *for all* $i > 1$, *then*

$$c_f^* = \inf_{s>0} \ln(\lambda_1(s))/s. \tag{14.91}$$

This proposition applies to the example above, where all blocks are of size one with $\lambda_1(0) = R_1 + a_1 > 1$ and $\lambda_2(0) = 1/R_2 < 1$. Hence, the speed at which N_1 spreads into the steady state $(0, 1)$ is given by the linearization at the unstable state. The proposition applies more generally when one invading species spreads into a steady state of finitely many other species that is stable in the absence of the invader. Stability in the absence of the invader implies $\lambda_i < 1$ for $i > 1$. This insight is relevant for many spreading phenomena, but it requires the cooperative structure. We observed the same result numerically for the invasion of a predator into its prey (Sect. 14.6), but the preceding proposition does not apply to that situation.

As a final aspect of our cooperative system (14.85), we illustrate how one species can overtake another and increase the other species' spread rate. The setup is as before, so that species 2 in isolation spreads faster than species 1 in isolation, but species 1 now has a head start of 50 space units. The plots in Fig. 14.15 show the front location of species 1 (thick) and species 2 (thin). Initially, species 1 spreads at its single-species speed $c_1^{\min} = \sigma_1\sqrt{2\ln(R_1)}$, indicated by the thick dashed line. Species 2 spreads into species 1 at the speed given by the linearization at $(1, 0)$, namely $\sigma_2\sqrt{2\ln(R_2 + a_2)}$, indicated by the thin dashed line. When the two fronts meet, species 2 slows down to its single-species speed $c_2^{\min} = \sigma_2\sqrt{2\ln(R_2)}$. The first species speeds up. When the positive effect from species 2 is small ($a_1 = 0.1$), species 1 spreads at the speed given by the linearization at $(0, 1)$, which is $\sigma_1\sqrt{2\ln(R_1 + a_1)}$. When the positive effect is large ($a_1 = 0.3$), the speed of the linearization at $(0, 1)$ predicts a speed faster than that of species 2, which obviously cannot be sustained ahead of the front of species 2. Hence, both spread at the same rate.

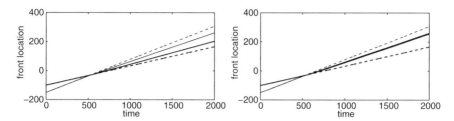

Fig. 14.15 Two scenarios of spread in the cooperating species model (see text for explanation). Parameters are $R_1 = R_2 = 1.5$, $a_2 = 0.2$, $b_1 = 0.4$, $b_2 = 0.3$, $\sigma_1^2 = \sigma_2^2 = 0.05$, and $a_1 = 0.1$ (left) versus $a_2 = 0.3$ (right). Both kernels are Gaussian kernels.

Anomalous Spreading Speed

We use a discrete-time version of the invasion model of two types (or morphs) from Elliott and Cornell (2012) to explain and illustrate the ideas and mechanisms behind an anomalous spreading speed. Elliott and Cornell (2012) assume that individuals in a population come in two types that may differ in their growth rate and dispersal ability. Mutations lead to changes in type. Offspring of type i are of type $j \neq i$ with probability μ_i and remain type i with probability $1 - \mu_i$. A linear model that describes this system is

$$N_{1,t+1}(x) = (1 - \mu_1) \int_{-\infty}^{\infty} K_1(x - y) R_1 N_{1,t}(y) dy + \mu_2 \int_{-\infty}^{\infty} K_2(x - y) R_2 N_{2,t}(y) dy,$$

$$N_{2,t+1}(x) = \mu_1 \int_{-\infty}^{\infty} K_1(x - y) R_1 N_{1,t}(y) dy + (1 - \mu_2) \int_{-\infty}^{\infty} K_2(x - y) R_2 N_{2,t}(y) dy,$$

$$(14.92)$$

where $R_i > 1$ describe growth and K_i dispersal of offspring from type i. The subsequent results do not change qualitatively when dispersal behavior is determined by the type of the offspring instead of the parent. To obtain a system with bounded solutions, we can truncate the linear IDE at some positive density without affecting the spreading behavior (Weinberger et al. 2007).

Here, we analyze the system in the cases where the second type does not mutate ($\mu_2 = 0$), so that the system is in Frobenius form. In the next section, we will study the case where the mutation rates are equal ($\mu_1 = \mu_2$). To allow for explicit calculations, we shall assume Gaussian dispersal kernels with variances σ_i^2 and mean zero. The respective moment-generating functions are $M_i(s) = \exp(\sigma_i^2 s^2 / 2)$.

When $\mu_2 = 0$, the system is in Frobenius form (14.89), so that the results from Weinberger et al. (2007) apply. We shall assume that μ_1 is small enough that $R_1(1 - \mu_1) = \tilde{R}_1 > 1$. The diagonal elements of matrix $\mathbf{H}(s)$ are $\lambda_1(s) = \tilde{R}_1 M_1(s)$ and $\lambda_2 = R_2 M_2(s)$. We define functions

$$c_i(s) = \frac{1}{s} \ln(\lambda_i(s)).$$

$$(14.93)$$

Unless the dispersal kernels are degenerate, these functions are convex and have unique minima $\tilde{c}_i$, the component spreading speeds, at critical values $\tilde{s}_i < \infty$. For Gaussian kernels, the explicit expressions are

$$\tilde{s}_1 = \sqrt{2\ln(\tilde{R}_1)/\sigma_1^2} \quad \text{and} \quad \tilde{s}_2 = \sqrt{2\ln(R_2)/\sigma_2^2}, \tag{14.94}$$

with

$$\tilde{c}_1 = \sqrt{2\sigma_1^2 \ln(\tilde{R}_1)} \quad \text{and} \quad \tilde{c}_2 = \sqrt{2\sigma_2^2 \ln(R_2)}. \tag{14.95}$$

The system has an anomalous spreading speed if the fastest speed is faster than the component speeds, i.e., if $c_f^* > \max(\tilde{c}_i)$.

Weinberger et al. (2007) give an upper bound, $c^{(u)}$, for c_f^* as

$$c^{(u)} = \inf_{s_1 \geq s_2 > 0} \{\max[c_1(s_1), c_2(s_2)]\}. \tag{14.96}$$

Note that the infimum is taken only over $s_1 \geq s_2 > 0$ and not $s_1, s_2 > 0$. If we define $C_1(s) = \tilde{c}_1$ for $0 < s \leq \tilde{s}_1$ and $C_1(s) = c_1(s)$ for $s > \tilde{s}_1$, then this expression can be simplified to

$$c^{(u)} = \inf_{s>0} \{\max[C_1(s), c_2(s)]\}. \tag{14.97}$$

The larger of the two component spreading speeds is $\max\{\inf(c_1(s)), \inf(c_2(s))\}$, so that the question becomes when the infimum of the maxima is larger than the maximum of the infima. The following proposition gives exact conditions for the upper bound to be larger than the component speeds.

Proposition 14.7 *We have* $c^{(u)} > \max\{\tilde{c}_i\}$ *if and only if the following three conditions are satisfied:*

1. $\tilde{s}_1 < \tilde{s}_2$,
2. $c_1(\tilde{s}_1) < c_2(\tilde{s}_1)$,
3. $c_1(\tilde{s}_2) > c_2(\tilde{s}_2)$.

If the conditions are satisfied, $c^{(u)}$ is given by the unique value $c_1(\bar{s}) = c_2(\bar{s})$ with $\tilde{s}_1 < \bar{s} < \tilde{s}_2$.

Proof It is clear that if the three conditions are satisfied, then $c^{(u)}$ has the value given; see Fig. 14.16 for illustration. We show that the three conditions are also necessary. Assume that $\tilde{s}_1 \geq \tilde{s}_2$. Then

$$c^{(u)} = \inf_{s>0} \{\max(C_1(s), c_2(s))\} \leq \max(C_1(s_2), c_2(s_2)) = \max\{\tilde{c}_i\}. \tag{14.98}$$

Fig. 14.16 Illustration of the case where $c^{(u)} > \max\{\tilde{c}_i\}$ in Proposition 14.7. The dash-dot curve denotes $C_1(s)$, and the star on it indicates $(\tilde{s}_1, \tilde{c}_1)$. The dashed curve denotes $c_2(s)$, and the star is $(\tilde{s}_2, \tilde{c}_2)$. The solid curve is the maximum of the two, and the star on it is its minimum, $c^{(u)}$. Parameters in this plot are $R_1 = 1.5$, $R_2 = 2$, $\mu_1 = 0.1$, $\sigma_1^2 = 0.05$, and $\sigma_2^2 = 0.02$.

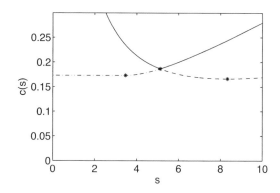

Hence, $\tilde{s}_1 < \tilde{s}_2$ is necessary. Now assume that the first condition is satisfied but the second is violated. Then $c^{(u)} \leq c_1(\tilde{s}_1) = \tilde{c}_1$, which is the opposite of the claim. The necessity of the third condition follows by the same reasoning.   □

In the case of Gaussian kernels, we can calculate all the quantities involved and reach further biological insights. The intersection point $c_1(\bar{s}) = c_2(\bar{s})$ is characterized by

$$2(\ln(\tilde{R}_1) - \ln(R_2)) = \bar{s}^2(\sigma_2^2 - \sigma_1^2). \tag{14.99}$$

Hence, the expressions $(\ln(\tilde{R}_1) - \ln(R_2))$ and $(\sigma_2^2 - \sigma_1^2)$ must have the same sign. The biological implication is that an anomalous spreading speed can happen only if there is a trade-off between reproduction and dispersal ability: the type that has the higher "self-reproductive" rate has to have the lower dispersal ability. By self-reproductive rate we mean the number of offspring that is of the same type as the parent (i.e., $\tilde{R}_1$ or R_2).

But we can be more precise. Condition $\tilde{s}_1 < \tilde{s}_2$ is equivalent to

$$\frac{\ln(\tilde{R}_1)}{\ln(R_2)} < \frac{\sigma_1^2}{\sigma_2^2}. \tag{14.100}$$

The third condition in Proposition 14.7 is equivalent to

$$\frac{\ln(\tilde{R}_1)}{\ln(R_2)} > 1 - \frac{\sigma_2^2}{\sigma_1^2}. \tag{14.101}$$

Combined with (14.100), we find $2 - \sigma_1^2/\sigma_2^2 < \sigma_1^2/\sigma_2^2$, which implies $\sigma_1^2 > \sigma_2^2$.

Biologically speaking, an anomalous spreading speed can arise only if the type that has the higher dispersal ability (and lower self-reproduction) is the one that produces mutations.

Proposition 14.7 gives sufficient conditions for there to be no anomalous spreading speed, but only necessary conditions for there to be one, since $c^{(u)}$ is only an upper bound for c_f^*. Weinberger et al. (2007) prove the existence of an anomalous spreading speed in a reaction–diffusion system; Holzer (2014) gives another such example. System (14.92) with $\mu_2 = 0$ is somewhat related to the model by Weinberger et al. (2007). We show that an anomalous spreading speed exists in a special case.

We rewrite system (14.92) with $*$ for convolutions and with $\mu_2 = 0$ as

$$N_{1,t+1} = (1 - \mu_1) R_1 K_1 * N_{1,t},$$
$$N_{2,t+1} = \mu_1 R_1 K_1 * N_{1,t} + R_2 K_2 * N_{2,t}. \tag{14.102}$$

We assume Gaussian dispersal kernels, denoted by $G(x; \sigma_i^2)$, with variance $\sigma_1^2 > \sigma_2^2$, and set $1 < R_1(1 - \mu_1) < R_2$, as required by the proposition. Starting with localized initial conditions $N_{i,0}(x) = \delta(x)$, we can solve the system explicitly as

$$N_{1,t} = (1 - \mu_1)^t R_1^t G(x; t\sigma_1^2),$$

$$N_{2,t} = R_2^t G(x; t\sigma_2^2) + \mu_1 \sum_{j=1}^{t} (1 - \mu_1)^{t-j-1} R_1^{t-j} R_2^j G(x; (t - j)\sigma_1^2 + j\sigma_2^2).$$
$$\tag{14.103}$$

Since all the terms in the sum are positive, $N_{2,t}$ is bounded below by each of them individually, i.e.,

$$N_{2,t}(x) \geq \mu_1(1 - \mu_1)^{t-j-1} R_1^{t-j} R_2^j G(x; (t - j)\sigma_1^2 + j\sigma_2^2), \quad 0 \leq j \leq t. \tag{14.104}$$

For each of these terms, we can calculate the location x_t of the level set $N_{2,t} = \tilde{N}$ as in Sect. 5.2, namely

$$x_t^2 = 2\tilde{\sigma}^2 \left[\ln\left(\mu_1(1 - \mu_1)^{t-j-1} R_1^{t-j} R_2^j \right) - \ln(\tilde{N}\sqrt{2\pi\tilde{\sigma}^2}) \right], \tag{14.105}$$

with $\tilde{\sigma}^2 = (t - j)\sigma_1^2 + j\sigma_2^2$.

Now we choose $t = 2\tau$ to be even and choose $j = \tau$. Then the expression becomes

$$x_{2\tau}^2 = 2\tau(\sigma_1^2 + \sigma_2^2) \left[\ln\left(\mu_1(1 - \mu_1)^{\tau-1} R_1^\tau R_2^\tau \right) - \ln(\tilde{N}\sqrt{2\pi\tilde{\sigma}^2}) \right]. \tag{14.106}$$

The asymptotic speed of a level set is obtained from

$$\frac{x_{2\tau}^2}{4\tau^2} = \frac{\sigma_1^2 + \sigma_2^2}{2} \left[\ln(\tilde{R}_1 R_2) + \frac{1}{\tau} \ln(\mu_1/(1 - \mu_1)) - \frac{1}{\tau} \ln\left(\tilde{N}\sqrt{2\pi\tau(\sigma_1^2 + \sigma_2^2)} \right) \right], \tag{14.107}$$

where $\tilde{R}_1 = R_1(1 - \mu_1)$ as before. In the limit as $\tau \to \infty$, we find

$$\frac{x_{2\tau}^2}{4\tau^2} \to \frac{1}{2}(\sigma_1^2 + \sigma_2^2) \ln(\tilde{R}_1 R_2) =: c^2. \tag{14.108}$$

We want to show $c > \max\{\tilde{c}_1, \tilde{c}_2\}$ with $\tilde{c}_i$ as in (14.95). We make the simplifying assumption that $\tilde{c}_1 = \tilde{c}_2$. We substitute $\sigma_1^2 \ln(\tilde{R}_1) = \sigma_2^2 \ln(R_2)$ and find

$$c^2 = \frac{1}{2}\sigma_1^2 \ln(\tilde{R}_1)\left(1 + \frac{\sigma_2^2}{\sigma_1^2}\right)\left(1 + \frac{\sigma_1^2}{\sigma_2^2}\right). \tag{14.109}$$

After some algebra, we find that $c > \tilde{c}_1 = \tilde{c}_2$ if

$$\left(1 + \frac{\sigma_2^2}{\sigma_1^2}\right)\left(1 + \frac{\sigma_1^2}{\sigma_2^2}\right) > 4. \tag{14.110}$$

This inequality is equivalent to $z + 1/z > 2$, which is true for all positive $z \neq 1$. We summarize these results as follows.

Proposition 14.8 *Consider system (14.102) with Gaussian dispersal kernels with variances $\sigma_1^2 > \sigma_2^2$, mutation probability $\mu_1 \in (0, 1)$, and growth rates $R_2 > R_1(1 - \mu_1) > 1$. Assume that the component spreading speeds, $\tilde{c}_i$, are equal. Then the fastest speed, c_f^*, is bounded below by c from (14.109). In particular, the system has an anomalous spreading speed since $c_f^* \geq c > \tilde{c}_i$.*

We illustrate these results in Fig. 14.17. The plots on the left show the density of type 2 spreading in the absence of type 1. The theoretical and numerical spread rates are $\tilde{c}_2 = 0.173$ and $c_2^{num} = 0.1716$ (top) and $\tilde{c}_2 = 0.1533$ and $c_2^{num} = 0.1511$ (bottom). The plots on the right show both types spreading simultaneously. The top plots illustrate the increased, anomalous spread rate when the conditions in Proposition 14.8 hold, in particular $R_2 > \tilde{R}_1$ and $\sigma_1^2 > \sigma_2^2$. Type 2 spreads faster than type 1 and faster than in the absence of type 1. We find $c_2^{num} = 0.1883$ and $c_1^{num} = 0.17$, whereas the upper bound for the anomalous speed is $c^{(u)} = 0.1886$ and equals the lower bound from (14.109). The bottom plots show that if $\tilde{R}_1 > R_2$ and $\sigma_1^2 < \sigma_2^2$, then the two types spread at (almost) the same rate, and that rate is the same as for type 2 in isolation. As expected, the numerical results deviate slightly. They give $c_2^{num} = 0.1522$ and $c_1^{num} = 0.1514$.

Anomalous Spread with Bidirectional Mutation

Elliott and Cornell (2012) assumed that mutation between types happens in both directions, not just in one like in the preceding example. They found that, under

Fig. 14.17 Illustration of Proposition 14.8 (see text for explanation). Density of type 1 is plotted dashed, density of type 2 solid. Parameters for top plots are $R_1 = 1.5$, $R_2 = 2$, $\mu_1 = 0.1$, and $\sigma_1^2 = 0.05$. Parameters for bottom plots are $R_1 = 2$, $R_2 = 1.5$, $\mu_1 = 0.1$, and $\sigma_1^2 = 0.01$. Parameter σ_2^2 is determined by the condition $\sigma_1^2 \ln(\tilde{R}_1) = \sigma_2^2 \ln(R_2)$. Profiles are plotted every 30 time steps. Plots on the left have initial conditions for type 1 equal to zero. All other initial conditions are characteristic functions on $\{x < 0\}$. The linear equation is truncated so that $N_{i,t} \leq 1$.

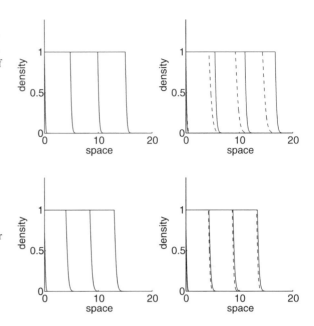

certain conditions, the spread of both types together was faster than for each type in isolation and in the absence of mutation. They called this effect an anomalous spreading speed, but the phenomenon is slightly different from the one found by Weinberger et al. (2007). The reaction–diffusion model by Elliott and Cornell (2012) is irreducible, there are no semi-trivial steady states, and there are no proper component spreading speeds. Nonetheless, the effect and the mathematics behind it are related to the phenomenon from Weinberger et al. (2007). We reproduce some of the results by Elliott and Cornell (2012) in our IDE model and explicitly calculate the spread rate in some special cases.

We consider system (14.92) again, but now with $\mu_1, \mu_2 \in [0, 1]$. As before, we could truncate the linear system at some positive density (and will do so for numerical simulations) or think of the linear system as the linearization of some appropriate nonlinear system at the trivial state. Neither modification will change the spreading speed from that of the linear system.

When $\mu_1 = \mu_2 = 0$, the system decouples into two equations, and there are two spreading speeds, $\tilde{c}_i$. Under the usual conditions (see Chap. 5), these speeds are given by

$$\tilde{c}_i = \inf_{s>0} c_i(s) = \inf_{s>0} \frac{1}{s} \ln(R_i M_i(s)). \qquad (14.111)$$

With Gaussian kernels as above, we have $\tilde{c}_i = \sigma_i \sqrt{2 \ln(R_i)}$.

When $\mu_1, \mu_2 \in (0, 1)$, linear system (14.92) is irreducible so that the results from Lui (1989a) apply; see Sect. 13.8. In particular, the system has a single spreading speed given by

$$c^* = \inf_{s>0} c(s) = \inf_{s>0} \frac{1}{s} \ln(\lambda(s)), \qquad (14.112)$$

where $\lambda(s)$ is the dominant eigenvalue of

$$\mathbf{H}(s) = \begin{bmatrix} (1 - \mu_1) R_1 M_1(s) & \mu_2 R_2 M_2(s) \\ \mu_1 R_1 M_1(s) & (1 - \mu_2) R_2 M_2(s) \end{bmatrix}. \qquad (14.113)$$

We can explicitly calculate the spreading speed for this system in the special case where the spreading speeds in isolation and the mutation rates are identical.

Proposition 14.9 *Assume that K_i are Gaussian dispersal kernels with moment-generating functions $M_i(s) = \exp(\sigma_i^2 s^2/2)$ and that $R_i > 1$, $R_1 \neq R_2$. Assume also that $\tilde{c}_1 = \tilde{c}_2$. Finally, assume that $0 < \mu_1 = \mu_2 =: \mu < 1$. Then we have the following properties of λ and c^*:*

1. *If there is some s with $R_1 M_1(s) = R_2 M_s(s)$, then $\lambda(s) = R_i M_i(s)$.*
2. *The curves $c_i(s) = \ln(R_i M_i(s))/s$ intersect exactly once for $s > 0$. The intersection occurs at*

$$s^* = \sqrt{\frac{2 \ln(R_1/R_2)}{\sigma_2^2 - \sigma_1^2}} = \sqrt{\frac{2 \ln(R_1)}{\sigma_2^2}} = \sqrt{\frac{2 \ln(R_2)}{\sigma_1^2}}. \qquad (14.114)$$

3. *We have $\lambda(s^*) = R_1 \exp\left(\sigma_1^2 \ln(R_1)/\sigma_2^2\right)$.*
4. *The derivative of $c(s)$ vanishes at s^*; i.e.,*

$$\frac{d}{ds} \frac{\ln(\lambda(s))}{s}\Big|_{s=s^*} = 0. \qquad (14.115)$$

5. *The spreading speed is given by*

$$c^* = \frac{\lambda(s^*)}{s^*} = \sqrt{\frac{\sigma_2^2 \ln(R_1)}{2}} \left(1 + \frac{\sigma_1^2}{\sigma_2^2}\right) = \frac{1}{2} \left(\sqrt{\frac{\sigma_2^2}{\sigma_1^2}} + \sqrt{\frac{\sigma_1^2}{\sigma_2^2}}\right) \tilde{c}_1. \qquad (14.116)$$

6. *If $\sigma_1^2 \neq \sigma_2^2$, then $c^* > \tilde{c}_1 = \tilde{c}_2$.*

Proof Most of the claims are straightforward calculations; only the fourth claim is somewhat tedious. Not all claims use all conditions. We give illustrations in Fig. 14.18. The condition $\tilde{c}_1 = \tilde{c}_2$ implies

$$\sigma_1^2 \ln(R_1) = \sigma_2^2 \ln(R_2). \qquad (14.117)$$

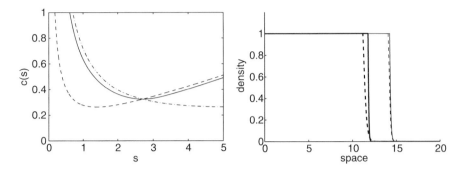

Fig. 14.18 Left: The dispersion relation for types 1 (dash-dot) and 2 (dashed) in the absence of mutation ($\mu = 0$), and for the joint speed with mutation ($\mu > 0$, solid). **Right:** Density profile of type 1 (solid) and type 2 (dashed) in the absence of mutation (thick), and joint spread with mutation (thin) after 200 generations. Parameters are $R_1 = 2$, $R_2 = 1.2$, $\sigma_1^2 = 0.05$. Initial conditions are the characteristic function on $\{x < 0\}$. Densities were truncated at one.

The explicit expression for λ is

$$\lambda = \frac{1}{2}\left((1-\mu)(R_1 M_1 + R_2 M_2) + \sqrt{(1-\mu)^2(R_1 M_1 - R_2 M_2)^2 + 4\mu^2 R_1 R_2 M_1 M_2} \right).$$

The first claim follows immediately. The second and third claims are simple calculations. For the fourth claim, we need to show

$$\lambda'(s^*)s^* = \lambda(s^*)\ln(\lambda(s^*)). \tag{14.118}$$

Calculating the derivative and substituting the condition $R_1 M_1(s^*) = R_2 M_2(s^*)$ results in

$$\lambda'(s^*) = \frac{1}{2}(R_1 M_1'(s^*) + R_2 M_2'(s^*)). \tag{14.119}$$

Differentiating M_i gives $M_i' = \sigma_i^2 s M_i$. Substituting, we find

$$\lambda'(s^*)s^* = \frac{(s^*)^2}{2}(\sigma_1^2 + \sigma_2^2) R_1 M_1(s^*) = \frac{(s^*)^2}{2}(\sigma_1^2 + \sigma_2^2)\lambda(s^*). \tag{14.120}$$

Hence, it remains to show that $(s^*)^2(\sigma_1^2 + \sigma_2^2) = 2\ln(\lambda(s^*))$. Substituting the expression for s^*, we calculate

$$2\ln(\lambda(s^*)) = 2\ln(R_1 M_1(s^*)) = 2\ln(R_1) + \sigma_1^2(s^*)^2 = \sigma_2^2(s^*)^2 + \sigma_1^2(s^*)^2. \tag{14.121}$$

Lui (1989a) showed that $\lambda(s)/s$ is a convex function. Its only critical point is at s^*. This has to be a minimum. Hence, $c^* = \lambda(s^*)/s^*$, which gives the expression in

(14.116). The function $f(z) = (z + z^{-1})/2$, $z > 0$, has its unique global minimum $f = 1$ at $z = 1$. Hence, whenever $\sigma_1^2 \neq \sigma_2^2$, we have $c^* > \tilde{c}_1 = \tilde{c}_2$. □

The plots in Fig. 14.18 illustrate the increase in the spread rate when there is mutation between the two types. For the parameters chosen, we have $c^* \approx 1.23\tilde{c}_1$ according to (14.116). The explicit calculation of c^* is possible only under the quite restrictive assumptions $\tilde{c}_1 = \tilde{c}_2$ and $\mu_1 = \mu_2$. It is clear from the construction (and by continuity) that the phenomenon of an increased spreading speed with mutation is not restricted to this case. Perhaps surprisingly, the joint spreading speed is independent of the mutation rate(s) when $0 < \mu_1 = \mu_2 < 1$. A population of two types, one a better disperser, the other with a higher growth rate, can spread faster than each of the types could on its own, hence the term "anomalous."

We take a final look at the degenerate case when $\mu_1 = \mu_2 = 1$. Each type produces only the other type. The system of equations can be reduced to a single equation over two generations, namely

$$N_{i,t+2} = R_1 R_2 K_1 * K_2 * N_{i,t} . \tag{14.122}$$

If we write c for the speed per generation, then the dispersion relation over two generations is given by

$$e^{2sc} = R_1 R_2 M_1(s) M_2(s) .$$

For Gaussian kernels with variances σ_i^2, we can find the minimum value of c explicitly as

$$c^* = \inf_{s>0} \frac{1}{2s} \ln \left(R_1 R_2 e^{(\sigma_1^2 + \sigma_2^2)s^2/2} \right) = \sqrt{2 \frac{\sigma_1^2 + \sigma_2^2}{2} \ln(\sqrt{R_1 R_2})} .$$

Hence, we find the speed by taking the arithmetic mean of the variances and the geometric mean of the growth rates. If we impose condition (14.117) as in Proposition 14.9, we arrive at the same expression for c^* as in (14.116).

This degenerate case of complete mutation is not biologically relevant since mutation rates are typically very small. It has, however, an alternative interpretation in terms of temporal variation. Suppose a population has growth rate R_1 and dispersal kernel K_1 in odd generations and growth rate R_2 and dispersal kernel K_2 in even generations. Then the linear equation for the population density between two generations is the same as in (14.122). Suppose that there is a trade-off between growth and dispersal between the 2 years. Then the population can spread faster when individuals choose to alternate between high and low growth rates (and corresponding low and high dispersal) than when they behave the same way in every generation. We consider temporal variation in more detail in Chap. 16.

14.8   Spread of Competing Species

Competition has long been recognized as a major force shaping ecological communities. Accordingly, theoretical models for competition have been studied for a long time. We distinguish between exploitative competition and interference competition. In the former, competitors affect each other indirectly through their depletion of a common limiting resource. Models for this case must include resource dynamics and, hence, are variants and extensions of the consumer–resource models that we discussed in Sect. 14.3. In the latter, consumers interact directly, e.g., via aggression. In models for this case, the growth function of one species decreases with the presence of the other species. We shall consider only the latter case here.

We can derive competition models from single-species models by including the density of competing species in the per capita growth rate of a given species. For example, if $G(N)$ denotes the (decreasing) per capita growth rate of species N in isolation and $\tilde{N}$ the density of a competing species, we write $G(N + \alpha \tilde{N})$ for the per capita growth rate of N in the presence of the competitor. If $\alpha > 1$, then a competitor individual has a stronger influence on the growth function than a conspecific; if $\alpha < 1$, then the competitor's effect is weaker. When we apply this construction to the scaled Beverton–Holt growth function, we obtain the two-species competition model

$$
\begin{aligned}
N_{1,t+1} &= \frac{R_1 N_{1,t}}{1 + (R_1 - 1)(N_{1,t} + \alpha N_{2,t})}, \\
N_{2,t+1} &= \frac{R_2 N_{2,t}}{1 + (R_2 - 1)(N_{2,t} + \beta N_{1,t})},
\end{aligned}
\tag{14.123}
$$

where $R_i > 1$ are the maximum growth rates and α, β the intraspecific competition coefficients.

The dynamics of this model are equivalent to those of the classical Lotka–Volterra competition model (Kot 2001). The trivial state $(0, 0)$ is unstable. The semi-trivial state $(1, 0)$ is stable if $\beta > 1$ and unstable if $\beta < 1$. Similarly, $(0, 1)$ is stable (unstable) if $\alpha > 1$ ($\alpha < 1$). The coexistence state,

$$
(N_1^*, N_2^*) = \left(\frac{1 - \alpha}{1 - \alpha \beta}, \frac{1 - \beta}{1 - \alpha \beta} \right),
\tag{14.124}
$$

is biologically meaningful only if either $0 < \alpha, \beta < 1$ or $\alpha, \beta > 1$. It is stable in the former case and unstable in the latter. Hence, we have four scenarios, depending on the interaction parameters. If $0 < \alpha, \beta < 1$, the coexistence state is globally stable. The competitors have a relatively weak influence on each other. If $0 < \alpha < 1 < \beta$, then $(1, 0)$ is globally stable. The first species competitively excludes the second. The case $0 < \beta < 1 < \alpha$ is analogous. If $\alpha, \beta > 1$, both semi-trivial states are locally stable and coexistence is unstable. This case is sometimes called "founder control."

The question of how and to what extent spatial dispersal can allow competing species to coexist if local dynamics predict competitive exclusion has been studied with many different modeling approaches but received only marginal attention within the IDE framework. Lutscher (2008) shows how coexistence can arise on a bounded domain if the competitively superior species disperses sufficiently more than the inferior species. High dispersal leads to high loss through the boundary, which reduces population density at the boundary. Near the boundary, competition pressure on the inferior species is weak so that its population can grow there. If it disperses only a little, sufficiently many of its offspring stay within the domain, so that the population can persist.

The question of spread of competing species has received much more attention, in particular in the context of the spread of invasive organisms. For example, Allen et al. (1996) and Hart and Gardner (1997) model the spread of competing plants with IDE systems. Okubo et al. (1989) analyze the spread of gray squirrels in the presence of red squirrels in Great Britain with a reaction–diffusion system; for a more systematic treatment of invasions, see Lewis et al. (2016). Biologically, we would like to know how the presence of a competitor affects the spread rate of a species, in particular, if it can slow the spread of an invading species. Mathematically, we are interested in asymptotic spreading speeds and traveling waves. Since there are so many steady states of the nonspatial model, there are many possibilities for traveling waves to connect them. The most obvious one, the connection from the trivial state to the coexistence state (if it is stable), turns out to be surprisingly difficult to study. Somewhat simpler is the question of one species invading an established competitor and either coexisting with it or replacing it. In the case of founder control, there could be bistable fronts connecting the two stable semi-trivial states.

Lewis et al. (2002) study the following spatial version of (14.123); see also Li et al. (2005) and Weinberger et al. (2007):

$$
\begin{aligned}
N_{1,t+1}(x) &= \int_{-\infty}^{\infty} K_1(x-y) \frac{R_1 N_{1,t}(y)}{1+(R_1-1)(N_{1,t}(y)+\alpha N_{2,t}(x))}\,dy\,, \\
N_{2,t+1}(x) &= \int_{-\infty}^{\infty} K_2(x-y) \frac{R_2 N_{2,t}(y)}{1+(R_2-1)(N_{2,t}(y)+\beta N_{1,t}(y))}\,dy\,,
\end{aligned}
\tag{14.125}
$$

where K_i is the dispersal kernel of species i. They consider the situation where $\alpha < 1$, so that the state $(0, 1)$ is unstable to invasion from species N_1. They then study the existence of a spreading speed and the validity of the linear conjecture for this invasion. Their approach is to transform the system into a cooperative system by the change of variables $(N_1, N_2) \mapsto (U_1, U_2) = (N_1, 1 - N_2)$ and then use the theory for cooperative systems (Sect. 14.7).

In the new variables, the system reads

$$U_{1,t+1}(x) = \int_{-\infty}^{\infty} K_1(x-y) \frac{R_1 U_{1,t}(y)}{1+(R_1-1)(N_{1,t}(y)+\alpha(1-U_{2,t}(x)))} dy,$$

$$U_{2,t+1}(x) = \int_{-\infty}^{\infty} K_2(x-y) \frac{U_{2,t}(y)+\alpha(R_2-1)U_{1,t}(y)}{1+(R_2-1)(\beta N_{1,t}(y)+1-U_{2,t}(y))} dy.$$

$$(14.126)$$

Since the original system is confined to the unit square (i.e., $0 \leq N_1, N_2 \leq 1$), the same is true for the new system. In that range, the new system is monotone. Biologically, we could say that if one species competes with another, then it "cooperates with the absence of the other." Under the change of variables, $(0,0)$ becomes $(0,1)$, $(0,1)$ becomes $(0,0)$, and $(1,0)$ becomes $(1,1)$. The coexistence state, (U_1^*, U_2^*), is inside the unit square as long as $\beta < 1$. Hence, if $\beta < 1$, then $(0,0)$ is unstable and (U_1^*, U_2^*) is stable, the system is cooperative in the ordered rectangle spanned by these two points, and there is no other steady state in that rectangle. If $\beta > 1$, then the point $(1,1)$ is stable, and there is only one boundary equilibrium, $(1,0)$, inside the square spanned by $(0,0)$ and $(1,1)$. In both cases, the theory for cooperative systems guarantees the existence of a joint asymptotic spreading speed of U_1 and U_2 (Lewis et al. 2002). Hence, if U_1, U_2 are initially confined to some bounded set, they will eventually spread at the same speed, c^*. In terms of the original variables, if N_1 is initially confined to some bounded set and N_2 is present only *outside* some bounded set, then the spread of N_1 and the retreat of N_2 at the invasion front of N_1 will eventually occur at that speed. We illustrate these scenarios in Fig. 14.19. The top row shows how two competitors coexist in the wake of the invasion ($\beta < 1$); the bottom row shows replacement of N_2 by N_1 ($\beta > 1$).

We aim to calculate the asymptotic speed of N_1 (or U_1 and U_2, respectively). We linearize the nonspatial version of (14.126) at $(0,0)$ and find the Jacobian matrix

$$B = \begin{bmatrix} \frac{R_1}{1+\alpha(R_1-1)} & 0 \\ * & 1/R_2 \end{bmatrix}. \qquad (14.127)$$

Since the matrix is lower triangular, its eigenvalues are independent of the expression denoted by $*$. We follow the same steps as in (13.30) and define the component spreading speeds c_i as in (14.90). Since $1/R_2 < 1$, speed c_2 is negative. Therefore, our candidate for the spreading speed is

$$\hat{c} = \inf_{s>0} \frac{1}{s} \ln \left(\frac{R_1}{1+\alpha(R_1-1)} M_1(s) \right), \qquad (14.128)$$

where M_1 is the moment-generating function of dispersal kernel K_1. Just as in the preceding sections, an anomalous spreading speed could arise, so that additional conditions are required before we can conclude that the linear conjecture holds, i.e., that $\hat{c} = c^*$.

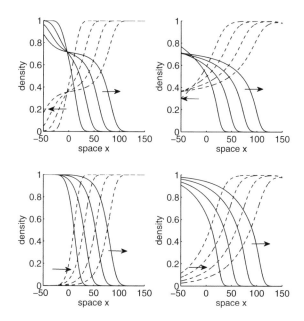

Fig. 14.19 Four possible outcomes of one species (solid) invading an established competitor (dashed). **Top row:** Weak competition leads to coexistence in the wake of the invasion ($\beta = 0.9$). **Bottom row:** Strong competition leads to replacement ($\beta = 1.1$). **Left column:** The invading species spreads at the linearized speed ($\sigma_2^2 = 1$). **Right column:** The invading species spreads faster than the linearization predicts ($\sigma_2^2 = 50$). Parameters are $R_1 = R_2 = 1.2$, $\alpha = 0.8$, $\sigma_1^2 = 1$. Initially, the invading species is present for $x < 0$ and the established species for $x > 0$. Profiles are plotted every 100 time steps.

Theorem 14.1 (Theorem 3.1 in Lewis et al. 2002) *Assume that K_i are symmetric and continuous and have finite moment-generating functions M_i. Assume that $R_i > 1$, $0 < \alpha < 1$, and $\beta > 0$. Let $s^* < \infty$ be the value where (14.128) attains its minimum. If*

$$\frac{R_1}{1 + \alpha(R_1 - 1)} M_1(s^*) \geq \frac{1 + (R_2 - 1)\max(\alpha\beta, 1)}{R_2} M_2(s^*), \qquad (14.129)$$

then the spreading speed c^* of (14.126) is linearly determined, i.e., it is equal to $\hat{c}$ in (14.128).

When we choose Gaussian dispersal kernels with mean zero and variances σ_i^2, we can calculate the condition of the theorem explicitly. The minimum of $c_1(s)$ occurs at

$$s^* = \frac{1}{\sigma_1} \sqrt{2 \ln\left(\frac{R_1}{1 + \alpha(R_1 - 1)}\right)}. \qquad (14.130)$$

Condition (14.129) becomes

$$\left(\frac{R_1}{1+\alpha(R_1-1)}\right)^{2-(\sigma_2^2/\sigma_1^2)} \geq \frac{1+(R_2-1)\max(\alpha\beta,1)}{R_2} . \tag{14.131}$$

The simulations in Fig. 14.19 show the difference between spread at the linearly determined speed when the above condition is satisfied (left column) and the faster, anomalous speed when the condition is not satisfied (right column). For the chosen parameter values, we have $\hat{c} = 0.26$. The numerically observed speeds are $c^{num} = 0.25$ (left column), $c^{num} = 0.267$ (top right), and $c^{num} = 0.28$ (bottom right).

How can the spread speed of one species depend on the dispersal behavior of its competitor? When the established competitor disperses only short distances, its population density will show a relatively steep transition from near carrying capacity to near zero at the edges of its current range. The leading edge of the invading species is inside the competitor's range, where the competitor is near carrying capacity, so that the invading species experiences the full force of competition. If, however, the competitor disperses a long distance, its population density declines more slowly near the edge of its range. The invading species does not experience the full force of the competition and is able to invade faster.

The scenario of one species spreading into a competing species is certainly the most relevant in the context of invasion biology. But what happens if two competing species are released simultaneously and at the same location into previously unoccupied habitat? Generically, one of the two will have the larger spreading speed in isolation. We expect that species to spread and occupy the habitat more quickly, setting the state for the previous scenario: the slower-spreading species (in isolation) will find itself competing against an (almost) established species. If it can invade, its spreading speed in the presence of the competitor will typically be slower than in isolation and at least as fast as the linearization at the semi-trivial state predicts, but not necessarily determined by it (see above). Proofs of these statements with necessary conditions for this qualitative behavior can be found in Lin et al. (2011). We illustrate these two phases of co-invasion in Fig. 14.20.

A number of researchers have studied various aspects of traveling waves in competition systems. Li et al. (2005) prove that under certain conditions, the spreading speed in a cooperative system can be characterized as the slowest speed of traveling waves. This result carries over to the two-species competition model here for the scenario where one species invades another. Li (2009) shows the existence of traveling waves where two competing species invade previously unoccupied habitat for all speeds higher than some threshold. The behavior in the wake of the wave is left open. Lin et al. (2011) prove the existence of a wave connecting the zero state to the coexistence state under certain conditions and with Gaussian dispersal kernels. This wave exists for all speeds faster than the single-species spreading speeds and converges to zero exponentially. Wang and Castillo-Chavez (2012) use the Ricker model instead of the Beverton–Holt model for local population dynamics. The resulting nonmonotone competition model is known as the Hassell and Comins

Fig. 14.20 Two phases of the co-invasion of competing species; see text for detailed explanation. The faster-spreading species (solid) slows the slower-spreading species (dash-dot in isolation, dashed in competition). The setup is as in Fig. 14.19, but initial conditions are characteristic functions on $\{x < 0\}$ for both species and $\sigma_2^2 = 2$. Densities are plotted every 50 generations.

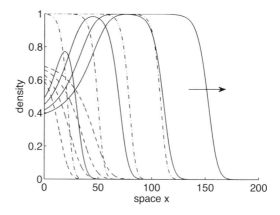

model. They show the existence of spreading speeds and of a traveling wave of invasion and replacement. Li and Li (2012b) also study the spatial Hassell and Comins model of competition; they prove the existence of a traveling wave for joint invasion. Li and Li (2012a) study the asymptotic behavior and prove uniqueness of traveling waves in the spatial Hassell and Comins model. Pan and Yang (2014) consider a competition model based on the logistic single-species growth model. Pan and Lin (2014) study traveling waves in a spatial Hassell and Comins model with history; i.e., the competition terms may depend on all past generation densities. Lin (2015) considers traveling waves in multi-species, multi-generation models. Li et al. (2016b) use Gaussian dispersal kernels and the interaction model in (14.125) to study asymptotic behavior and uniqueness of traveling waves. The case of bistable fronts, connecting the locally stable semi-trivial equilibria if the coexistence state is unstable, is studied by Zhang and Zhao (2012).

14.9   Further Reading

Predator–Prey Systems

Predator–prey or host–parasitoid systems are fascinating to study because of the multitude and diversity of patterns and phenomena that they support. One aspect that we have not touched upon is the study of spreading phenomena with nonstationary dynamics in the wake of a wave. Since local predator–prey dynamics can show sustained oscillations or chaotic behavior, we can ask how such dynamics spread spatially. Sherratt et al. (1997) compare the patterns in the wake of a (one-dimensional) invasion in four different model structures: reaction–diffusion equations, IDEs, coupled map lattices, and cellular automata. They find that all model structures support complex patterns in the wake, and they conclude that these phenomena are robust to model choice.

Legaspi et al. (1998) study a complex (discrete-space) IDE simulation model of hosts and parasitoids with three developmental stages each, in two dimensions, via numerical simulations. Their model considers the life-cycle structure of the boll weevil (*Anthonomus grandis Boheman*) and its exotic parasitoid *Catolaccus grandis* (Burks). They consider management options under scenarios such as dominant wind direction and temperature-dependent reproduction.

Allen et al. (2001) consider a discrete-space host–parasitoid system in one, two, and three spatial dimensions. They find complex patterns such as spiral waves and scroll waves. They explore the stabilizing effects of spatial structure by varying model parameters across the domain, thereby coupling patches with stable and unstable local dynamics, and they obtain globally stable densities under some conditions. They propose the use of transfer functions to study stabilizing aspects analytically.

White and White (2005) clarify the relationship between discrete-space models, known as coupled map lattices but also referred to as discrete-space IDEs by Allen et al. (2001), and (continuous-space) IDEs with regard to the respective pattern-formation conditions. Neubert et al. (2002) relate transient dynamics of the local model to pattern-formation conditions of the IDE model.

IDEs are used to model the transition from hunter–gatherer to farmer societies in the neolithic period. Fort et al. (2008) argue that the modeling framework of IDEs is better suited to this application than reaction–diffusion equations because they can naturally include the fact that parents typically move with their children and not away from them. The interaction between hunter–gatherers and farmers is different from a typical predator–prey model in that each population can persist on its own. Their interaction, however, typically leads to population growth for the farmers at the expense of the hunter–gatherers and therefore constitutes a consumer–resource relationship. Fort et al. (2008) formulate their model in two spatial dimensions and calculate an expected spreading speed based on the assumption of linear determinacy. They also estimate the time of coexistence of the two populations during the replacement process by estimating the maximum slope of the invasion front and inferring its width from it. Fort (2012) considers two processes of spatial spread in the neolithic transition: the movement of individuals (demic diffusion) and the movement of ideas (cultural diffusion). He estimates their relative impact on the speed of the invasion front from data.

Wright and Hastings (2007) study how mechanisms that are very different from the ones in Sect. 14.4 can generate stable spatial patterns in host–parasitoid IDEs. They use singular perturbation theory to give conditions for abrupt spatial transitions between two "steady states" of the local system. They find that the local dynamics must exhibit an Allee effect of the host as well as an instability of the coexistence equilibrium for such patterns to exist. The singular perturbation arises from considering host dispersal on a very small scale.

Aydogmus et al. (2017) present an analytical framework to derive the so-called Stuart–Landau equations for a single nonlocal IDE. We expect that these techniques carry over to the case of coupled IDEs and could be employed to analyze the amplitude of patterns in dispersal-driven instabilities.

Predator and prey affect not only each other's population dynamics but poten-tially also each other's dispersal behavior. Dwyer and Morris (2006) present an IDE simulation model for density-dependent dispersal of predator and prey; see Chap. 17. Another important aspect of predator–prey dynamics is landscape structure. Hughes et al. (2015) consider a host–parasitoid IDE model in fragmented landscapes of different characteristics. We devote Chap. 15 to questions of landscape heterogeneity and consider their model in that context. Finally, Gouhier et al. (2010) derive and implement a model of an aquatic ecosystem of mussels and their succession. The model shows locally oscillatory behavior driven by disturbance and regrowth. Gouhier et al. (2010) simulate the coupled model (a discrete-space IDE) and observe the emergence of synchrony of these local cycles at scales much larger than the dispersal scale of the organisms.

Cooperative Systems

The theory in Weinberger et al. (2002, 2007) covers more general systems than what we presented here. Weinberger et al. (2002) provide constructions to deal with the case where the equation has intermediate steady states and prove that there is a single spreading speed. Theorem 3.1 in Weinberger et al. (2002) also shows that for the linear conjecture to be true, the linear boundedness condition $Q[\mathbf{N}] \leq Q'[0]\mathbf{N}$ needs to hold only in the direction of the dominant eigenvalue and not for all $\mathbf{N}$; compare Theorem 13.3. Weinberger et al. (2007) discuss an error in the proof of Lemma 2.3 in Weinberger et al. (2002) and give additional hypotheses that can be imposed for the results to hold.

Castillo-Chavez et al. (2013) review the literature on spreading speeds, linear determinacy, and traveling waves in IDEs and reaction–diffusion equations for scalar and vector-valued models with and without monotonicity.

In a "delayed" IDE, reproduction in generation t depends on the densities in generations t and $t - 1$, so that

$$N_{t+1}(x) = \int_{-\infty}^{\infty} K(x - y) F(N_t(y), N_{t-1}(y)) \mathrm{d}y . \qquad (14.132)$$

This equation can be written as a system for $N_{1,t} = N_t$ and $N_{2,t} = N_{t-1}$. Under some conditions on F, the resulting system is cooperative. Lin and Li (2010) show the existence of a spreading speed and establish that this speed is the slowest speed for which a monotone traveling wave solution exists. They also show that wave fronts are stable. Pan and Lin (2011) analyze the same equation but allow for more general growth functions. In particular, under the assumption that the growth function be "locally monotone," they obtain the existence of a spreading speed, and the existence and nonexistence of monotone traveling waves, as well as their stability.

Ecological theory distinguishes between facultative and obligate mutualisms. In the former, each species can grow independently, whereas in the latter, each species requires the latter to grow. All the examples presented here treat facultative mutualisms. Obligate mutualisms typically exhibit strong Allee-like effects, as has been investigated in a nonspatial model for plant–pollinator systems by Lutscher and Iljon (2013).

Competitive Systems

Lin (1995) studies a spatial genetics model for the dynamics of an advantageous gene replacing an inferior gene. The model is an extension of a model in Weinberger (1982) in that the densities of both genotypes are described, whereas the earlier model had assumed that the total population density is constant, so that the density of one genotype is simply the complement of the other. Even though the resulting model in Lin (1995) is not monotone, the author proves the existence of traveling waves and shows that the asymptotic spreading speed is characterized as the slowest traveling wave speed.

Allen et al. (1996) derive a competition model for two flowering plant types. They assume that the "relative total yield," i.e., the sum of the yield of the two species, is constant. Hence, as the yield of one species is increased, the other has to decrease. Furthermore, the relative yield of each species is proportional to the relative amount of seeds produced. For homogeneous solutions, they find exactly the same four scenarios as we did in our competition model. Numerically, they also show the same invasion–coexistence and invasion–replacement behavior as we did in Fig. 14.19.

The competition model by Hart and Gardner (1997) is based on "lottery competition," where the density of offspring is determined by the ratio of the seeds produced. This choice allows the authors to collapse the two equations into a single one. For that single equation, the authors find the exact spreading speed (based on the same arguments as in Chap. 5). They also derive a linear approximation of that speed.

Carrillo et al. (2009) also consider a genetics model for two types: one that reproduces sexually, while the other reproduces asexually. They show that under some conditions, the two genotypes can coexist in an invasion front.

Kanary et al. (2014) study the competition between two variants of the green crab, an invading species on the east coast of North America. They include pseudo-age-structure and dispersal bias in their model and describe the feasibility or failure of management options.

Ramanantoanina et al. (2014) (see Ramanantoanina et al. (2015) for a correction) explore how the presence of different dispersal types affects the speed of spread of an invasive species. They assume that individuals differ only in their dispersal behavior so that the system effectively becomes a competition system of these different types. They observe that if the type with the greatest dispersal ability is initially present only at very low density, then the overall population begins spreading slowly but speeds up as the type with the greatest dispersal ability becomes more prevalent. The population will eventually be spatially sorted with the types with higher dispersal ability ahead of those with lower dispersal ability.

.

Chapter 15
Spatial Variation

Abstract Most landscapes are heterogeneous at many spatial scales. Heterogeneity can reflect natural resource distribution (e.g., nutrients or temperature) or human activity (e.g., harvesting or agricultural land use). All of these factors may affect the local growth rate of individuals and their interactions. These effects can be modeled by an explicit dependence of the growth function on spatial location. Spatial heterogeneity can also affect individual dispersal patterns, either passively (e.g., dispersal barriers, wind direction) or actively (e.g., search for resources, avoidance of dangers). Including spatially varying dispersal patterns in IDEs and analyzing the resulting population dynamics poses numerous challenges. In previous chapters, we concentrated on two kinds of landscapes: a homogeneous infinite landscape and a single (homogeneous) bounded region. In this chapter, we present several approaches to include more realistic spatial heterogeneity in IDEs. We begin with models where only the growth function is affected by spatial heterogeneity and later present ways to include heterogeneity in movement models and dispersal kernels.

15.1 Habitat Quality Function

Latore et al. (1999) propose using a "habitat quality function" to model the effect of habitat heterogeneity on population persistence. They also introduce the notion of a "critical habitat-size," which generalizes that of the critical patch-size from Chap. 3. We present some of their ideas here, in particular a special case that can be solved explicitly and that we will extend later to include temporal variation as well; see Chap. 16 and Zhou and Fagan (2017).

The habitat quality function, H, assigns to each location a number between zero and one as the "quality" of the habitat at that location. It can be interpreted as the probability that an offspring that arrives there will become established. The IDE for the population density from one year to the next is then a slight modification of (2.1), namely

$$N_{t+1}(x) = H(x) \int_{-\infty}^{\infty} K(x - y) F(N_t(y)) \mathrm{d}y . \tag{15.1}$$

© Springer Nature Switzerland AG 2019

F. Lutscher, *Integrodifference Equations in Spatial Ecology*, Interdisciplinary Applied Mathematics 49, https://doi.org/10.1007/978-3-030-29294-2_15

The growth function models population growth under ideal conditions; the dispersal kernel remains independent of habitat quality. If habitat quality affects production rather than establishment of offspring, we write the habitat quality function at location y inside the integral (Zhou and Fagan 2017).

When H is constant, we have an infinite homogeneous habitat. When H is the "uniform habitat quality function" (Latore et al. 1999), i.e., the characteristic function of an interval, the model is equivalent to the one for the critical patch-size in Chap. 3. When the integral of H is finite, it can be interpreted as the total habitat quality. The critical habitat-size can then be defined as the minimal total habitat quality that allows a population to persist.

Latore et al. (1999) propose several different habitat quality functions and compare persistence conditions for constant total habitat quality. We present two of their examples. The "Gaussian habitat quality function" is given by

$$H(x) = \exp\left(-\frac{x^2}{2\rho^2}\right). \tag{15.2}$$

It models a single high-quality region in an infinite landscape without abrupt edges. Parameter ρ^2 measures the spread of the habitat. The total habitat quality equals $\sqrt{2\pi\rho^2}$. For the "periodic habitat quality function," we divide the interval $[0, L]$ into $2m$ subintervals, which alternate between habitat quality one (good) and zero (bad); i.e.,

$$H(x) = \begin{cases} 1 & \text{if } x \in \left[\frac{(n-1)L}{m}, \frac{(2n-1)L}{2m}\right], \quad n = 1, \ldots, m, \\ 0, & \text{otherwise.} \end{cases} \tag{15.3}$$

The total habitat quality equals $L/2$, independent of m.

As in Chap. 3, we assume that the growth function has no Allee effect so that we can study population persistence by determining the stability of the trivial steady state by linearization. As usual, we say that the population persists if the trivial solution is unstable. In the case of a Gaussian dispersal kernel and the Gaussian habitat quality function, we can find explicit solutions of the linear equation and determine persistence conditions analytically. For the periodic case, we use a numerical scheme.

Gaussian Habitat Quality Function

We derive an explicit solution for the linear equation with Gaussian habitat quality

$$N_{t+1}(x) = Re^{-\frac{x^2}{2\rho^2}} \int_{-\infty}^{\infty} K_G(x - y; \sigma^2) N_t(y) \mathrm{d}y, \tag{15.4}$$

where $R = F'(0)$ is the linearized growth rate and K_G is the Gaussian kernel (2.25) with mean zero and variance σ^2.

If $N_0(x) \equiv \bar{N}_0$ is a constant, then $N_1(x) = R\bar{N}_0 H(x)$; i.e., N_1 is a multiple of a Gaussian. We claim that if N_t is a multiple of a Gaussian, then so is N_{t+1}.

Proposition 15.1 *The linear IDE in (15.4) has a solution of the form*

$$N_t(x) = A_t \exp\left(-\frac{x^2}{2v_t^2}\right), \tag{15.5}$$

with

$$v_{t+1}^2 = \frac{\rho^2(\sigma^2 + v_t^2)}{\rho^2 + \sigma^2 + v_t^2} \quad \text{and} \quad A_{t+1} = A_t R \sqrt{\frac{v_t^2}{\sigma^2 + v_t^2}}. \tag{15.6}$$

Proof We write $N_t(x) = A_t \sqrt{2\pi v_t^2} K_G(x; v_t^2)$. Then the convolution integral in (15.4) is the convolution of two Gaussian kernels with zero mean, which results in a Gaussian kernel where the two variances are simply added. Hence, we have

$$N_{t+1}(x) = Re^{-\frac{x^2}{2\rho^2}} A_t \sqrt{2\pi v_t^2} K_G(x; v_t^2 + \sigma^2) \tag{15.7}$$

$$= R A_t \sqrt{\frac{v_t^2}{v_t^2 + \sigma^2}} \exp\left(-\frac{x^2}{2\rho^2} - \frac{x^2}{2(v_t^2 + \sigma^2)}\right) \tag{15.8}$$

$$= R A_t \sqrt{\frac{v_t^2}{v_t^2 + \sigma^2}} \exp\left(-\frac{x^2}{2}\left(\frac{v_t^2 + \sigma^2 + \rho^2}{\rho^2(v_t^2 + \sigma^2)}\right)\right). \tag{15.9}$$

This proves the claim.   □

Thanks to the proposition, studying the behavior of (certain) solutions of the IDE in (15.4) reduces to studying the two-dimensional difference equations for v_t^2 and A_t. Moreover, the equation for v_t^2 decouples and forms a monotone and bounded iteration. In particular, it converges to the stable fixed point

$$\hat{v}^2 = \frac{\sigma^2}{2}\left(\sqrt{1 + 4\frac{\rho^2}{\sigma^2}} - 1\right). \tag{15.10}$$

Eventually, then, the iteration for A_t is

$$A_{t+1} \approx A_t R \sqrt{\frac{\hat{v}^2}{\sigma^2 + \hat{v}^2}}. \tag{15.11}$$

Hence, for large t, A_t will increase if

$$R > \sqrt{1 + \frac{\sigma^2}{\widehat{v}^2}} \quad \text{or} \quad \widehat{v}^2 > \frac{\sigma^2}{R^2 - 1}. \tag{15.12}$$

Substituting the expression for $\widehat{v}^2$ from above gives us the persistence conditions in terms of the original parameters as

$$\rho^2 > \rho_*^2 = \frac{\sigma^2 R^2}{(R^2 - 1)^2}. \tag{15.13}$$

We summarize these calculations in the following result.

Proposition 15.2 *If the habitat quality function is given by (15.2) and the dispersal kernel is a Gaussian kernel with variance σ^2, then the population in (15.1) can persist provided (15.13) holds. Therefore, the critical habitat-size is $\sqrt{2\pi\rho_*^2}$.*

The left panel in Fig. 15.1 shows the Gaussian and the uniform habitat quality functions with the same total habitat quality. The right panel compares the critical habitat-size as a function of growth rate for these two habitat quality functions. For any given growth rate, the required habitat size is smaller in the uniform case than in the Gaussian case, although the difference between the two decreases with increasing growth rate. Hence, population persistence is more likely when the resources are concentrated spatially than when they are spread out.

To numerically calculate the critical habitat-size for the uniform habitat quality function, Latore et al. (1999) suggest an algorithm inspired by the power method for matrices (see Chap. 8). The idea is to iterate the linear IDE, starting from a very small population density, for a large number of time steps and then compare whether the density grows or declines between the final two time steps. If the density declines, the length is below the critical length; if it increases, it is above.

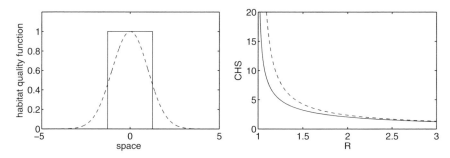

Fig. 15.1 Left: Illustration of the uniform (solid) and Gaussian (dashed) habitat quality functions. **Right:** Comparison of the corresponding critical habitat-sizes (CHS) for the uniform (solid) and Gaussian (dashed) case. The dashed curve is given by $\sqrt{2\pi\rho_*^2}$, with ρ_*^2 as in (15.13). The solid curve is obtained numerically as described in the text.

Alternatively, we can discretize the interval (scaled to unity), write a matrix for the discretized dispersal kernel (with scaled variance), and use standard computational routines to calculate the dominant eigenvalue, $\lambda = \lambda(L)$, for the linear integral operator; see Sect. 8.4 for details. Then we use the relationship $\lambda(L)R = 1$ to find the critical value of L for a given value of R.

Periodic Habitat Quality Function

Latore et al. (1999) use their periodic habitat quality function (15.3) to evaluate how the critical habitat-size depends on habitat fragmentation. More specifically, they compare the critical habitat-size under different combinations of growth rates and number of good (and bad) patches. By construction, the total habitat quality is exactly half of the total habitat length, independent of the number of patches into which the habitat is divided (left panels, Fig. 15.2).

The right panel in Fig. 15.2 demonstrates that the critical habitat-size increases as fragmentation (i.e., the number of patches) increases. It also demonstrates that the increase is fairly moderate and levels off when R is larger but that it is almost linear when R is smaller. To understand the reasons behind this difference, we note that as the habitat size, L, increases, so does the distance between two adjacent good patches, which is $L/(2m)$. The variance of the dispersal kernel, however, is fixed. Therefore, if L is relatively small, a large number of individuals disperse from one good patch to the next (e.g., the bottom left panel in Fig. 15.2). If L is relatively large, then the amount of dispersal between adjacent patches is very small (e.g., the top left panel in Fig. 15.2). When R is small, the habitat needs to be large, and consequently the gaps between patches are large. Because the dispersal

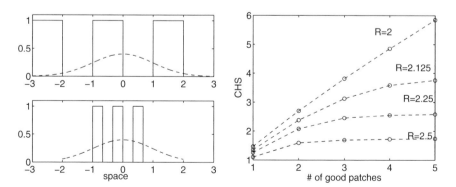

Fig. 15.2 Left: Illustration of the periodic habitat quality function (15.3) with $m = 3$ good and bad patches each and the Gaussian dispersal kernel with variance one (dashed). The habitat has length $L = 6$ in the top panel and length $L = 2$ in the bottom panel. **Right:** Comparison of critical habitat-sizes (CHS) for increasing fragmentation $m = 1, \ldots, 5$ and different growth rates, as indicated. The number of patches is discrete; the dashed lines serve only for visual structure.

between patches is small, each patch operates essentially as an isolated patch, and the population can persist if it can persist on each single patch. Hence, the critical habitat-size grows almost linearly with the number of patches.

This aspect of the work by Latore et al. (1999) is closely related to our discussion of habitat fragmentation in Sect. 3.5. There, we chose a sinusoidally varying linear growth rate with constant integral. We evaluated the dominant eigenvalue of the linearized IDE with the cosine kernel. We found that the dominant eigenvalue is a decreasing function of the number of peaks of the growth rate; i.e., population growth is highest when the resources are concentrated in one peak. The main difference between that setup and the one here is that the total length of the habitat was fixed. As the number of peaks increased, the width of each peak decreased, allowing increased dispersal between adjacent peaks, rather than the decreased dispersal that occurred, when the habitat was more fragmented.

The habitat quality functions by Latore et al. (1999) and the concept of a critical habitat-size provide a very flexible framework for exploring various aspects of habitat heterogeneity and population persistence. From a mathematical point of view, the framework is only partly satisfying since almost all results require numerical simulation. It turns out that by extending the periodic habitat quality function to the entire infinite line, we obtain a model that is more accessible for analyzing persistence conditions and even spreading phenomena. We consider this scenario in the next section.

15.2 Persistence in an Infinite Periodic Habitat

The systematic study of population dynamics on infinite fragmented habitats began with the pioneering work by Shigesada et al. (1986) for reaction–diffusion equations. While it may seem counterintuitive that an infinite domain could be simpler to study than a finite domain, the additional assumption of periodicity reduces the problem on the infinite domain to one on a bounded domain with a simple structure. We can calculate many of the interesting quantities explicitly in special cases and thereby gain important ecological insights. The study of IDEs on infinite periodic landscapes began with the work of Van Kirk (1995) and has since been expanded considerably (Van Kirk and Lewis 1997; Weinberger 2002; Robbins 2004; Kawasaki and Shigesada 2007; Weinberger et al. 2008; Dewhirst and Lutscher 2009; Musgrave and Lutscher 2014a,b).

All of these works study some aspects of the population dynamics of the IDE

$$N_{t+1}(x) = Q[N_t](x) = \int_{-\infty}^{\infty} K(x, y) F(N_t(y), y) \mathrm{d}y \,, \tag{15.14}$$

where the growth function is assumed to be periodic with some positive period L in the second variable; i.e., $F(N, y) = F(N, y + L)$ for all $y \in \mathbb{R}$. The corresponding periodicity condition for the dispersal kernel is

$$K(x, y) = K(x + L, y + L) \quad \text{for all} \quad x, y \in \mathbb{R}.\tag{15.15}$$

For explicit calculations, we consider a binary periodic growth function: each period consists of one "good" patch of length L_1 and one "bad" patch of length L_2, with $L_1 + L_2 = L$. The growth function is spatially constant within each patch; i.e.,

$$F(N, y) = \begin{cases} F_1(N), & y \text{ in good patch,} \\ F_2(N), & y \text{ in bad patch.} \end{cases}\tag{15.16}$$

We also refer to a patch with growth function F_i as type i. For the linear growth function $F_i(N) = R_i N$, we have a good patch if $R_i > 1$ and a bad patch if $R_i < 0$. In the ecological literature, these patch types are also called source and sink habitat, respectively. We shall use this terminology in the analysis below. For a nonlinear growth function, the property $F_i(N) < N$ for all $N \geq 0$ certainly describes a bad patch. A good patch requires $F_i(N) > N$ for some $N \geq 0$. But what exactly constitutes a good patch and how to distinguish between "good" and "better" in general is difficult and depends on many model aspects (e.g., whether there is overcompensation or an Allee effect).

We first study persistence conditions in a mixed source–sink landscape and ask which combination of good and bad patches allows the population to persist. As was the case for the critical patch-size, the answer will depend on the dispersal behavior of individuals. We begin with an explicit calculation in a special case that closely mimics the procedure in Chap. 3.

The Linear Equation with the Laplace Kernel

We assume that the growth function is linear in N and piecewise constant in space; i.e., $F(N, x) = R(x)N$ and $R(x) = R_i$ in patch type i. We choose the Laplace kernel (2.27) with parameter a. In particular, we assume that the dispersal behavior is independent of landscape quality. This assumption will be relaxed in Sect. 15.5. Equation (15.14) becomes

$$N_{t+1}(x) = \int_{-\infty}^{\infty} \frac{a}{2} e^{-a|x-y|} R(y) N_t(y) dy,\tag{15.17}$$

which can also be interpreted as the linearization of (15.14) at $N = 0$ with $R(x) = \partial F / \partial N(0, x)$. We are interested in the corresponding eigenvalue problem

$$\lambda \phi(x) = \int_{-\infty}^{\infty} \frac{a}{2} e^{-a|x-y|} R(y) \phi(y) dy,\tag{15.18}$$

and in particular in the dominant eigenvalue. For now, we simply assume that such a dominant eigenvalue exists; we return to the problem below. As before, we say that the population persists if there is at least one eigenvalue $|\lambda| > 1$, and we say that the population goes extinct if all eigenvalues satisfy $|\lambda| < 1$. Since we integrate with respect to a continuous function, we may assume ϕ to be continuous even though $R\phi$ is not. We will, in fact, assume that ϕ is smooth. We return to this question in Sect. 15.5, where we study more complicated dispersal behavior.

From a biological point of view, we can interpret the dominant eigenvalue λ as the overall population growth rate, whereas $R(x)$ is the local population growth rate. From this interpretation, it is clear that λ should be bounded above and below by the maximum and minimum of R, respectively. Mathematically, we can show that the dominant eigenvalue is monotone in R; i.e., if $R(x) \geq \tilde{R}(x)$, then the dominant eigenvalue corresponding to R is greater than or equal to that corresponding to $\tilde{R}$.

As in Chap. 3, we can differentiate Eq. (15.18) twice with respect to x and substitute to arrive at the second-order differential equation

$$\phi'' = a^2 \left(1 - \frac{R(x)}{\lambda} \right) \phi . \tag{15.19}$$

This periodic equation is a form of Hill's equation and therefore possesses a countable set of real eigenvalues, $1/\lambda$ (Magnus and Winkler 1979).

For explicit calculations, we choose the interval $[-L_1/2, L_1/2]$ as a good patch of length L_1 and the adjacent interval $[L_1/2, L_1/2+L_2]$ as a bad patch of length L_2. With these intervals as building blocks, the landscape is L-periodic; see Fig. 15.3. With this notation, and under the assumption that the eigenfunction is L-periodic, (15.19) can be written as the pair

$$\phi'' = a^2 \left(1 - R_1/\lambda \right) \phi, \quad x \in [-L_1/2, L_1/2],$$
$$\phi'' = a^2 \left(1 - R_2/\lambda \right) \phi, \quad x \in [L_1/2, L_1/2 + L_2], \tag{15.20}$$

Fig. 15.3 Schematic illustration of an infinite periodic landscape with two patch types (dashed). A periodic function (solid) that is also invariant under the transformation $x \mapsto -x$ must have zero slope at the midpoint of each patch.

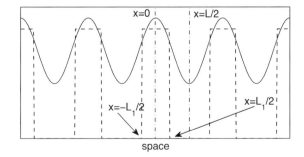

where ϕ is smooth and L-periodic. Since (15.19) is also symmetric, i.e., invariant under the transformation $x \mapsto -x$, the derivative of ϕ must be zero at the midpoint of each of the intervals; see Fig. 15.3. Hence, we can reduce problem (15.20) to

$$\phi'' = a^2 \left(1 - R_1/\lambda\right) \phi, \quad x \in [0, L_1/2],$$

$$\phi'' = a^2 \left(1 - R_2/\lambda\right) \phi, \quad x \in [L_1/2, L/2], \tag{15.21}$$

with no-flux boundary conditions $\phi'(0) = \phi'(L/2) = 0$ and smooth matching conditions

$$\lim_{x \searrow \frac{L_1}{2}} \phi(x) = \lim_{x \nearrow \frac{L_1}{2}} \phi(x) \quad \text{and} \quad \lim_{x \searrow \frac{L_1}{2}} \phi'(x) = \lim_{x \nearrow \frac{L_1}{2}} \phi'(x) \tag{15.22}$$

at $L_1/2$.

By the previous consideration, the dominant eigenvalue, λ, is bounded by $R_2 < \lambda < R_1$. Hence, we can write the general solution of the first equation as

$$\phi(x) = A_1 \cos(v_1 x) + B_1 \sin(v_1 x), \quad \text{with} \quad -v_1^2 = a^2 \left(1 - \frac{R_1}{\lambda}\right), \tag{15.23}$$

for $x \in [0, L_1/2]$. For the second equation, we obtain hyperbolic rather than trigonometric functions, and we choose to center them at the right-hand end of the interval; i.e.,

$$\phi(x) = A_2 \cosh\left(v_2 \left(\frac{L}{2} - x\right)\right) + B_2 \sin\left(v_2 \left(\frac{L}{2} - x\right)\right) \tag{15.24}$$

for $x \in [L_1/2, L/2]$ with $v_2^2 = a^2(1 - R_2/\lambda)$. The no-flux boundary conditions at $x = 0$ and $x = L/2$ force $B_1, B_2 = 0$. The matching conditions at $x = L_1/2$ give conditions for A_i. Continuity of ϕ requires

$$A_1 \cos\left(v_1 \frac{L_1}{2}\right) = A_2 \cosh\left(v_2 \frac{L_2}{2}\right), \tag{15.25}$$

whereas continuity of ϕ' requires

$$v_1 A_1 \sin\left(v_1 \frac{L_1}{2}\right) = v_2 A_2 \sinh\left(v_2 \frac{L_2}{2}\right). \tag{15.26}$$

These two conditions form a linear system for A_i. For a nontrivial solution, the determinant of the coefficient matrix must vanish. This condition gives us the relation

$$\nu_1 \tan\left(\frac{\nu_1 L_1}{2}\right) - \nu_2 \tanh\left(\frac{\nu_2 L_2}{2}\right) = 0 \,. \tag{15.27}$$

This relation is the analogue for infinitely many patches of (3.13) for a single patch. Just as in Chap. 3, we can numerically calculate λ if all other parameters are given. Alternatively, we can explicitly calculate the persistence boundary by setting $\lambda = 1$ and solving for L_1 to obtain the threshold

$$L_1^* = \frac{2}{a\sqrt{R_1 - 1}} \arctan\left[\sqrt{\frac{1 - R_2}{R_1 - 1}} \tanh\left(\frac{aL_2\sqrt{1 - R_2}}{2}\right)\right]. \tag{15.28}$$

The population can persist if $L_1 > L_1^*$. The relation between L_1^* and the other parameters is illustrated in Fig. 15.4. It increases with the size of bad patches, L_2, and decreases with the growth rate R_2. In the limit as $L_2 \to \infty$, the critical size approaches

$$L_{1,\infty} = \frac{2}{a\sqrt{R_1 - 1}} \arctan\left[\sqrt{\frac{1 - R_2}{R_1 - 1}}\right], \tag{15.29}$$

which is the critical size of an isolated patch in an infinite but not completely hostile environment. In the limit of a hostile environment ($R_2 \to 0$), we recover the critical patch-size of a single patch in (3.14). Hence, if the bad patches in an infinite landscape become completely hostile and very long, then each good patch acts as an isolated single patch. Of course, we can also solve (15.27) for L_2 to determine the maximal length of a bad patch that allows for population persistence. To obtain results for the ratio of good to bad patches, we scale space before solving the equations.

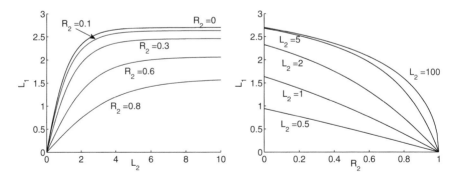

Fig. 15.4 Persistence threshold L_1^* from (15.28) in an infinite periodic landscape. **Left:** L_1^* increases with the length of bad patches and approaches the asymptote $L_{1,\infty}$ as $L_2 \to \infty$. **Right:** L_1^* decreases with R_2 and becomes zero when $R_2 = 1$, i.e., when bad patches become good patches.

We note that whenever lengths L_i occur in (15.27), they are multiplied by inverse mean dispersal distance a. As in Chap. 3, we can scale the equations in (15.21) to reduce the number of parameters. We set $\lambda = 1$ to study the persistence boundary. By setting $y = 2x/L$, we obtain

$$
\begin{aligned}
\phi'' &= \tfrac{a^2 L^2}{4}\,(1 - R_1)\,\phi\,, & y \in [0, p]\,, \\
\phi'' &= \tfrac{a^2 L^2}{4}\,(1 - R_2)\,\phi\,, & y \in [p, 1]\,,
\end{aligned}
\tag{15.30}
$$

where $p = L_1/L$ denotes the fraction of good habitat. Van Kirk (1995) found a clever substitution to reduce the number of parameters even further. By choosing

$$
\mathscr{H}^2 = \frac{a^2 L^2}{4}\sqrt{(R_1 - 1)(1 - R_2)} \quad \text{and} \quad q^2 = \sqrt{\frac{R_1 - 1}{1 - R_2}}\,,
\tag{15.31}
$$

we reduce the problem to

$$
\begin{aligned}
\phi'' &= -\mathscr{H}^2 q^2 \phi\,, & y \in [0, p]\,, \\
\phi'' &= \mathscr{H}^2 \phi/q^2\,, & y \in [p, 1]\,,
\end{aligned}
\tag{15.32}
$$

with only three parameters. We interpret $\mathscr{H}$ as the heterogeneity of the habitat relative to dispersal ability. When organisms are highly mobile, a and therefore also $\mathscr{H}$ are very small. Similarly, when R_i are both close to unity, then $\mathscr{H}$ is small. Parameter q is the ratio of the quality of good patches to the quality of bad patches. It measures habitat quality independent of dispersal ability.

Since (15.32) is of the same form as (15.21), the condition for nontrivial solutions is of the same form as (15.27), namely

$$
\tanh\!\left(\frac{\mathscr{H}}{q}(1 - p)\right) = q^2 \tan(\mathscr{H}qp)\,.
\tag{15.33}
$$

We illustrate the persistence conditions in terms of p and q in Fig. 15.5. As relative heterogeneity increases, a smaller percentage of good patches is required for persistence. In particular, in a fragmented landscape consisting of sources and sinks, a population is more likely to persist if its mean dispersal distance is smaller. While a small dispersal distance is potentially detrimental for individuals within a bad patch, it is crucially beneficial for individuals in good patches. When the mean dispersal distance is large compared to the period, individuals experience only some average growth rate, which might be insufficient for persistence. We present some ideas of spatial averaging in Sect. 15.4. Here, we show that a population may persist in a heterogeneous landscape even if the spatially averaged growth rate is less than unity. This calculation follows the ideas of Van Kirk and Lewis (1997).

We solve (15.19) for $R(x)$ as

$$
R(x) = \lambda\left(1 - \frac{1}{a^2}\frac{\phi''(x)}{\phi(x)}\right)\,.
\tag{15.34}
$$

Fig. 15.5 The minimum percentage of good habitat, p, required for persistence in a given landscape quality q with fragmentation level $\mathscr{H} = 1$ (dashed), $\mathscr{H} = 2$ (dash-dot), and $\mathscr{H} = 5$ (solid) according to (15.33). Higher quality and increasing levels of (relative) heterogeneity require a smaller percentage of good habitat.

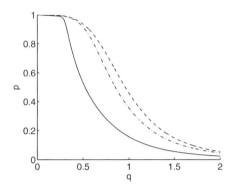

Averaging over one period on each side gives

$$\frac{1}{L}\int_0^L R(x)\mathrm{d}x = \lambda\left(1 - \frac{1}{a^2 L}\int_0^L \frac{\phi''(x)}{\phi(x)}\mathrm{d}x\right). \tag{15.35}$$

Integrating by parts on the right-hand side leads to

$$\int_0^L \frac{\phi''(x)}{\phi(x)}\mathrm{d}x = \frac{\phi'(x)}{\phi(x)}\Big|_0^L + \int_0^L \frac{(\phi'(x))^2}{(\phi(x))^2}\mathrm{d}x. \tag{15.36}$$

By periodicity, the boundary term vanishes; the remaining integral is positive. Hence, the factor multiplying λ in (15.35) is less than unity. Therefore, we find

$$\lambda > \frac{1}{L}\int_0^L R(x)\mathrm{d}x. \tag{15.37}$$

Hence, the overall population growth rate is greater than the average of the local growth rates. In particular, if the average is only slightly below unity, the overall growth rate can still be above unity.

General Persistence Conditions

In this section, we outline how to proceed in the general case, when the kernel is not the Laplace kernel, and we justify the existence of a dominant eigenvalue that determines persistence in the previous calculations. This section follows the ideas from Van Kirk (1995) that were generalized in Musgrave (2013) and Musgrave and Lutscher (2014b). We begin with operator Q from (15.14). A simple calculation shows that under the periodicity assumption (15.15), this operator maps L-periodic functions into L-periodic functions.

Lemma 15.1 *Let Q be defined by (15.14) and assume that (15.15) holds. Then Q leaves the subspace of L-periodic functions on* $\mathbb{R}$ *invariant.*

Proof Let N be L-periodic. Then

$$
\begin{aligned}
Q[N](x) &= \int K(x, y)F(N(y), y)\mathrm{d}y \\
&= \int K(x+L, y+L)F(N(y+L), y+L)\mathrm{d}y \\
&= \int K(x+L, \hat{y})F(N(\hat{y}), \hat{y})\mathrm{d}\hat{y} = Q[N](x+L), \quad (15.38)
\end{aligned}
$$

where $\hat{y} = y + L$ results from a change of variables.   □

The dynamics of the IDE on the space of all bounded functions on $\mathbb{R}$ are certainly not identical to those on the periodic functions. However, as long as we restrict our investigation to fixed points, we may reduce the spatial domain to a single period.

Lemma 15.2 *If N is an L-periodic fixed point of the operator Q in (15.14) and if (15.15) holds, then N is precisely the L-periodic extension of a fixed point of the operator* Q_L, *defined by*

$$
Q_L[N](x) = \int_0^L \widehat{K}(x, y)F(N(y), y)\mathrm{d}y, \quad (15.39)
$$

where

$$
\widehat{K}(x, y) = \sum_{m\in\mathbb{Z}} K(x, y+mL), \quad x, y \in [0, L], \quad (15.40)
$$

on the space of bounded functions on $[0, L]$. *Conversely, if N is a fixed point of* Q_L, *then its L-periodic extension to* $\mathbb{R}$ *is a fixed point of Q.*

Proof The calculation is straightforward. We split the integral and use Tonelli's theorem so that the definition of $\widehat{K}$ emerges:

$$
\begin{aligned}
Q[N](x) &= \int_{-\infty}^{\infty} K(x, y)F(N(y), y)\mathrm{d}y \\
&= \sum_{m\in\mathbb{Z}} \int_{mL}^{(m+1)L} K(x, y)F(N(y), y)\mathrm{d}y \\
&= \sum_{m\in\mathbb{Z}} \int_0^L K(x, z+mL)F(N(z+mL), z+mL)\mathrm{d}z \\
&= \int_0^L \sum_{m\in\mathbb{Z}} K(x, z+mL)F(N(z), z)\mathrm{d}z = Q_L[N](x). \quad (15.41)
\end{aligned}
$$

This proves the claim.   □

We can now enlarge the function space of Q_L to the space of square-integrable functions on $[0, L]$ and obtain results similar to those in Chap. 13. In particular, we get the existence of a dominant eigenvalue.

Proposition 15.3 *Let the growth function be nonnegative, differentiable in both variables, and L-periodic in the second variable. Let the dispersal kernel be positive and continuous and satisfy the periodicity condition (15.15). Then operator Q_L defined in (15.39) is completely continuous on $L^2[0, L]$. It is Fréchet differentiable at zero, and its derivative is the completely continuous operator*

$$Q'_L[0]\phi(x) = \int_0^L \widehat{K}(x, y)R(y)\phi(y)\mathrm{d}y, \tag{15.42}$$

where $R(x) = \partial F/\partial N(0, x)$.

Under some additional conditions on the growth function, similar to those in Chap. 13, the Fréchet derivative is a superpositive operator and has a dominant eigenvalue with positive eigenfunction. Hence, the stability of the zero solution for Q_L is determined by this dominant eigenvalue, and consequently the persistence of the population described by Q_L is as well. What remains to see is that the behavior of Q_L on the bounded domain also determines the behavior of Q on the unbounded domain. The proof of the following theorem can be found in Sect. 4.3 of Musgrave (2013).

Theorem 15.1 *The trivial solution for Q is stable (locally asymptotically stable) if and only if the trivial solution is stable (locally asymptotically stable) for Q_L.*

If the growth function is also concave down in N for every x, then we obtain the existence of a positive steady state and upward convergence toward this state from low densities if the zero state is locally unstable. This statement follows from the same techniques as in Sect. 13.4. The existence of steady states will be an important ingredient when we consider spreading phenomena in the next section.

15.3   Spread in an Infinite Periodic Habitat

We expect that a species that can grow locally from low density and that disperses to surrounding areas will spread spatially. In a homogeneous landscape, this spread may occur in the form of a traveling wave with constant speed, as we saw in Chaps. 5, 6, and 13. When landscape quality varies, we expect the rate of spatial spread to increase where population growth is high and to slow down where population growth is low, at least if dispersal is independent of landscape quality. If dispersal behavior depends on landscape quality, the relationship between local growth conditions and spread rate could be more complicated. In any case, we cannot expect a population front to move at a constant speed. However, if landscape quality varies periodically in space, we can hope that while the local speed would vary within each period, there could still be an analogue to the asymptotic spreading speed (Chap. 5) that emerges on the scale of the spatial period. Simulations (see below) and theoretical results (Weinberger 2002; Weinberger et al. 2008) show that this is indeed the case, at least under certain conditions.

We begin with a numerical exploration in the spirit of the preceding section. We simulate the model

$$N_{t+1}(x) = \int_{-\infty}^{\infty} \frac{a}{2} e^{-a|x-y|} F(N_t(y), y) \mathrm{d}y , \tag{15.43}$$

where F is piecewise constant with respect to the second argument as in (15.16). We choose F_1 to be the scaled Beverton–Holt function (2.13) and $F_2 = 0$. For simplicity, we set $L = 1$ so that $L_1 = p \leq 1$ represents the percentage of source habitat. We obtain the critical percentage for persistence from formula (15.33). For the parameter values chosen in Fig. 15.6, this critical value is $p^* \approx 0.38$. When $p < p^*$, the initial population can spread from its initial release location only to a few neighboring source patches and dies out quickly. When $p > p^*$, the population spreads. The population density is near zero ahead of the front and exhibits a periodic pattern behind the front.

It is tempting to try to define a speed for this spreading process by considering the time that it takes the profile to move by one period. In reaction–diffusion models for spatially periodic habitats, this approach is valid (Shigesada et al. 1986). In our setting, however, we are limited to discrete time points, and the population density in one generation may never be an exact translation of a previous generation. If it were, speed and landscape period would have to be rationally related. Instead, Kawasaki and Shigesada (2007) define an "instantaneous" and an "averaged" speed.

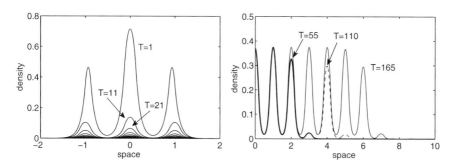

Fig. 15.6 Simulation of IDE (15.43) with Beverton–Holt growth in good patches and hostile bad patches. **Left:** When $L_1 = 0.25$, persistence condition (15.33) is not satisfied and the population declines to zero. **Right:** When $L_1 = 0.7$, the persistence condition is satisfied and the population spreads (only the positive x-direction is shown). Parameters are $R = 1.5$, $a = 10$, and $L = 1$. The initial profile is $N_0 = \chi_{[-1,1]}$. The density is plotted every 10 generations (left plot) and every 55 generations (right plot).

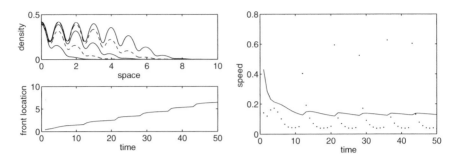

Fig. 15.7 **Left:** Spreading population front, plotted every 10 generations (top), and corresponding front location (bottom). **Right:** Instantaneous (dots) and average (curve) speed. Dispersal kernel and growth function are as in Fig. 15.6. The initial density is positive for all $x < 0$. Parameters are $R = 1.5$, $a = 5$, and $L_1 = 0.7$.

The location of the front, denoted as x_t, is defined as the rightmost point where $N_t(x_t) = \tilde{N}$ for some threshold value $\tilde{N}$; compare Sect. 5.2 and Fig. 5.1. The instantaneous speed from generation t to generation $t + 1$ is then $x_{t+1} - x_t$. The average speed up to generation t is the mean of the instantaneous speeds up to that time; i.e.,

$$c_t = \frac{1}{t} \sum_{t'=1}^{t} (x_{t'+1} - x_{t'}) = \frac{x_t}{t}.\tag{15.44}$$

This definition assumes that the initial population front is at the origin so that $x_0 = 0$. In Fig. 15.7, we plot snapshots of a spreading population and the location of its front as a function of time, as well as the instantaneous and the average speed. From the simulations it appears that the limit $\bar{c} = \lim_{t \to \infty} c_t$ exists. This would be a candidate for the asymptotic spreading speed in periodic landscapes.

Heuristic Calculation of the Spreading Speed

We begin with the explicit example from Kawasaki and Shigesada (2007), who study the linear IDE with Laplace dispersal kernel as in (15.17) and with piecewise-constant growth function $R(x) = R_i \geq 0$ in patch type i. By scaling space, we may set the dispersal parameter $a = 1$ to simplify notation. Even though we cannot expect densities at subsequent time steps to be spatial translations of one another, it turns out to be helpful to emulate the ansatz from the continuous-time case and set $N_t = \exp(-s(x - ct))\phi(x)$ with some L-periodic function ϕ (Shigesada et al. 1986; Kawasaki and Shigesada 2007). Substituting this ansatz into (15.17), we obtain the relation

$$e^{sc} N_t(x) = \int_{-\infty}^{\infty} \frac{1}{2} e^{-|x-y|} R(y) N_t(y) dy . \tag{15.45}$$

As we have seen several times already, the Laplace kernel allows us to turn the integral equation into a differential equation. By repeated differentiation, we obtain

$$N_t''(x) = (1 - e^{-sc} R(x)) N_t(x) . \tag{15.46}$$

After substituting our ansatz, we find the corresponding equation for ϕ as

$$\phi'' - 2s\phi' + (s^2 - 1 + e^{-sc} R(x))\phi = 0 . \tag{15.47}$$

As in the previous section, we need to find conditions for the existence of a nontrivial eigenfunction ϕ. Since ϕ needs to be L-periodic, we have again a Hill's equation as in (15.19). Unlike in the previous case, the equation here is not symmetric with respect to $x \mapsto -x$. Hence, we cannot reduce the equation to half a domain period as we did in the preceding section. In this case, notation is simplified by choosing good and bad patches in the L-periodic landscape to be located at $[-L_1, 0]$ and $[0, L_2]$. Hence, we study the equation

$$\phi'' - 2s\phi' + (s^2 - 1 + e^{-sc} R_i)\phi = 0 \quad \text{for} \quad \begin{cases} x \in [-L_1, 0), \ i = 1 , \\ x \in [0, L_2), \quad i = 2 , \end{cases} \tag{15.48}$$

and periodicity conditions

$$\lim_{x \to 0^+} \phi(x) = \lim_{x \to 0^-} \phi(x) , \quad \lim_{x \to -L_1^+} \phi(x) = \lim_{x \to L_2^-} \phi(x) , \tag{15.49}$$

and similarly for ϕ'. Superscripts $\pm$ denote one-sided limits from the right and left, respectively.

We write the solution of the characteristic equation of (15.48) as

$$s \pm q_i = s \pm \sqrt{1 - e^{-sc} R_i} \tag{15.50}$$

and the general solution of the equation as

$$\phi(x) = \begin{cases} A_1 e^{(s+q_1)x} + B_1 e^{(s-q_1)x} , & x \in [-L_1, 0) , \\ A_2 e^{(s+q_2)x} + B_2 e^{(s-q_2)x} , & x \in [0, L_2) . \end{cases} \tag{15.51}$$

The periodic matching conditions give us relations between the coefficients. For example, at $x = 0$ we require $A_1 + B_1 = A_2 + B_2$. Altogether, we can write the conditions in matrix form as

$$\begin{bmatrix} 1 & 1 & -1 & -1 \\ e^{-(s+q_1)L_1} & e^{-(s-q_1)L_1} & -e^{-(s+q_2)L_2} & -e^{-(s-q_2)L_2} \\ q_1 & -q_1 & -q_2 & q_2 \\ q_1 e^{-(s+q_1)L_1} & -q_1 e^{-(s-q_1)L_1} & -q_2 e^{-(s+q_2)L_2} & q_2 e^{-(s-q_2)L_2} \end{bmatrix} \begin{bmatrix} A_1 \\ B_1 \\ A_2 \\ B_2 \end{bmatrix} = 0 .$$

(15.52)

As in the previous section, we are looking for a nonzero solution, which requires the determinant of the matrix to vanish. This requirement establishes a relation between the speed, c, and the steepness, s, of the profile. Tedious computations give this *dispersion relation* as (Kawasaki and Shigesada 2007)

$$\cosh(sL) = \cosh(q_1 L_1)\cosh(q_2 L_2) + \frac{q_1^2 + q_2^2}{2q_1 q_2}\sinh(q_1 L_1)\sinh(q_2 L_2) ,$$

(15.53)

where speed c is implicit in the definition of q_i in (15.50).

The dispersion relation defines a function $c = c(s)$. If c is real valued for some positive s, then it is a candidate for the frontal speed. As in the homogeneous landscape, we expect that the smallest such speed,

$$\hat{c} = \min_{s>0} c(s) ,$$

(15.54)

is the one that is actually realized from compactly supported initial data; see Chap. 5 and formula (5.17). The theory below will confirm this expectation.

Figure 15.8 shows the dispersion relation as well as the speed of spread according to (15.54) as a function of the length of a good patch. When the good patch is so small that persistence is impossible according to the condition in (15.33), the speed is zero. When persistence is possible, the speed is positive, and it is an increasing function of the length of good patches.

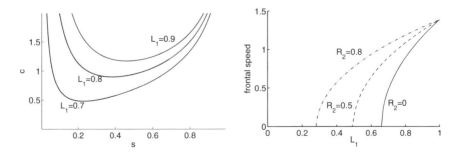

Fig. 15.8 Left: The dispersion relation for the frontal speed in a periodic landscape according to (15.53). **Right:** The minimal speed from (15.54) as a function of the length of good patches. Parameters are $R_1 = 1.5$, $R_2 = 0$, and $L = 1$, unless otherwise noted.

Spread Speed Theory

The first results on the existence of an asymptotic spreading speed and periodic traveling waves for IDEs with spatial periodicity are due to Weinberger (2002), who generalizes his earlier work in Weinberger (1982). The field has since expanded considerably (Weinberger et al. 2008; Liang and Zhao 2010). More and earlier results exist for continuous-time models with periodic spatial variation. We refer the reader to Berestycki and Hamel (2002) and the references therein.

It is not surprising that we have to distinguish cases according to whether the dynamics are monotone, nonmonotone, monostable, bistable, or their various combinations, just as we did in homogeneous landscapes; see Chaps. 5, 6, and 11. It does, however, take a bit longer to properly formulate these conditions when the landscape is heterogeneous. Instead of (pointwise) conditions on the growth function, we now require conditions on the IDE operator on one period of the landscape. We formulate the simplest case of monotone, monostable dynamics here, adapted from Weinberger (2002), and illustrate other cases with simulations and references to the literature.

We consider the spatially periodic IDE

$$N_{t+1}(x) = Q[N_t](x) = \int_{-\infty}^{\infty} K(x, y) F(N_t(y), y) \mathrm{d}y , \qquad (15.55)$$

where $F(\cdot, y + L) = F(\cdot, y)$ and $K(x + L, y + L) = K(x, y)$. Under these assumptions, Q commutes with translations of multiples of L.

Since we always assume that the growth function is zero when the density is zero, Q also has zero as a steady state. We now assume that there is a unique positive periodic steady state N^* of Q; i.e., $Q[N^*] = N^*$. We define the set

$$\mathscr{C}_{N^*} = \{N | N \text{ is continuous on } \mathbb{R} \text{ and } 0 \leq N \leq N^*\} . \qquad (15.56)$$

With this notation, we formulate the theorem in the monotone monostable case.

Theorem 15.2 (Weinberger 2002) *Assume that Q has the following properties:*

(i) $Q[N(\cdot - L)](x) = Q[N](x - L)$.
(ii) $Q[0] = 0$, *and there is exactly one positive L-periodic equilibrium* $Q[N^*] = N^*$.
(iii) Q *leaves* $\mathscr{C}_{N^*}$ *invariant and is order preserving; i.e., if* $0 \leq N(x) \leq \tilde{N}(x) \leq N^*(x)$, *then* $0 \leq Q[N](x) \leq Q[\tilde{N}](x) \leq N^*(x)$.
(iv) *0 is unstable and N^* is stable in the following sense: if N_0 is L-periodic and* $0 < N_0 \leq N^*$, *then the recursion $N_{t+1} = Q[N_t]$ converges to N^* uniformly.*
(v) Q *is continuous on* $\mathscr{C}_{N^*}$ *with respect to uniform convergence on bounded sets.*
(vi) Q *is compact on* $\mathscr{C}_{N^*}$.

Then Q has a spreading speed $c^* \in (-\infty, \infty]$ in the following sense:

1. *If $c^* < \infty$, $0 \le N_0 \le N^*$, and $N_0(x) = 0$ for $x \ge 0$, then for every $c > c^*$,*

$$\limsup_{t \to \infty} \left[\sup_{x > ct} N_t(x) \right] = 0 \,. \tag{15.57}$$

2. *If $0 \le N_0 \le N^*$ and there is a constant $k > 0$ such that $N_0(x) > 0$ for $x < -k$, then for every $c < c^*$,*

$$\lim_{t \to \infty} \left[\sup_{x < ct} (N^*(x) - N_t(x)) \right] = 0 \,. \tag{15.58}$$

The theorem says that if the initial density is zero on the positive half-line and positive on some negative half-line, then an observer who travels faster than the asymptotic speed will eventually see no population, while an observer who travels slower than this speed will eventually see the population at its positive periodic steady state.

The theorem as stated considers only one direction. The corresponding results in the opposite direction follow from the coordinate change $x \mapsto -x$. The speeds in the two directions could differ if the dispersal process is not symmetric. In particular, if there is a strong directional bias in the dispersal process, then the spreading speed could be negative; see Sect. 12.2.

The definition of the spreading speed here also differs slightly from the one in Definition 5.1 in that the initial condition here is not compactly supported. It is this version of the definition that extends to the case with Allee effect (see below). However, the statement also holds for compactly supported initial data, as stated in Theorems 2.2 and 2.3 by Weinberger (2002). The theory there is formulated more generally for directional spreading speeds and "ray speeds" in any spatial dimension, which makes the statement of the theorem more tedious.

Under the concavity conditions discussed at the end of Sect. 15.2, the assumptions of uniqueness of the positive steady state and of upward convergence in Theorem 15.2 are satisfied. Hence, we obtain the existence of a spreading speed, but the theorem does not give a concrete formula for it. If Q has a linearization at zero and satisfies certain properties, such a formula exists.

Theorem 15.3 (Weinberger 2002) *Suppose that Q satisfies the conditions of Theorem 15.2. Suppose also that Q has a linearization at zero, denoted by $Q'[0] = Q'_0$, that has the following properties:*

(i) *$Q[N] \le Q'_0[N]$ for all $0 \le N \le N^*$.*
(ii) *$Q[N] \ge (1 - \delta) Q'_0[N]$ for some small $\delta > 0$ when $N \ge 0$ is small.*
(iii) *Q'_0 is L-periodic and strongly order preserving, and $Q'_0[e^{s|x|}]$ exists for all s.*
(iv) *There is a positive, L-periodic function $\underline{N}$ such that $Q'_0[\underline{N}] > \underline{N}$.*
(v) *The truncated operator $N \mapsto \min\{Q'_0[N], \underline{N}\}$ satisfies the assumptions in Theorem 15.2.*

Then the asymptotic spreading speed of Q can be calculated as

$$c^* = \hat{c} = \inf_{s>0} \left(\frac{1}{s} \lambda(s) \right), \qquad (15.59)$$

where λ is the dominant eigenvalue of the equation

$$\lambda \phi(y) = e^{sy} Q_0'[e^{-sx} \phi(x)](y). \qquad (15.60)$$

In particular, the spreading speed is linearly determined.

The above two theorems are stated for continuous functions, but their validity has been extended to the case of lower semicontinuous functions in Weinberger et al. (2008). In particular, the statements apply to the scenario of piecewise-constant growth functions that we used to motivate and illustrate the results. We close this section by considering appropriate analogues to traveling waves (see Chaps. 5 and 10), known as "traveling periodic waves" (Shigesada et al. 1986), "periodic traveling waves" (Weinberger 2002), or "pulsating waves" (Berestycki and Hamel 2002); see Fig. 15.6 (right plot) and Fig. 15.7 (top left plot).

Definition 15.1 A solution, N_t, of IDE (15.55) is called a *traveling periodic wave* of speed c if it has the form $N_t(x) = W(x - ct, x)$, where $W(s, x)$ has the following properties:

1. For each s, the function $W(x + s, x)$ is continuous in x.
2. For each s, the function $W(s, x)$ is L-periodic in x.
3. For each s, the function $W(s, x)$ is nonincreasing in s.
4. $W(-\infty, x) = N^*(x)$ and $W(\infty, x) = 0$.

Theorem 15.4 (Weinberger 2002) *Under the assumptions of Theorem 15.2, IDE (15.55) has a traveling periodic wave of speed c if and only if $c \geq c^*$.*

Overcompensation

The theory in the preceding section requires the operator Q to be monotone, which means that it does not apply to models with overcompensation, such as the Ricker or the logistic growth function. In a homogeneous landscape, we can still define a spreading speed when the growth function shows overcompensation and show that this speed is still linearly determined under some conditions (Chap. 5). Weinberger et al. (2008) extend both the theory for the monotone periodic case in Weinberger (2002) and the nonmonotone homogeneous case in Hsu and Zhao (2008) and Li et al. (2009), to the nonmonotone periodic case. In particular, the construction of monotone operators $Q^{\pm}$ following (5.35) that form an upper and lower bound of the operator Q is still valid. Furthermore, the spreading speed is linearly determined under the conditions that we can expect from the homogeneous case. We illustrate the ideas and results via numerical simulations.

We consider the IDE with Ricker growth function and spatially periodic, piecewise-constant parameter function

$$N_{t+1}(x) = Q[N_t](x) = \int_{-\infty}^{\infty} \frac{a}{2} e^{-a|x-y|} R(y) N_t(y) e^{-N_t(y)} dy, \qquad (15.61)$$

where $R(x) = R_i$ as in (15.16). The monotone upper and lower bounds defined in (5.35) for the growth function can be written explicitly as

$$F^+(N) = \begin{cases} N R e^{-N}, & N \le 1, \\ R/e, & N > 1, \end{cases} \quad \text{and} \quad F^-(N) = \begin{cases} N R e^{-N}, & N \le R/e, \\ e^{-R/e} R^2/e, & N > R/e. \end{cases}$$
$$(15.62)$$

We denote by $N_\pm^*$ the positive steady states of the corresponding operators $Q^\pm$ and by N^* a positive steady state of Q. Proving the uniqueness of the positive steady state for Q in general is quite difficult and depends on properties of the dispersal kernel (Weinberger et al. 2008).

When parameters R_i are small enough, Q is monotone for functions bounded between zero and $N^*(x)$, and $N_-^* = N_+^* = N^*$. Hence, the theory from the preceding section applies. There is an asymptotic spreading speed, it is given by the linearization at zero, and there is a traveling periodic wave for each speed that is at least as large as the asymptotic speed. The top panel in Fig. 15.9 illustrates how the invading front very quickly approaches the periodic steady state.

When at least one of the R_i is large enough, Q is not monotone. Three ordered steady states, $N_-^* < N^* < N_+^*$, emerge. The density of an invading front approaches the steady state N^*, and it stays below N^* at all times (middle panel of Fig. 15.9). When the growth rate(s) increase even further, the density at the front overshoots the steady-state density, then dips below it again, and eventually approaches it (bottom panel of Fig. 15.9)

With even higher growth rate(s), N^* becomes unstable in what seems to be a flip bifurcation. This behavior is expected since the nonspatial model also has such a bifurcation; however, there have been no analytical results on the periodic case so far. Weinberger et al. (2008) show numerical examples of a two-cycle and, eventually, chaotic behavior. We saw in Chap. 11 that in a homogeneous landscape, a spreading population may exhibit two phases during the invasion process with overcompensation. At first, a nonmonotone wave spreads with the asymptotic spreading speed and leaves in its wake the (unstable) steady-state profile. Then, a secondary wave connects this state with the (stable) two-cycle profile, but at a slower speed. We present some numerical evidence that the same phenomenon arises in the periodic landscape as well.

Figure 15.10 shows that the approach to the (unstable) steady state takes many spatial periods, during which the density varies somewhat randomly between periods (top panel). The middle and lower panels show the density of the spreading population at two subsequent generations. We recognize the secondary wave that replaces the (unstable) steady-state profile with the periodic two-cycle profile. This wave moves at a much slower speed than the first profile. By generation 200, it has

Fig. 15.9 Numerical simulation of IDE (15.61) with spatially periodic piecewise-constant Ricker growth function. The thick curve shows the density of the invading population front. The dashed and dash-dot curves correspond to the upper and lower steady states, N_+^* and N_-^*. The thin curve is the steady state N^*. Dispersal parameters are $L = 1$, $L_1 = 0.7$, and $a = 5$. Growth parameters are $R_2 = 0.9$ and $R_1 = 1.5$ (top), $R_1 = 5$ (middle), and $R_1 = 7$ (bottom). The initial condition for the invasion front is unity for $x < -10$ and zero for $x \geq -10$.

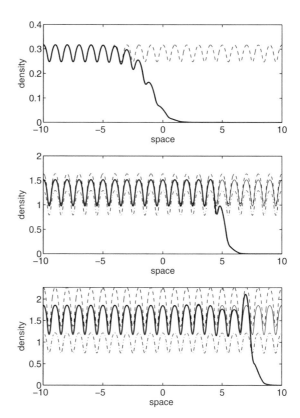

barely reached the location $x = 0$, whereas the first profile passed that point in generation 25.

The steady states $N_\pm^*$, N^* in all our numerical examples have the shape of sinusoidal functions but can have much more complicated profiles in general. Weinberger et al. (2008) show examples where N^* has two maxima in each spatial period, located near the boundary of the good patch and not in the center. We saw similar shapes emerge as steady states on a single patch with Ricker dynamics in Fig. 4.5. Much more numerical and analytical work is necessary to understand the possible shape of periodic steady-state profiles and the convergence of spreading profiles toward them.

When the periodic steady state is unstable, the following modification helps visualize it numerically (Weinberger et al. 2008). Instead of IDE (15.14), we iterate

$$N_{t+1}(x) = (1 - \rho)N_t(x) + \rho Q[n_t](x) \tag{15.63}$$

for some $0 < \rho < 1$. Clearly, the steady states of the two equations are the same. If ϕ is an eigenfunction of Q' with eigenvalue λ, then ϕ is an eigenfunction of the

Fig. 15.10 Numerical simulation of IDE (15.61) with spatially periodic piecewise-constant Ricker growth function. The setup and parameters are as in Fig. 15.9, except here $R_1 = 9.5$. The positive steady state is unstable. **Top:** The first front at generation 25, where the density increases from zero to the unstable state over many periods. **Middle and bottom:** The second front at generations 199 and 200, respectively, where the positive unstable state is replaced by a two-cycle that moves at a much slower speed.

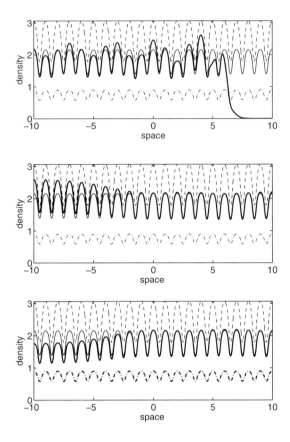

modified IDE with eigenvalue $\rho\lambda$. Hence, by choosing ρ small enough, we can stabilize the steady state. However, convergence to this state may be very slow.

Allee Effect

In a homogeneous landscape, a population with Allee effect may spread or retreat in the form of a traveling wave; see Chap. 6. For specifically chosen parameter values, the population may also form a standing wave (cline); see Theorem 6.1. In a heterogeneous landscape, new phenomena arise; e.g., a "standing periodic wave" (a traveling periodic wave with speed zero) occurs for a much larger set of parameters. There are only a few publications that study population spread with Allee effect in heterogeneous landscapes (Weinberger 2002; Dewhirst and Lutscher 2009; Musgrave et al. 2015). We summarize and illustrate their most important findings here.

The analytical results by Weinberger (2002) include the case of an Allee effect to the extent possible. In particular, the following version of Theorem 15.2 clarifies the existence of a spreading speed.

Theorem 15.5 (Weinberger 2002) *Assume that Q satisfies the assumptions of Theorem 15.2, except that assumptions (ii) and (iv) are replaced by, respectively, the following:*

(ii') *$Q[0] = 0$, and there are exactly two positive L-periodic equilibria $0 \leq N_a^* < N^*$.*

(iv') *N_a^* is unstable and N^* is stable in the following sense: if N_0 is L-periodic and $N_a^* < N_0 \leq N^*$, then the recursion $N_{t+1} = Q[N_t]$ converges to N^* uniformly.*

Then there exists a spreading speed $c^* \in (-\infty, \infty]$ of Q in the following sense:

1. *If $c^* < \infty$, $0 \leq N_0 \leq N^*$, and $N_0(x) = 0$ for $x \geq 0$, then for every $c > c^*$,*

$$\limsup_{t \to \infty} \left[\sup_{x > ct} (N_t(x) - N_a^*) \right] \leq 0. \tag{15.64}$$

2. *If $0 \leq N_0 \leq N^*$ and there is a constant $k > 0$ such that $N_0(x) > N_a^*$ for $x < -k$, then for every $c < c^*$,*

$$\lim_{t \to \infty} \left[\sup_{x < ct} (N^*(x) - N_t(x)) \right] = 0. \tag{15.65}$$

Steady state N_a^* corresponds to a spatially periodic Allee threshold, but it is not the pointwise Allee threshold of the growth function, because it includes the effects of dispersal. There are no particular assumptions of the theorem with regard to the stability of the zero state. For that reason, the claim is not that the population density approaches zero far ahead of the front, but only that it remains below the spatial Allee threshold. Likewise, since the spreading speed can be negative, it is in general not sufficient for the population to exceed the spatial Allee threshold on a bounded set.

We illustrate how a decrease in the available amount of good habitat can lead from a spreading population to a stalled population and eventually a retreating population. Our model equation is IDE (15.14) with a growth function that is periodic and piecewise constant in space, as in (15.16). As the simplest meaningful case, we choose F to be the Allee growth function (2.22) with $\gamma = 2$ in (good) patches of type 1, and set $F = 0$ in (bad) patches of type 2. We use the Laplace dispersal kernel and scale the period to $L = 1$ so that L_1 denotes the percentage of good habitat.

The plots in Fig. 15.11 show that the population can spread when the fraction of good habitat is large enough (top panel), but stalls (middle panel) or retreats (bottom panel) if the fraction is too small. The population density is plotted every 20 generations. In the middle panel, there is no visible change between generations.

Fig. 15.11 Numerical
simulation of IDE (15.61)
with Allee growth function in
good patches and hostile bad
patches. **Top:** The population
spreads when $L_1 = 0.9$.
Middle: The population stalls
when $L_1 = 0.8$. **Bottom:** The
population retreats when
$L_1 = 0.65$. The dashed curve
indicates the steady state
(invading front) with growth
function F^+; see text for
details. Parameters are $L = 1$,
$R = 6$, and $a = 5$. The
population density is plotted
for $t = 20, 40, 60, 80, 100$. In
the middle panel, these five
curves are practically
indistinguishable. The initial
condition is zero for $x > 0$
and one for $x < 0$.

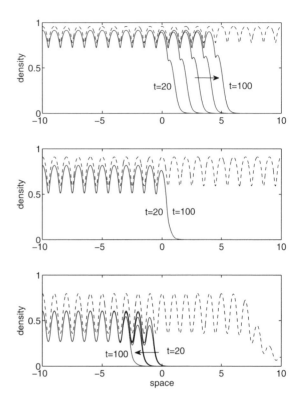

We observe a standing periodic wave, or cline, which is a nonperiodic steady state
of Q. The bottom panel shows the slow retreat. The dashed curves in Fig. 15.11
indicate the solution at generation 100 with the monotone and concave upper bound
F^+ of the Allee function F; see (6.31). For the two upper panels, this curve is close
to the steady state. In the lower panel, the population spreads to the right with growth
function F^+ but retreats to the left with Allee growth function F. When the fraction
of good habitat is very small, the population will decline uniformly in space (plot
not shown).

The standing periodic wave arises when (1) the distance from one good patch to
the next is so large that the density of individuals that can arrive in that patch does
not exceed the Allee threshold, and (2) the population can persist on an isolated
single patch (from a sufficiently high initial density). We calculate these thresholds
in a particular case. In a good patch of length L_1, we choose the piecewise-constant
growth function from (6.1) with Allee threshold N_a; in a bad patch of length L_2,
we set the growth function to zero. We choose the Laplace dispersal kernel with
parameter a. We consider the good and bad patches in an L-periodic landscape to
be located at $[-L_1, 0)$ and $[0, L_2)$, respectively.

We find conditions under which a population that is established for $x < 0$ will
persist. By the growth function, the pre-dispersal density is one in occupied patches

and zero in bad patches. If the post-dispersal density at zero exceeds the Allee threshold, the population will not retreat. More formally, we set $N(x) = 1$ for $x \in \cup_{n=0}^{\infty}[-nL - L_1, -nL)$ and zero elsewhere. The condition for persistence is then $Q[N](0) > N_a$. We can calculate this condition explicitly as

$$Q[N](0) = \sum_{n=0}^{\infty} \int_{-nL-L_1}^{-nL} \frac{a}{2} e^{ay} dy$$

$$= \frac{1}{2} \sum_{n=0}^{\infty} \left(e^{-anL} - e^{-anL-aL_1} \right) = \frac{1 - e^{-aL_1}}{2(1 - e^{-aL})}. \qquad (15.66)$$

Hence, $Q[N](0) > N_a$ if and only if

$$L_1 > L_1^* = -\frac{1}{a} \ln \left(1 - 2N_a \left(1 - e^{-aL} \right) \right). \qquad (15.67)$$

As $L \to \infty$, we obtain the persistence threshold for a single isolated patch, which is exactly the critical spatial extent that we calculated in Chap. 6 after formula (6.8).

With the same setup, the population can spread if $Q[N](L_2) > N_a$. This condition can also be evaluated explicitly to get $L_2 < L - L_1^{**}$ with

$$L_1^{**} = \frac{1}{a} \ln \left(1 + 2N_a \left(e^{aL} - 1 \right) \right). \qquad (15.68)$$

Finally, we calculate conditions under which the positive periodic state exists and is stable. In our setup, we require $Q[1](0) > N_a$; i.e., if the pre-dispersal density is equal to one on all good patches, then the next-generation density must exceed the Allee threshold at the boundary of a good patch. The calculations give

$$Q[N](0) = \frac{\sinh(aL_1)}{1 - e^{-aL}}, \quad \text{so that} \quad L_1^{***} = \frac{1}{a} \arcsin \left(N_a \left(1 - e^{-aL} \right) \right)$$

$$(15.69)$$

is the threshold value for the positive periodic state.

We illustrate the three threshold values for L_1 in Fig. 15.12. Below the solid curve, there is no positive periodic state. The population collapses. Between the solid and the dashed curves, the positive state exists, but a population that only occupies the region $x < 0$ will retreat. Between the dashed and dash-dot curves, the population will persist in a standing wave but not spread. Finally, above the dash-dot curve, the population will spread.

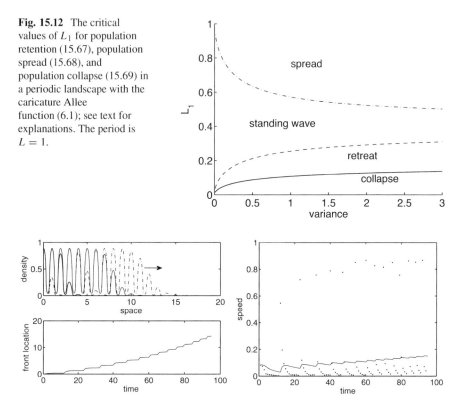

Fig. 15.12 The critical values of L_1 for population retention (15.67), population spread (15.68), and population collapse (15.69) in a periodic landscape with the caricature Allee function (6.1); see text for explanations. The period is $L = 1$.

Fig. 15.13 Accelerating front in the IDE with a (heavy tailed) exponential square root kernel (5.37) with $a = 10$ in a fragmented landscape; compare Fig. 15.7. The growth function is the Beverton–Holt function with $R = 1.5$ in good patches and $R = 0$ in bad patches. Other parameters are $L_1 = 0.6$ and $L = 1$. The density in the top left panel is plotted every 20 generations.

Heavy-Tailed Kernels

In Chap. 5, we saw that when the dispersal kernel is not exponentially bounded (heavy tailed), we may observe accelerating invasion fronts in a homogeneous landscape. There are currently no systematic studies on how landscape heterogeneity may interact with heavy-tailed kernels. Simulations by Dewhirst and Lutscher (2009) indicate that while fragmentation slows the overall speed, it does not prevent the advancing front from accelerating.

Figure 15.13 illustrates such an accelerating front with the exponential square root kernel (5.37). The front location grows faster than linearly in time, and the average frontal speed appears to follow a linear increase.

15.4   Approximations of Spread

The theory so far gives implicit expressions for persistence conditions and spread rates in a special case (namely using the Laplace kernel and a piecewise-constant periodic growth function); for the general case, it gives abstract results, which can be tricky to calculate even numerically. Several authors considered approximations to simplify the calculations involved with periodic, heterogeneous habitats. We present two of these approaches in some detail.

Large Dispersal Scale

When the scale of dispersal is significantly larger than the scale of habitat heterogeneity, it seems plausible that the population dynamics depend not so much on the exact spatial landscape configuration as on some suitably averaged quantities. For reaction–diffusion equations, spatial averaging ("homogenization") is a well-developed and highly useful tool (Othmer 1983; Pavliotis and Stuart 2008). The first application to IDEs was developed by Dewhirst and Lutscher (2009). They use the linear IDE

$$N_{t+1}(x) = \int_{-\infty}^{\infty} K(x-y; y)R(y)N_t(y)dy, \tag{15.70}$$

where the second argument in K indicates that kernel parameters such as the variance may depend on the location of origin of the dispersing organisms. Various mechanisms could create such a dependence. For example, if individuals grow up in a resource-poor environment, they simply may not have the energy to disperse long distances, so that the variance of their dispersal kernel could be small. On the other hand, individuals from resource-poor locations may try to disperse as far as they possibly can to increase their chances of ending up in resource-rich locations. In that case, the variance of their dispersal kernel could be large in resource-poor areas.

We assume that R is L-periodic and that K is L-periodic in its second argument. As before, we construct an exponential ansatz for a periodic traveling wave, $N_{t+1}(x) = N_t(x-c) = \exp(-s(x-c))\phi(x)$, with a periodic function ϕ. With this ansatz, we obtain the eigenvalue problem

$$e^{sc}\phi(x) = \int_{-\infty}^{\infty} K(x-y; y)e^{s(x-y)}R(y)\phi(y)dy \tag{15.71}$$

as before. The scaling $x = Lw$ and $y = Lz$ leads to

$$e^{sc}\hat{\phi}(w) = \int_{-\infty}^{\infty} L\hat{K}(L(w-z); z)e^{sL(w-z)}\hat{R}(z)\hat{\phi}(z)dz, \tag{15.72}$$

where $\hat{\phi}(z) = \phi(Lz)$, $\hat{R}(z) = R(Lz)$, and $\hat{K}(\cdot; z) = K(\cdot; Lz)$. Since all functions are now 1-periodic, we split the integral into a sum of integrals of length one. The equation becomes

$$e^{sc}\hat{\phi}(w) = \int_0^1 \left[\sum_{n=-\infty}^{\infty} L\hat{K}(L(w - z - n); z)e^{sL(w-z-n)} \right] \hat{R}(z)\hat{\phi}(z)dz .$$

$$(15.73)$$

The sum in brackets is the Riemann-sum approximation of the integral

$$\sum_{n=-\infty}^{\infty} L\hat{K}(L(w - z - n); z)e^{sL(w-z-n)} \approx \int_{-\infty}^{\infty} \hat{K}(v; z)e^{sv}dv = M(s; z) .$$

$$(15.74)$$

Hence, in the limit as $L \to 0$, we obtain the moment-generating function. Then we can write (15.73) as

$$e^{sc}\hat{\phi}(w) = \int_0^1 M(s; z)R(z)\hat{\phi}(z)dz .$$ $$(15.75)$$

The right-hand side is now independent of w, so that the left-hand side has to be independent as well. But then $\hat{\phi}$ is a constant and can be canceled from the equation. We are left with the relation

$$e^{sc} \approx \int_0^1 \hat{R}(z)M(s; z)dz .$$ $$(15.76)$$

Hence, we arrive at the minimal speed of a traveling wave as

$$\hat{c} \approx \inf_{s>0} \frac{1}{s} \ln \left(\int_0^1 \hat{R}(z)M(s; z)dz \right) .$$ $$(15.77)$$

For the case of a piecewise-constant growth function of two periodically alternating values, the formula becomes

$$\hat{c} \approx \inf_{s>0} \frac{1}{s} \ln \left(p R_1 M_1(s) + (1 - p)R_2 M_2(s) \right) ,$$ $$(15.78)$$

where p is the proportion of landscape where $\hat{R}(z) = R_1$ and M_i denotes the moment-generating function of the dispersal kernel from patch type i.

To compare this approximate explicit formula with the exact implicit relation in (15.53), we choose $K(x - y; y)$ to be a (scaled) Laplace kernel ($a = 1$), independent of the second argument. Then $M_1(s) = M_2(s) = (1 - s^2)^{-1}$, so that the preceding formula gives

$$\hat{c} \approx \inf_{s>0} \frac{1}{s} \ln \left(\frac{pR_1 + (1-p)R_2}{1-s^2} \right).$$ (15.79)

We can derive the same expression from the exact dispersion relation in (15.53) as follows. When L is small, then so are $L_{1,2}$. We expand the hyperbolic functions as $\sinh(x) \approx x$ and $\cosh(x) \approx 1 + x^2/2$. We obtain

$$1 + \frac{(sL)^2}{2} \approx \left(1 + \frac{(q_1 L_1^2)}{2} \right) \left(1 + \frac{(q_2 L_2^2)}{2} \right) + \frac{q_1^2 + q_2^2}{2} L_1 L_2.$$ (15.80)

After expanding, this expression becomes

$$e^{-sc}[R_1 L_1^2 + R_2 L_2^2 + (R_1 + R_2)L_1 L_2] \approx (1-s^2)L^2 + O(L_1^2 L_2^2).$$ (15.81)

Since the $O(L_1^2 L_2^2)$-term is small compared to all the quadratic terms, we neglect it. Then we divide by L^2 and sort to get

$$e^{sc} = \frac{R_1 L_1 + R_2 L_2}{L(1-s^2)},$$ (15.82)

which, with $p = L_1/L$ and $1 - p = L_2/L$, leads to the expression in (15.79).

The formula in (15.79) also provides a rule of thumb for the minimal fraction of good habitat required for population persistence and spread. The numerator in the expression must exceed unity for $\hat{c}$ to be positive. If we assume that $R_1 > 1 > R_2$, then the minimal percentage of type-1 habitat for persistence and spread is

$$p^* = \frac{1 - R_2}{R_1 - R_2}.$$ (15.83)

Comparison with numerical simulations reveals that this approximation works very well for the Gaussian kernel, even if the dispersal scale is comparable to the landscape scale; see Fig. 15.14, left panel. For the Laplace kernel, the approximation is less accurate near the persistence threshold, p^*, if the dispersal is relatively large; see Fig. 15.14, right panel.

Collingham and Huntley (2000) simulated the spread rate of tree species on a discrete lattice representing a fragmented landscape of favorable and hostile patches. They found a minimum fraction of 10–40% for the persistence and spread of the population with a growth rate of R between 1.02 and 1.09 in favorable patches. With these values, the persistence threshold (15.83) is almost 100%, which is clearly not realistic. However, Collingham and Huntley (2000) simulated only long-distance dispersal between lattice points; most individuals did not disperse outside their patch. To match their simulations, we need to include a proportion of sessile individuals as in Sect. 12.4, e.g., model (12.54) on $\Omega = \mathbb{R}$, but with spatial heterogeneity. The linearized model is

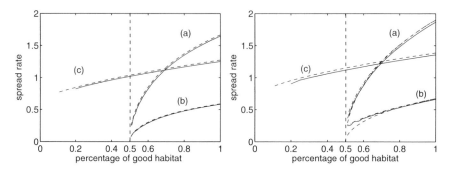

Fig. 15.14 Comparison of the spreading speed obtained from averaging in (15.78) (dashed) with simulation results (solid) for the Gaussian kernel (left panel) and with the exact value from (15.53) (dash-dot) for the Laplace kernel (right panel). Parameters are (a) $R_1 = 2$, $R_2 = 0$, and $\sigma_{1,2}^2 = 2$; (b) $R_1 = 2$, $R_2 = 0$, and $\sigma_{1,2}^2 = 0.25$; and (c) $R_{1,2} = 1.5$, $\sigma_1^2 = 2$, and $\sigma_2^2 = 0.25$. Plots adapted from Dewhirst and Lutscher (2009).

$$N_{t+1}(x) = gR(x)N_t(x) + (1-g)\int_{-\infty}^{\infty} K(x-y; y)R(y)N_t(y)\mathrm{d}y. \tag{15.84}$$

We can still construct the exponential ansatz for a traveling periodic wave and the Riemann-sum approximation to arrive at the analogous equation to (15.75), namely

$$e^{sc}\hat{\phi}(w) = g\hat{R}(w)\hat{\phi}(w) + (1-g)\int_0^1 M(s; z)\hat{R}(z)\hat{\phi}(z)\mathrm{d}z. \tag{15.85}$$

However, the previous argument that $\hat{\phi}$ has to be a constant does not apply at this point. Under the assumption that $\hat{R}$ is piecewise constant (and that $\hat{R}_2 = 0$ for simplicity), Dewhirst and Lutscher (2009) proceed as follows. They write $\hat{\phi}_1(w) = \hat{\phi}(w)$ for $0 \le w < p = L_1/L$ and $\hat{\phi}_2(w) = \hat{\phi}(w)$ for $p \le w < p = 1$. These two functions satisfy

$$e^{sc}\hat{\phi}_1(w) = g\hat{R}_1\hat{\phi}_1(w) + (1-g)\hat{R}_1\int_0^p M(s; z)\hat{\phi}_1(z)\mathrm{d}z,$$

$$e^{sc}\hat{\phi}_2(w) = (1-g)\hat{R}_1\int_0^p M(s; z)\hat{\phi}_1(z)\mathrm{d}z. \tag{15.86}$$

The first equation decouples and can be written as

$$(e^{sc} - g\hat{R}_1)\hat{\phi}_1(w) = (1-g)\hat{R}_1\int_0^p M_1(s)\hat{\phi}_1(z)\mathrm{d}z, \tag{15.87}$$

where M_1 is the moment-generating function of K_1 on favorable patches. Now the previous argument applies: $\hat{\phi}_1(w)$ is constant and can be canceled from the equation. What remains can be solved for c to find

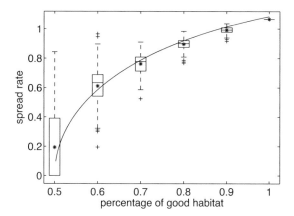

Fig. 15.15 Each box and whisker plot represents spread-rate statistics for 100 simulations in randomly generated landscapes. The curve indicates the theoretical value obtained by the averaging formula (15.78). Parameters are $R_1 = 1.8$ and $R_2 = 0.2$, and dispersal kernels are Gaussians with $\sigma_1^2 = 1$ and $\sigma_2^2 = 3$. Plot adapted from Dewhirst and Lutscher (2009).

$$\hat{c} = \inf_{s>0} \frac{1}{s} \ln(g R_1 + (1 - g) p R_1 M_1(s)), \tag{15.88}$$

which implies a persistence threshold of $p^* = \frac{1-gR_1}{(1-g)R_1}$. With $R_1 = 1.04$ and $g = 0.95$, this expression yields a reasonable threshold of $p^* \approx 23\%$; see Appendix 3 in Dewhirst and Lutscher (2009) for more details.

Dewhirst and Lutscher (2009) also present a heuristic argument to apply their averaging method to the model with the caricature Allee effect from (6.1). Since the final formula in (15.78) contains only the percentage of each of the two habitat types, it could potentially also apply to (certain) randomly varying landscapes. Dewhirst and Lutscher (2009) test this idea by simulations and find that the formula makes a reasonable prediction of the mean over many simulation runs. For example, Fig. 15.15 shows the mean, median, percentile, and outliers of 100 simulation runs on randomly generated landscapes together with the curve given by the homogenization approximation.

Small Landscape Variation

Gilbert et al. (2014a) consider a different small parameter, namely the spatial variation in landscape quality and dispersal behavior. Their work is for stage-structured models; we present the main idea for the simpler scalar case. We also start with the linearized L-periodic equation (15.70), but instead of assuming that L is small we assume that K and R vary only slightly around their means. More precisely, let us denote by $\sigma(x)$ the L-periodic function that indicates how a parameter in the dispersal kernel (e.g., the variance) depends on spatial location. We define the averages

$$\overline{R} = \frac{1}{L} \int_0^L R(x)\mathrm{d}x \quad \text{and} \quad \overline{K}(x) = \frac{1}{L} \int_0^L K(x; y)\mathrm{d}y, \tag{15.89}$$

and similarly for $\overline{\sigma}$. The main assumption that the functions differ only slightly from their averages can be expressed as

$$R(x) = \overline{R}(1 + \epsilon R_1(x) + O(\epsilon^2)), \quad \sigma(x) = \overline{\sigma}(1 + \epsilon\sigma_1(x) + O(\epsilon^2)),$$

$$K(x; y) = \overline{K}(x)(1 + \epsilon\sigma_1(y) + O(\epsilon^2)) \tag{15.90}$$

for some small $\epsilon > 0$ and functions R_1 and σ_1 with zero mean. We continue as above by forming the ansatz of a traveling periodic wave with speed $c = c(s, \epsilon)$ and arrive at eigenvalue equation (15.71), which we write in operator form as

$$e^{sc(s,\epsilon)}\phi(x) = \mathcal{K}[\phi](x) = \int_{-\infty}^{\infty} K(x - y; y)e^{s(x-y)}R(y)\phi(y)\mathrm{d}y \tag{15.91}$$

$$= \int_0^L \sum_m K(x - y + mL; y)e^{s(x-y+mL)}R(y)\phi(y)\mathrm{d}y.$$

The latter expression uses the L-periodicity of the functions involved, which guarantees that we can always work on a bounded spatial domain; see (15.40).

When we substitute the expansions for R and K as above, we can write $\mathcal{K} = \mathcal{K}_0 + \epsilon\mathcal{K}_1 + O(\epsilon^2)$, where

$$\mathcal{K}_0[\psi](x) = \int_{-\infty}^{\infty} \overline{K}(x - y)e^{s(x-y)}\overline{R}\psi(y)\mathrm{d}y \quad \text{and}$$

$$\mathcal{K}_1[\psi](x) = \int_{-\infty}^{\infty} \overline{K}(x - y)e^{s(x-y)}\overline{R}[\sigma_1(y) + R_1(y)]\psi(y)\mathrm{d}y. \tag{15.92}$$

On the left-hand side of (15.91), we substitute the expansions

$$c = c(s, \epsilon) = c_0(s) + c_1(s)\epsilon + O(\epsilon^2) \quad \text{and} \quad \phi(x) = \phi_0 + \epsilon\phi_1(x) + O(\epsilon^2). \tag{15.93}$$

Since ϕ is an eigenfunction, its average, ϕ_0, is a scalable constant.

Comparing like terms in the expansion, we find to lowest order the known formula

$$c_0(s) = \frac{1}{s} \ln(\overline{R}\,\overline{M}(s)), \tag{15.94}$$

where $\overline{M}$ is the moment-generating function of $\overline{K}$. The terms of order ϵ give the equation

$$e^{sc_0}\phi_1 - \mathcal{K}_0[\phi_1] = \mathcal{K}_1[\phi_0] - e^{sc_0}c_1\phi_0. \tag{15.95}$$

We denote by ϕ_0^* the (constant) eigenfunction of the adjoint operator $\mathcal{K}_0^*$ to the eigenvalue e^{sc_0} with respect to the inner product on the set of square-integrable functions on $[0, L]$. Taking the inner product of the left-hand side of (15.95), we find

$$\langle \phi_0^*, (e^{sc_0}\phi_1 - \mathcal{K}_0[\phi_1]) \rangle = \langle (e^{sc_0}\phi_0^* - \mathcal{K}_0^*[\phi_0^*]), \phi_1 \rangle = 0. \tag{15.96}$$

Hence, the inner product of ϕ_0^* with the right-hand side of (15.95) is also zero. Then we find an expression for c_1 as

$$e^{sc_0}c_1 = \frac{\langle \phi_0^*, \mathcal{K}_1[\phi_0] \rangle}{\langle \phi_0^*, \phi_0 \rangle}. \tag{15.97}$$

We claim that this expression is zero. Since ϕ_0 and ϕ_0^* are constants, we may set them to unity. Then we have to evaluate

$$\int_0^L \int_0^L \sum_m K_0(x - y - mL)e^{s(x-y-mL)}\overline{R}[\sigma_1(y) + R_1(y)]\mathrm{d}y\mathrm{d}x. \tag{15.98}$$

The change of variables $z = y$ and $w = x - z$ gives

$$\int_{-\infty}^{\infty} \sum_m K_0(w - mL)e^{s(w-mL)} \int_{-w}^{L-w} \overline{R}[\sigma_1(z) + R_1(z)]\mathrm{d}z\mathrm{d}w. \tag{15.99}$$

By construction, the integrals of σ_1 and R_1 over one period vanish. Hence, the entire expression is zero. Under some additional technical assumptions that guarantee that the minimum of c with respect to s is attained for finite s (see Appendix B in Gilbert et al. 2014a for details), we find that the dispersion relation is given by

$$c(s, \epsilon) \approx c_0(s) + O(\epsilon^2). \tag{15.100}$$

In other words, if the landscape variation is of order ϵ, then the speed obtained from the averaged equation is an order-ϵ^2 approximation to the true speed.

Some landscapes show forms of heterogeneity (or fragmentation) that are not covered by either of the two approaches to averaging. These landscapes have strong heterogeneity on a scale that is comparable to or larger than the dispersal scale. An example could be certain wildflowers that find good habitat only on the fringes of agricultural fields and forests. Their good habitat patches are far apart, and the difference between "good" and "bad" patches in terms of population reproduction is large. The consideration of such sparse landscapes led Gilbert et al. (2014b) to develop a different approximation technique for scalar and stage-structured IDE models. They reduce the continuous source–sink habitat to a discrete lattice model by averaging dispersal on each patch (similar to the ideas behind the dispersal success approximation in Chap. 9). Then they use a discrete version of Theorem 5.1 (also proved by Weinberger 1982) to show the existence of a spreading speed and the validity of the linearization formula.

15.5   Spatially Varying Dispersal Behavior

So far, we have assumed that only the growth function depends on spatial location. However, individuals also adapt their dispersal behavior to local conditions. Foraging theory postulates that individuals should move faster in unsuitable habitat and more slowly in suitable habitat. A meta-study by Crone et al. (2019) confirms this prediction empirically for many different taxa. If given a choice, individuals should also preferentially move toward suitable habitat. Empirical results that confirm such habitat preference for several taxa are listed in Maciel and Lutscher (2013). Including these and potentially other aspects of behavior in dispersal kernels poses challenges for modeling and analysis. Several authors begin with a random-walk model (Chap. 7), include spatial variation in the diffusion and settling parameters, derive corresponding dispersal kernels, and analyze the resulting dynamics (Van Kirk 1995; Van Kirk and Lewis 1997, 1999; Powell and Zimmermann 2004; Robbins 2004; Musgrave 2013; Musgrave and Lutscher 2014a,b; Musgrave et al. 2015). Neupane and Powell (2015) also include temporal variability in the settling process. We present some aspects of the work by Musgrave (2013), which generalizes many of the results preceding it, and give an application to invasive forest insects (Lutscher and Musgrave 2017).

Dispersal Kernels in Patchy Landscapes

We generalize random-walk models (Chap. 7) to patchy landscapes. As in Sects. 15.2 and 15.3, we consider a one-dimensional landscape divided into patches of two types. In a patch (interval) of type i, we write a diffusion and settling equation with patch-dependent parameters for the density of a dispersing organism as (compare (7.5))

$$\frac{\partial u}{\partial t} = D_i \frac{\partial^2 u}{\partial x^2} - \alpha_i u \,. \tag{15.101}$$

We neglect dispersal-related mortality here (see Sect. 12.1), but it is included in the model by Musgrave (2013). We need to define matching conditions for the density at the interface between two adjacent patches. Our approach follows Ovaskainen and Cornell (2003) and Maciel and Lutscher (2013), who include patch preference in the movement model. For simplicity, we consider a single interface, located at $x = 0$, with a patch of type 1 to the right and a patch of type 2 to the left. We denote by Δ_t the time step and by Δ_i the space step in patch type i; compare Chap. 7. We

Fig. 15.16 Schematic
illustration of the
random-walk probabilities at
an interface between two
patch types.

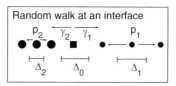

set $\Delta_0 = (\Delta_1 + \Delta_2)/2$. An individual inside a patch of type i moves left or right
with equal probability p_i. An individual at the interface chooses to move into patch
type i with probability γ_i, where $\gamma_1 + \gamma_2 \leq 1$; see Fig. 15.16.

The master equations for the probability of the individual's location near the
interface are

$$u(t + \Delta_t, 0)\Delta_0 = (1 - \gamma_1 - \gamma_2)u(t, 0)\Delta_0 + \frac{p_1}{2}u(t, x + \Delta_1)\Delta_1$$

$$+ \frac{p_2}{2}u(t, \Delta_2)\Delta_2,$$

$$u(t + \Delta_t, \Delta_1)\Delta_1 = (1 - p_1)u(t, \Delta_1)\Delta_1 + \gamma_1 u(t, 0)\Delta_0$$

$$+ \frac{p_1}{2}u(t, 2\Delta_1)\Delta_1,\tag{15.102}$$

$$u(t + \Delta_t, -\Delta_2)\Delta_2 = (1 - p_2)u(t, -\Delta_2)\Delta_2 + \gamma_2 u(t, 0)\Delta_0 + \frac{p_2}{2}u(t, -2\Delta_2)\Delta_2.$$

We expand the terms on the left-hand side in a Taylor series in Δ_t and the terms
containing $2\Delta_i$ in a Taylor series in Δ_i. Then we multiply the second equation by
γ_1 and the third by γ_2 and substitute the first equation. Eventually, we find

$$\gamma_1 p_2 \Delta_2 u(t, -\Delta_2) = \gamma_2 p_1 \Delta_1 u(t, \Delta_1) + O(\Delta^2),\tag{15.103}$$

where Δ^2 stands for the product of any two of Δ_1, Δ_2, and Δ_t. If we now assume
that $\Delta_1 = \Delta_2$, we multiply both sides by Δ_1/Δ_t, take the parabolic limit (7.6), and
arrive at

$$\gamma_1 D_2 u(t, 0^-) = \gamma_2 D_1 u(t, 0^+).\tag{15.104}$$

Here, D_i are the diffusion coefficients in patch type i, and $0^\pm$ indicate the one-sided
limits from the right and left, respectively. Different assumptions lead to slightly
different formulations, but we only consider (15.104) here (denoted Case M in
Musgrave (2013)). In a similar way, we can derive the continuity condition for the
flux as

$$D_2 \frac{\partial u}{\partial x}(t, 0^-) = D_1 \frac{\partial u}{\partial x}(t, 0^+);\tag{15.105}$$

see Appendix A in Maciel and Lutscher (2013) for details.

We write the diffusion-settling equation in a patchy periodic landscape with two types of patches as

$$\frac{\partial}{\partial t}u(t,x) = \frac{\partial^2}{\partial x^2}(D(x)u(t,x)) - \alpha(x)u(t,x) \qquad (15.106)$$

with piecewise-constant functions $D(x) = D_i$ and $\alpha(x) = \alpha_i$ on patch type i. At each interface, we impose matching conditions (15.104) and (15.105) or the corresponding conditions with the one-sided limits interchanged when the patch of type 1 (2) is on the left (right). To obtain a dispersal kernel, we impose the initial condition $u(0,x) = \delta(x - y)$ and denote the corresponding solution as $u(t,x;y)$. Then the dispersal kernel is defined as in (7.7), except that the settling rate now depends on space; i.e.,

$$K(x,y) = \int_0^\infty \alpha(x)u(t,x)dt . \qquad (15.107)$$

Musgrave (2013) explicitly calculates the resulting dispersal kernel in several cases, e.g., when there are only one or two interfaces in the entire landscape; see also Musgrave and Lutscher (2014a). In fact, for any finite number of patches, the kernel consists of a linear combination of exponential functions on each patch. The coefficients for these exponentials satisfy a linear system of equations, given by appropriate interface conditions for the kernel. These conditions are similar to conditions (15.104) and (15.105) but also contain the settling coefficients. Figure 15.17 shows one example of a kernel on a landscape consisting of three good patches (low diffusion and high settling) interspersed with and surrounded by bad patches (high diffusion and low settling). For comparison, we also plot the Laplace kernel on a homogeneously good landscape. Musgrave (2013) and Musgrave and Lutscher (2014a) also derive and analyze expressions for the dispersal success function, the average dispersal success, and the mean dispersal distance in patchy landscapes.

Fig. 15.17 Dispersal kernel in a patchy landscape with three good patches. Parameters are $D_1 = 0.1$, $D_2 = 0.3$, $\alpha_1 = 2$, $\alpha_2 = 1$, and $\gamma_1 = \gamma_2 = 1/2$. The dash-dot curve shows the Laplace kernel with parameters D_1 and α_1. Figure adapted from Lutscher and Musgrave (2017).

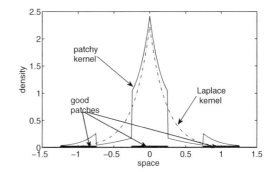

Persistence and Spread in Patchy Landscapes

To consider the population dynamic effects of patch-dependent dispersal behavior, we study the IDE

$$N_{t+1}(x) = \int_{-\infty}^{\infty} K(x, y) F(N_t(y), y) dy \qquad (15.108)$$

on an infinite landscape of two periodically alternating patch types. The growth function is piecewise defined as in (15.16), and the dispersal kernel is implicitly given in (15.107), derived from the random-walk model above. We assume that there is no Allee effect and study persistence conditions and spreading speeds based on the linearization at zero.

It turns out that all of the theoretical analysis and explicit calculations for persistence conditions in infinite periodic landscapes with homogeneous dispersal behavior (Sect. 15.2) can be generalized to the case of heterogeneous dispersal behavior here (Musgrave 2013; Musgrave and Lutscher 2014b). The key insight that allows explicit calculations is that the kernel is again the Green's function of a second-order differential operator. Therefore, the eigenvalue problem for stability of the trivial state can still be reduced to a pair of differential equations as in (15.21) but with different values for a in the different patches and with discontinuous matching conditions replacing the ones in (15.22). Specifically, the equations for eigenvalue λ and eigenfunction ϕ are

$$\phi'' = a_1^2 \left(1 - R_1/\lambda\right) \phi, \quad x \in [0, L_1/2], \qquad (15.109)$$
$$\phi'' = a_2^2 \left(1 - R_2/\lambda\right) \phi, \quad x \in [L_1/2, L/2],$$

with no-flux boundary conditions $\phi'(0) = \phi'(L/2) = 0$. We denoted $a_i = \sqrt{\alpha_i/D_i}$ and $R_i = F_i'(0)$ with $R_1 > 1 > R_2 \geq 0$. The interface matching conditions are

$$\gamma_2 \alpha_2 D_1 \phi\left(\frac{L_1^-}{2}\right) = \gamma_1 \alpha_1 D_2 \phi\left(\frac{L_1^+}{2}\right),$$

$$\alpha_2 D_1 \phi'\left(\frac{L_1^-}{2}\right) = \alpha_1 D_2 \phi'\left(\frac{L_1^+}{2}\right). \qquad (15.110)$$

Following the same steps as in Sect. 15.2, we can calculate the minimal size of good patches that allows the population to persist as

$$L_1^* = \frac{2}{a_1\sqrt{R_1 - 1}} \arctan\left[\frac{\gamma_2 a_2}{\gamma_1 a_1}\sqrt{\frac{1 - R_2}{R_1 - 1}} \tanh\left(\frac{a_2 L_2 \sqrt{1 - R_2}}{2}\right)\right]. \qquad (15.111)$$

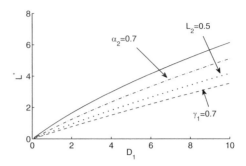

Fig. 15.18 Critical length of good patches in a periodic landscape according to (15.111). Default parameters for the solid curve are $D_2 = 1$, $L_2 = 1$, $\alpha_1 = \alpha_2 = 1$, $\gamma_1 = \gamma_2 = 0.5$, $R_1 = 2$, and $R_2 = 0$. Parameter values for other curves differ only in the parameter indicated, and $\gamma_2 = 1 - \gamma_1$.

In Fig. 15.18, we illustrate that the threshold length, L_1^*, increases with the movement rate in good patches, D_1 (solid curve). The biological explanation is the same as for the critical patch-size: as the movement rate increases, individuals are more likely to leave the good patches and die in bad patches. Varying only one parameter at a time, we observe that decreasing the length of bad patches, L_2 (dots); decreasing the settling rate in bad patches, α_2 (dash-dot); or increasing the preference for good patches, γ_1 (dashed), each decreases L_1^*. We note that increasing the movement rate in bad patches, D_2, decreases L_1^*; for illustrations, see Musgrave (2013) and Musgrave and Lutscher (2014b).

The explicit calculations of the minimal traveling periodic wave speed from Sect. 15.3 can be extended to the case of patch-dependent movement behavior (see below). The theoretical results by Weinberger (2002) cannot be applied here for two reasons. All previous theoretical work considers the density, N, to be continuous in space, but this is generally not the case when the landscape is patchy and the matching conditions are discontinuous. Also, numerical simulations reveal that even if the growth function is spatially homogeneous and monotone and has $N^* = 1$ as a stable fixed point, the solution operator may not leave the set of functions $\mathscr{C}_{N^*}$ in (15.56) invariant (Musgrave 2013). Nonetheless, numerical simulations suggest that the linear conjecture holds for IDE (15.108). We proceed under the assumption that it does.

Most of the steps leading to dispersion relation (15.53) between the speed and the steepness of a traveling periodic wave, $N_t(x) = e^{-s(x-ct)}\phi(x)$, can be extended to our case here, but it is helpful to make several substitutions; see (Musgrave 2013) for details. In the end, we arrive at the dispersion relation for a traveling periodic wave

$$\cosh(sL) = \cosh(q_1 L_1)\cosh(q_2 L_2) + \frac{q_1^2 + (\bar{\gamma}q_2)^2}{2\bar{\gamma}q_1 q_2}\sinh(q_1 L_1)\sinh(q_2 L_2),$$

$$(15.112)$$

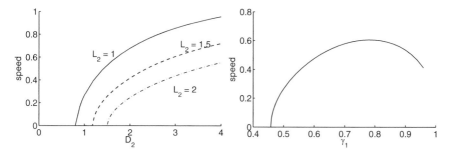

Fig. 15.19 Minimal speed of a traveling periodic wave according to (15.112). **Left:** Speed as a function of movement rate in bad patches, D_2, for three different lengths of bad patches. **Right:** Speed as a function of patch preference, γ_1, with $\gamma_2 = 1-\gamma_1$. Unless specified otherwise, parameter values are $L_1 = L_2 = 1$, $D_1 = D_2 = 1$, $\alpha_1 = \alpha_2 = 1$, $\gamma_1 = \gamma_2 = 0.5$, $R_1 = 2$, and $R_2 = 0$.

where

$$q_i^2 = \left(\frac{a_i}{D_i}\right)^2 \left(e^{-sc}R_i - 1\right) \quad \text{and} \quad \bar{\gamma} = \frac{\gamma_2}{\gamma_1}. \tag{15.113}$$

The plots in Fig. 15.19 illustrate how the minimal traveling periodic wave speed depends on model parameters. Above a certain threshold value, it is an increasing function of the movement rate in bad patches (left plot). The threshold value arises from the critical size of good patches (15.111). If an individual moves too slowly in bad habitat, it will settle there and not make it across to the next good habitat. The resulting population loss is too large for the population to persist and spread. The traveling periodic wave speed is a hump-shaped function of the preference for good patches (right plot). If individuals have a low preference for good patches (or even some preference for bad patches), the persistence condition in (15.111) is not satisfied, and the population will die out. If the preference for good patches is very high, only a very few individuals will leave a good patch, and those few will not be sufficient to establish a population on the next good patch, so that the population cannot spread.

Application to Invasive Forest Insects

Lutscher and Musgrave (2017) apply the ideas and results above to study the spread of the invasive emerald ash borer (*Agrilus planipennis* Fairmaire) in fragmented landscapes. Crone et al. (2019) use a slight modification of the dispersal model to evaluate the importance of low-quality habitat for the spread of the endangered Baltimore checkerspot butterfly (*Euphydryas phaeton*). Both studies emphasize the importance of bad (matrix) habitat for population persistence and spread and find the counterintuitive phenomenon that decreasing the extent of a species' good habitat could increase the species' spread rate.

Emerald ash borer is a wood-boring beetle whose larvae consume the phloem underneath the bark of ash trees, eventually killing the trees. In the attempt to control this and other forest insect pests, we would like to know how and to what extent silvicultural measures such as thinning or removal of host trees affect the spread rate of the insect. We expect that the spread rate decreases with decreasing resource density. We shall see that this reasoning may not apply when individuals can adapt their dispersal behavior and actively seek out areas of higher resource density or avoid areas of lower resource density.

Lutscher and Musgrave (2017) derive the following nonspatial model for emerald ash borer eggs (E_t) and ash tree phloem (P_t) at the beginning of generation t:

$$E_{t+1} = \frac{R}{2}\left(1 - e^{-P_t/\bar{P}}\right)sE_t, \qquad P_{t+1} = P_t e^{-wsE_t}. \qquad (15.114)$$

Here, s denotes the overwintering survival and hatching probability of eggs, the term in brackets is the phloem-dependent probability of maturation and emergence as adult moths, and R is the average number of eggs laid by a female moth. Parameter $\bar{P}$ measures the phloem requirement of larvae; the factor $1/2$ accounts for a 1:1 sex ratio. Phloem is depleted from year to year, where w corresponds to a consumption rate per larvae. Ash borers can kill a tree within 4–5 years, but trees require several decades before they are sufficiently mature for ash borers to lay eggs in. Because of this difference in time scales, our model considers only the first stage of the invasion of the ash borer and not the subsequent regrowth of the trees. The model shows very simple dynamics: from a typical initial condition of high phloem levels and low egg density, eggs initially increase each year while the phloem decreases. Once the phloem levels are low, eggs also decrease. Lutscher and Musgrave (2017) estimate parameter values from empirical data.

Since only the moths disperse, the spatial version of (15.114) is

$$E_{t+1}(x) = \int_{-\infty}^{\infty} K(x, y)\frac{R}{2}\left(1 - e^{-P_t(y)/\bar{P}}\right)sE_t(y)\mathrm{d}y,$$
$$P_{t+1}(x) = P_t(x)e^{-wsE_t(x)}, \qquad (15.115)$$

where K is the dispersal kernel of female ash borer moths. The model is somewhat degenerate since it has a continuum of steady states of the form $(0, P_0(y))$. When linearizing at such a state, the equation for E decouples and becomes

$$E_{t+1}(x) = \int_{-\infty}^{\infty} K(x, y)\frac{R}{2}\left(1 - e^{-P_0(y)/\bar{P}}\right)sE_t(y)\mathrm{d}y. \qquad (15.116)$$

The initial phloem density, $P_0(x)$, can be manipulated by, say, tree removal. We want to know how fast the insect population spreads, depending on the initial phloem configuration. We consider a homogeneous initial phloem level as the default and investigate how removing phloem changes the spread rate. We remove phloem in a "patchy" periodic way, so that the initial phloem level looks like the periodic patchy landscape in Fig. 15.2. The difference from the previous case is that the phloem is now a dynamic variable, whereas the landscape quality in Sect. 15.1 was fixed. Simulations of the nonlinear model reveal that a locally introduced population of ash borer forms two moving pulses, one in each direction, and leaves in its wake a landscape devoid of phloem (Lutscher and Musgrave 2017); see Fig. 15.20.

It is a priori not clear that the minimal traveling wave speed of linearized equation (15.116) will give the correct spreading speed. Musgrave and Lutscher (2014b) ran extensive numerical simulations to support the conjecture that it does. Furthermore, we can solve the equation for P and substitute the result into the equation for E to obtain the *delayed* equation (in the nonspatial case)

$$E_{t+1} = \frac{R}{2}\left(1 - \exp\left(-P_0/\bar{P}\exp\left(-ws\sum_{\tau=0}^{t-1}E_\tau\right)\right)\right) s E_t. \tag{15.117}$$

The function $F(z) = (1 - \exp(-\exp(-z)))\,z$ is bounded by its linearization at zero; i.e., $F(z) \leq F'(0)z$. Based on the results on delayed equations in homogeneous landscapes (e.g., Lin and Li 2010; see Sect. 14.9), it is plausible to assume that the linear conjecture holds for our case. Hence, we use dispersion relation (15.112) to calculate the minimal traveling periodic wave speed of the linearized equation and claim that this coincides with the spreading speed.

Figure 15.20 shows that the population spread rate in a heterogeneous landscape with *lower* overall resource availability can be *higher* than in homogeneously good habitat if individuals move faster in low-quality habitat. Alternatively or

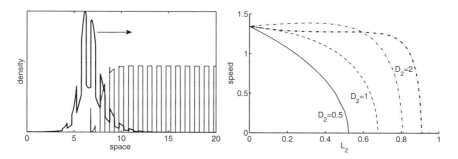

Fig. 15.20 Left: A "ragged" traveling pulse of ash borer (thick) moves into and consumes phloem (thin). Parameters as in Lutscher and Musgrave (2017). **Right:** Spread speed of (15.116) according to (15.112) as a function of the size of bad patches. When the diffusion rate in bad patches is high enough (dash-dot, thin), the spread rate can be higher in fragmented than in homogeneous landscapes. If, in addition, the preference for good patches is high (dash-dot, thick), the population can persist and spread at a very low resource density and the spread rate is essentially constant for a large range of L_2. Unless otherwise specified, parameters are $L_1 + L_2 = 1$, $D_1 = D_2 = 1$, $\alpha_1 = \alpha_2 = 1$, $\gamma_1 = \gamma_2 = 0.5$, $R_1 = 2$, and $R_2 = 0$.

additionally, a lower settling rate in bad patches can increase the spread rate. A higher preference for good patches can allow the population to persist at very low resource density. Together with high movement rates in bad patches, this mechanism can lead to high spread rates at a low, patchily distributed resource density.

When the population dynamics include an Allee effect, there are no simple formulas for persistence conditions or spread rates. Musgrave et al. (2015) study the effects of fragmentation and differential movement on Allee dynamics by numerical simulation. They also give some upper and lower bounds on the sizes of good and bad patches that ensure that the population will persist and spread, that it will persist but not spread, or that it will retreat.

Powell and Zimmermann (2004) study the spread of plant populations whose seeds are transported by animals, e.g., harvester ants moving wild ginger seeds and blue jays moving acorns. They also use the movement model (15.106) for the animal vector but without patch preference and with continuous matching conditions. Rather than deriving a dispersal kernel from the heterogeneous equation, they use homogenization at the level of the partial differential equation and obtain a dispersal kernel from the averaged equation. This kernel is a Laplace kernel with appropriately averaged diffusion coefficients and settling rates. Neupane and Powell (2015) extend the dispersal model by including a handling-time distribution of seeds by the disperser, in particular birds.

15.6 Further Reading

There are very few explicit examples of spatially varying dispersal kernels in heterogeneous landscapes. Weinberger et al. (2008) give the example

$$K(x, y) = [1 + (1/2)\cos(x)] \exp(|2x + \sin(x) - 2y - \sin(y)|) \tag{15.118}$$

for a spatially varying smooth kernel, but without any mechanistic interpretation. The approach by Mistro et al. (2005a) is different but conceptually related to our derivation of dispersal kernels based on random walks. They consider any dispersal kernel in a homogeneous landscape and modify it by a heuristically chosen attraction function (12.104) that reflects habitat quality; see Sect. 12.8.

Van Kirk and Lewis (1997) pioneered the approach of deriving dispersal kernels from diffusion equations in patchy landscapes with a simpler model than we presented in Sect. 15.5. They also extend the ideas of average dispersal success from Chap. 9 to an infinite periodic habitat. Lutscher and Lewis (2004) generalize their derivations to stage-structured models and derive patch models from continuous-space models. Botsford et al. (2001) and Lockwood et al. (2002) apply the dispersal success ideas to calculate the requirements for the size and spacing of marine reserves along a coastline to ensure the persistence of a harvested population.

Samia and Lutscher (2012) study a competition model in a patchy landscape. They show by simulation and by the average dispersal success approximation how the outcome of competition can vary depending on the percentage of habitat where one species has a competitive advantage. Hughes et al. (2015) study the effect of habitat fragmentation on host–parasitoid interaction and outbreak dynamics. They consider a two-dimensional landscape and measure fragmentation in terms of percentage of good habitat as well as habitat arrangement ("clumping"). They compare simulation outcomes with an approximation based on average dispersal success. Among other things, they find that the prey might benefit from habitat loss if prey individuals are less mobile than predators.

Westerberg and Wennergren (2003) discretize the landscape into patches and derive discrete dispersal kernels from leaving probabilities that depend on habitat quality. Lutscher (2008) studies spread rates in a patchy landscape with density-dependent dispersal. Marchetto et al. (2010) explore the effect of vegetation height on the spread of a thistle. Rather than modeling spatial variation explicitly, they measure wind speeds in different plant arrangements and use the WALD model to find appropriate dispersal kernels. Skelsey et al. (2010) study the effects of crop arrangement on disease spread of potato late blight and investigate which planting patterns could make the potatoes the least susceptible. Pittman et al. (2015) explore management options for the potential spread of a recently developed biofuel crop. They simulate a four-stage life-cycle model in a numerically generated neutral fractal landscape but do not consider heterogeneity in dispersal. Ramanantoanina and Hui (2016) simulate invasions in patchy landscapes with a mix of two dispersal kernels. One of their results is that the spread rate decreases as the period of the landscape increases.

On the analytical side, Ding and Liang (2015) prove the existence of spreading speeds in a spatially periodic landscape for homogeneous but otherwise very general kernels, e.g., kernels for which the resulting next-generation operator need not be compact. Wu and Zhao (2018) analyze a two-species competition system in a spatially periodic landscape. They allow a nonhomogeneous and not necessarily continuous dispersal kernel; they only require lower semicontinuity. They prove the existence of a spreading speed and traveling waves, as well as linear determinacy.

Pachepsky and Levine (2011) simulate population spread in a patchy landscape with stochastic growth dynamics. They find that density dependence could slow an invasion in a patchy landscape, in particular if individuals were assumed discrete. This behavior is in contrast to the linear conjecture, according to which the spread rate is determined by the low-density dynamics only.

Krkošek and Lewis (2010) study an aspect of population dynamics in heterogeneous landscapes that we have not touched on at all. They develop a theory of source–sink dynamics in IDEs, based on a spatial extension of R_0-theory. In nonspatial models, R_0 is a measure of the fitness of individuals in a population. In spatially explicit models, individual fitness may vary between locations. Reproductive output at one location has to be discounted by the probability that the offspring will land in a viable habitat. Krkošek and Lewis (2010) apply their theory to the competition of quagga and zebra mussels (*Dreissena*) and explain how the two competing species can coexist.

Li et al. (2015) consider a very different kind of habitat heterogeneity, namely a gradient habitat: The growth function is monotone nondecreasing in the spatial location and population density. The authors give conditions for the population to spread toward better habitat and to persist (or not) on the entire domain. In a subsequent paper, Li et al. (2016a) combine the gradient model with moving-habitat models and analyze conditions under which a population can persist in an expanding or contracting habitat.

Chapter 16
Temporal Variation

Abstract IDE models naturally allow a certain temporal variation within a generation since they divide each generation into separate growth and dispersal phases. However, so far we have assumed that the growth phases in all generations are identical and that the same holds for the dispersal phases. In realistic environments, external conditions in subsequent generations may vary substantially so that growth and dispersal behavior could differ. In this chapter, we present some theory on and examples of how to formulate and analyze IDEs with a periodically or randomly varying growth function and dispersal kernel. In the periodic case, much of the previous theory for temporally constant environments can be applied to the period map. In the random case, even the formulation of the problem requires substantially different tools from the theory of stochastic processes. We focus again on the two fundamental questions of population persistence and spread.

16.1 Nonspatial Models with Temporal Variation

We illustrate and explain some basic questions about how temporal variation affects population dynamics by using the simple nonspatial model from (2.3). We also introduce some terminology for subsequent spatial models. We denote the population density in year t by N_t and the temporally varying growth function by F_t. We study the dynamics of the equation

$$N_{t+1} = F_t(N_t) \tag{16.1}$$

when the environment varies periodically or randomly in time.

Periodic Variation

When the environment is periodic with period $T \in \mathbb{N}$, we can study the map of the Tth iteration

© Springer Nature Switzerland AG 2019
F. Lutscher, *Integrodifference Equations in Spatial Ecology*, Interdisciplinary Applied Mathematics 49, https://doi.org/10.1007/978-3-030-29294-2_16

$$N_{t+T} = G(N_t) = F_{t+T-1} \circ F_{t+T-2} \circ \cdots \circ F_{t+1} \circ F_t(N_t) \qquad (16.2)$$

with the usual techniques for discrete maps. The qualitative behavior of this map is independent of the choice of t.

For an example, we consider a two-periodic environment with growth functions

$$F_i(N) = \frac{R_i N}{1 + \kappa_i N}, \qquad i = 1, 2, \qquad (16.3)$$

from (2.11) with $R_i, \kappa_i > 0$. Since F_i are monotone increasing and concave down, $G = F_2 \circ F_1$ has the same properties. Therefore, the dynamics of (16.2) are determined by the local stability of the zero state. If the zero state is locally stable, then it is globally stable; if it is unstable, then there is a globally stable positive steady state; see Sect. 2.2. The linearization of G at $N = 0$ is given by

$$n_{t+2} = R_2 R_1 n_t, \qquad (16.4)$$

so that the zero state is unstable if and only if $R_2 R_1 > 1$. If this condition is satisfied, the positive steady state of G is given by

$$N^* = \frac{R_2 R_1 - 1}{\kappa_1 + \kappa_2 R_1}. \qquad (16.5)$$

The solutions of the original system, $N_{t+1} = F_t(N_t)$, will converge to zero if $R_2 R_1 < 1$ and will approach a positive stable two-cycle, (N_1^*, N_2^*), when the inequality is reversed. One of the two states of the two-cycle is given by N^* above and the other by the corresponding expression with all indices exchanged.

Stochastic Variation

When the environment varies randomly, the formulation of the equations and the terminology and techniques used to study them are based on stochastic processes and differ considerably from the deterministic theory covered in previous chapters. We refer to Allen (2010) or Meyn and Tweedie (2009) for a thorough introduction.

We begin with the linear model

$$N_{t+1} = R_t N_t, \qquad t = 0, 1, 2, \ldots, \qquad (16.6)$$

and follow the exposition by Lewis et al. (2016). We assume that R_t are finite positive random variables that are independent and identically distributed (iid) with finite positive expectation $E[R_t] = E[R_0]$ for all t. As usual, R_t is the per capita growth rate of the population, i.e., the average number of offspring that an individual has in the given year, t. Since R_t are random variables, so are N_t, and we can ask for their expectation, denoted by $E[N_t]$. Under the assumption that environmental conditions are independent of population density, we have

$$E[N_{t+1}] = E[R_t N_t] = E[R_t]E[N_t].\tag{16.7}$$

Hence, the expected population density satisfies a deterministic equation with growth rate $E[R_0]$. Its solution is explicitly given by

$$E[N_t] = E[N_0]E[R_0]^t = E[N_0]e^{t \ln E[R_0]}.\tag{16.8}$$

It will grow in time when the geometric growth rate satisfies $E[R_0] > 1$ or the arithmetic growth rate $\ln(E[R_0]) > 0$.

On the other hand, we can write the solution of (16.6) explicitly as

$$N_t = N_0 \prod_{j=0}^{t-1} R_j = N_0 \exp\left(t \frac{1}{t} \sum_{j=0}^{t-1} \ln R_j \right).\tag{16.9}$$

Hence, the expected arithmetic growth rate is $E[\ln(R_0)]$. The process will grow to infinity with probability one if $E[\ln(R_0)] > 0$ and decay to zero with probability one if the reverse inequality holds. Since the logarithm is a concave function, Jensen's inequality states that $E[\ln(R_0)] \leq \ln(E[R_0])$. Hence, it is possible that the expectation in (16.8) is predicted to grow, whereas the actual solution in (16.9) will decay to zero with probability one.

We present an example similar to the two-periodic example above. We assume that the growth rate is a Bernoulli random variable that assumes values R_1 and R_2 with probability p and $1 - p$, respectively. Clearly, the population can grow when $R_{1,2} > 1$. However, if we choose $R_1 > 1 > R_2 > 0$, we find two threshold probabilities. We have

$$\ln(E[R_0]) > 0 \quad \text{if and only if} \quad p > p^* = \frac{1 - R_2}{R_1 - R_2}\tag{16.10}$$

and

$$E[\ln(R_0)] > 0 \quad \text{if and only if} \quad p > p^{**} = \frac{-\ln(R_2)}{\ln(R_1) - \ln(R_2)}.\tag{16.11}$$

For example, choosing $R_1 = 2$ and $R_2 = 1/4$ gives $p^* = 3/7 < 1/2 < 2/3 = p^{**}$. For $p \in (p^*, p^{**})$, the expectation in (16.8) will grow but the solution in (16.9) will decay to zero with probability one.

The study of nonlinear stochastic processes is more complicated. It requires concepts and results that we cannot introduce in detail here; see, e.g., Allen (2010) or Meyn and Tweedie (2009). Instead, we briefly discuss the results on the

stochastic version of the Beverton–Holt equation from Ellner (1984). We present
their generalization to IDEs in Sect. 16.3.

Ellner (1984) considers model (16.1) with $F_t(N) = F(N, \alpha_t)$, where α_t
represents parameters in F that describe the random environment in year t. He
assumes that α_t are iid random variables. Then N_t is a homogeneous Markov
process. It turns out that if F has properties like the Beverton–Holt function, then
stability results similar to the deterministic case hold for the stochastic case. Of
course, instead of a stationary state, we now have a stationary distribution. More
precisely, we assume that F is differentiable, monotone increasing, concave, and
bounded for each possible random environment. Then the stochastic process N_t has
a stationary distribution, μ_*, independent of N_0. There are two possibilities. The
first is $\mu_*(\{0\}) = 1$, which means that the entire mass of the stationary distribution
is concentrated at $N = 0$. In this case, the process will die out with probability one.
The second possibility is $\mu_*(\{0\}) = 0$, which means that the stationary distribution
is supported in $(0, \infty)$. In this case, the process will persist with probability one.
The behavior of the process is decided by the linear process $n_{t+1} = F'(0, \alpha_t)n_t$. If
$E[\ln(F'(0, \alpha_t))] < 0$, the process will die out with probability one. If the inequality
is reversed, the process will persist. The results in Ellner (1984) are formulated for
more general growth functions.

16.2   The Gaussian Habitat Quality Model with Temporal Variation

We begin our study of the effects of temporal variation on population persistence in
a spatial model with an explicitly solvable model, namely the linear model in (15.4)
with Gaussian habitat quality function (Latore et al. 1999). Even with temporally
varying parameters, this model can be reduced to a two-dimensional difference
equation as in Proposition 15.1. The ideas and concepts from the preceding section
can then be applied to study the spatial problem as well.

Our model equation is

$$N_{t+1}(x) = R_t e^{-x^2/(2\rho_t^2)} \int_{-\infty}^{\infty} K_G(x - y; \sigma_t^2) N_t(y) \, dy \,, \tag{16.12}$$

where K_G is the Gaussian dispersal kernel. Parameters R_t, ρ_t^2, and σ_t^2 depend on
time. As before, R_t is the per capita reproduction rate in year t, ρ_t^2 measures the
extent of the habitat in year t, and σ_t^2 is the variance of the dispersal kernel in year
t. All parameters are assumed positive.

As in Proposition 15.1, model (16.12) has a solution of the form $N_t = A_t \exp(-\frac{x^2}{2v_t^2})$, where A_t and v_t^2 satisfy the difference equations

$$v_{t+1}^2 = F_t(v_t^2) := \frac{\rho_t^2(\sigma_t^2 + v_t^2)}{\rho_t^2 + \sigma_t^2 + v_t^2} \quad \text{and} \quad A_{t+1} = A_t R_t \sqrt{\frac{v_t^2}{\sigma_t^2 + v_t^2}}. \tag{16.13}$$

Function F_t is differentiable, monotone increasing, concave down, bounded, and positive for $v_t^2 > 0$.

Periodic Variation

When the environment is periodic, we can obtain explicit conditions for population persistence and thereby study trade-offs between "good" and "bad" years. We consider a two-periodic environment and denote the respective values of the parameters by $R_{1,2}$, $\rho_{1,2}^2$, and $\sigma_{1,2}^2$, as well as functions $F_{1,2}$. By the considerations in the preceding section, the iteration for v_t^2 converges to a stable two-cycle, (v_{1*}^2, v_{2*}^2). Here, v_{1*}^2 is the positive solution of the quadratic

$$\left(\rho_1^2 + \rho_2^2 + \sigma_2^2\right) v_{1*}^4 + \left(\rho_1^2\sigma_1^2 - \rho_2^2\sigma_2^2 + \rho_1^2\sigma_2^2 + \rho_2^2\sigma_1^2 + \sigma_1^2\sigma_2^2\right) v_{1*}^2$$
$$- \left(\rho_1^2\rho_2^2\sigma_1^2 + \rho_1^2\rho_2^2\sigma_2^2 + \rho_1^2\sigma_1^2\sigma_2^2\right) = 0 \tag{16.14}$$

and $v_{2*}^2 = F_1(v_{1*}^2)$. Hence, the iteration for A_t approaches the linear function

$$A_{t+1} = A_t R_j \sqrt{\frac{v_{j*}^2}{v_{j*}^2 + \sigma_j^2}}, \tag{16.15}$$

with $j = 1$ if t is odd and $j = 2$ if t is even.

According to the results from the preceding section, A_t will grow eventually if and only if

$$R_1 R_2 > \sqrt{\left(1 + \frac{\sigma_1^2}{v_{1*}^2}\right)\left(1 + \frac{\sigma_2^2}{v_{2*}^2}\right)}. \tag{16.16}$$

For a temporally constant habitat with $R_1 = R_2$, $\sigma_1 = \sigma_2$, and $\rho_1 = \rho_2$, this condition is just the persistence condition from (15.12).

We explore the persistence condition as follows. We express $\sigma_{1,2}^2 = \bar{\sigma}^2 \pm \epsilon_\sigma$ in terms of the mean, $\bar{\sigma}^2$, and deviation, ϵ_σ, and similarly for $\rho_{1,2}^2 = \bar{\rho}^2 \pm \epsilon_\rho$. Figure 16.1 shows the contour lines of the critical value of $R_1 R_2$ that guarantees persistence according to (16.16). We observe that variation in suitable habitat size alone ($\epsilon_\rho > 0$, $\epsilon_\sigma = 0$) requires a higher growth rate for persistence, whereas variation in dispersal distance only ($\epsilon_\rho = 0$, $\epsilon_\sigma > 0$) allows for a lower growth rate.

Fig. 16.1 Contour lines for persistence condition (16.16) with $\sigma_{1,2}^2 = 1 \pm \epsilon_\sigma$ and $\rho_{1,2}^2 = 2 \pm \epsilon_\rho$. The numbers on the contour lines indicate the threshold values. The persistence threshold in the absence of temporal variation is $R_1 R_2 = 2$.

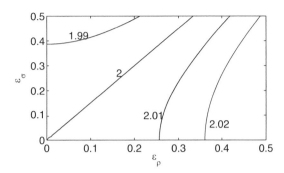

When both parameters vary simultaneously, the joint effect depends on the relative strength (variation) of the two.

Stochastic Variation

Now we assume that environmental conditions vary randomly, and we describe the corresponding growth rate, habitat size, and dispersal behavior by positive random variables R_t, ρ_t^2, and σ_t^2, each of which is assumed iid with positive finite expectation. According to Theorem 2.2 by Ellner (1984), v_t^2 converges to a stationary distribution, supported on $(0, \infty)$.

The equation for A_t can be solved explicitly as

$$A_t = A_0 \prod_{j=0}^{t-1} R_j \sqrt{\frac{v_j^2}{v_j^2 + \sigma_j^2}}, \tag{16.17}$$

which gives

$$\ln\left(\frac{A_t}{A_0}\right) = t\left(E[\ln(R_t)] + \frac{1}{2}E\left[\ln\left(\frac{v_t^2}{v_t^2 + \sigma_t^2}\right)\right]\right). \tag{16.18}$$

The population eventually grows with probability one if the term on the right-hand side is positive in the limit as $t \to \infty$. We can write this condition suggestively as

$$E[\ln(R_0)] > \frac{1}{2}E\left[\ln\left(1 + \frac{\sigma_0^2}{v_*^2}\right)\right], \tag{16.19}$$

where v_*^2 stands for a random variable whose distribution equals the stationary distribution of the variables v_t^2.

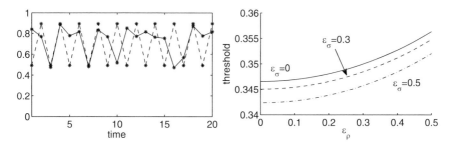

Fig. 16.2 Left: The expression $\ln(1 + \sigma_t^2/v_t^2)$ from (16.18) in the periodic case (dashed) and one realization of the corresponding stochastic case (solid). **Right:** Persistence threshold (16.19) as a function of ϵ_ρ for $\epsilon_\sigma = 0$ (solid), $\epsilon_\sigma = 0.3$ (dashed), and $\epsilon_\sigma = 0.5$ (dash-dot). The plot was obtained by simulating the stochastic process for up to 10,000 time steps and calculating the expectation numerically.

We can evaluate this condition numerically. We use the setup that most closely resembles the periodic model: ρ_t^2, σ_t^2 are Bernoulli random variables, where the two possible values $\rho_{1,2}^2$ and $\sigma_{1,2}^2$ appear with equal probability. We keep the mean of each variable fixed and vary the deviation. It turns out that the resulting persistence condition in (16.19) is exactly the same as the one in (16.16). In Fig. 16.2, we plot one realization of the process and compare it to the periodic case (left panel). We also plot the threshold condition from (16.19) as a function of ϵ_ρ, the deviation of ρ^2. Each curve increases with ϵ_ρ, indicating that persistence is harder to achieve as the variation in ρ^2 increases. However, for fixed ϵ_ρ, the threshold decreases with ϵ_σ, indicating that persistence is easier to achieve as the variation in σ^2 increases. The thresholds from the stochastic process and the periodic case are indistinguishable.

16.3 Persistence Under Temporal Variation

In this section, we present more formal and more general conditions for population persistence in temporally varying environments, extending the results on population persistence and existence of positive steady states from Chaps. 3 and 4. We focus on random environments but mention the corresponding results for periodic variation as well. We begin with the work by Hardin et al. (1988a), which can be considered a spatially explicit extension of the work by Ellner (1984).

Hardin et al. (1988a) formulate their model as

$$N_{t+1}(x) = Q_t[N_t](x) = \int_\Omega K(x, y) R(\alpha_t, y) F(N_t(y)) dy . \qquad (16.20)$$

Environmental variation affects the dynamics via the "fertility" function $R(\alpha_t, y)$, where $\alpha_t, t = 0, 1, 2, \ldots$ is a sequence of random variables. Density-dependent population limitation, modeled by F, is independent of time, as is dispersal, modeled by K. The habitat is a (fixed) compact subset $\Omega \subset \mathbb{R}$ with nonempty interior. The initial condition and subsequent population densities are random functions in $\mathscr{C}_+(\Omega)$, the positive cone of continuous functions on Ω.

Hardin et al. (1988a) make the following assumptions:

(V1)   Random variables α_t are iid from some index set $\mathscr{A}$.

(V2)   Dispersal kernel K is a continuous and strictly positive function on $\Omega \times \Omega$.

(V3)   For each $\alpha \in \mathscr{A}$, function $R(\alpha, x)$ is in $\mathscr{C}_+(\Omega)$, and there exist positive constants such that $0 < \underline{R} \leq R(\alpha, x) \leq \overline{R}$ for all $x \in \Omega$.

(V4)   Function F is continuous, nonnegative, and bounded. It is differentiable at zero, and $F(0) = 0$. Furthermore, F is nondecreasing, and $F(x)/x$ is strictly decreasing for $x > 0$.

The first assumption implies that temporal variations are uncorrelated. The second condition indicates that within one dispersal period, an individual can move from any location in the habitat to any other location. The third assumption excludes the possibility that the population dies out in a single year. The conditions on the density-dependent growth limitations imply that the per capita yield decreases with density. They are satisfied by a Beverton–Hold type function; see (2.11).

Theorem 16.1 (Theorem 4.2 in Hardin et al. 1988a) *Suppose that conditions (V1)–(V4) are satisfied for (16.20) and that $N_0 \neq 0$ with probability one. Then N_t converges in distribution to a stationary distribution μ_*, which is independent of N_0. Furthermore, we have either $\mu_*(\{0\}) = 0$ or $\mu_*(\{0\}) = 1$.*

Just as in the deterministic case and in the nonspatial stochastic case, the difference between extinction and persistence is given by the behavior of the process at small densities. We denote by $Q'_t[0]$ the Fréchet derivative of Q_t at zero.

Theorem 16.2 (Lemma 5.1 and Theorem 5.3 in Hardin et al. 1988a) *Suppose that conditions (V1)–(V4) are satisfied. Then the limit*

$$\lambda = \lim_{t \to \infty} \| Q'_t[0] \circ Q'_{t-1}[0] \circ \cdots \circ Q'_0[0] \|^{1/t} \tag{16.21}$$

exists with probability one. Furthermore, if $\lambda < 1$, then $\mu_*(\{0\}) = 1$ and $N_t \to 0$ with probability one. Alternatively, if $\lambda > 1$, then $\mu_*(\{0\}) = 0$.

Both of these results hold under somewhat weaker conditions and for more general processes than the ones in (16.20) (Hardin et al. 1988a).

Hardin et al. (1988b) prove the corresponding results in T-periodic environments. They study the operator

$$Q_t[N] = \int_\Omega K(x, y) F_t(n(y)) \, dy, \tag{16.22}$$

where K denotes a dispersal kernel; F_t models reproduction in year t with $F_{t+T} = F_t$; and Ω is a bounded domain, as above. They consider the existence of a positive fixed point for the period-T-map $\overline{Q}^T = Q_{T-1} \circ \cdots \circ Q_0$, as well as its local and global stability. Many of their results are contained in our Chap. 4, in particular Sect. 4.3 for global existence.

The persistence condition $\lambda > 1$ for the stochastic process is elegant theoretically but difficult to apply, even numerically. Jacobsen et al. (2015) present an equivalent condition that is computationally simpler to evaluate. Their model generalizes (16.20) in that it allows the dispersal kernel to vary in time. Their particular motivation was to study the effect of variable flow rates on the persistence of stream populations; see Sect. 12.2.

Consider the stochastic process

$$N_{t+1}(x) = Q_t[N_t](x) = \int_\Omega K_{\alpha_t}(x - y) F_{\alpha_t}(N_t(y)) dy, \qquad t = 0, 1, \ldots,$$
(16.23)

where α_t are iid random variables from some index set $\mathscr{A}$. We require the following generalizations of and additions to conditions (V1)–(V4):

(V2′) For each $\alpha \in \mathscr{A}$, K_α is a continuous function, and there exist constants such that $0 < \underline{K} \leq K_\alpha \leq \overline{K}$ for all α.

(V4′) For each $\alpha \in \mathscr{A}$, F_α is a nonnegative, continuous, and increasing function such that $F_\alpha(x)/x$ is decreasing and the right-sided limit $F_\alpha'(0)$ exists. Functions F_α are uniformly bounded by $m > 0$.

(V5) We have uniform limits $F_\alpha(x)/x \to F_\alpha'(0)$ as $x \to 0$ and uniform bounds $0 < \underline{F} \leq F_\alpha'(0) \leq \overline{F}$.

(V6) For $b = m\overline{K}|\Omega|$, there exists $\inf_{\alpha \in \mathscr{A}} F_\alpha(b) > 0$.

(V7) There exists $\alpha^* \in \mathscr{A}$ such that $Q_\alpha[N] \leq Q_{\alpha^*}[N]$ for all $\alpha \in \mathscr{A}$ and nonnegative, continuous functions N on Ω.

Theorem 16.3 (Theorems 1 and 2 in Jacobsen et al. 2015) *Assume that conditions (V2′), (V4′), and (V5)–(V7) are satisfied. Then Theorems 16.1 and 16.2 hold for (16.23). Furthermore, the limit*

$$\Lambda = \lim_{t \to \infty} \Lambda(t) = \lim_{t \to \infty} \left[\int_\Omega n_t(x) dx \right]^{1/t}$$
(16.24)

exists and is independent of n_0, where n_t is defined by $n_{t+1} = Q_t'[0]n_t$. Finally, $\Lambda = \lambda$.

The theory presented thus far considered temporal variation in growth and dispersal but assumed that the size and location of the habitat patch are fixed over time. There are many examples of natural habitats whose size and location vary within and between years. Wetlands are a particular example where surface area and depth vary according to rainfall and other climatic conditions. These observations motivate the study by Zhou and Fagan (2017), in which habitat size and location can vary with time. The authors implement a temporally varying habitat via a habitat quality function.

Zhou and Fagan (2017) analyze the model

$$N_{t+1}(x) = \int_{-\infty}^{\infty} K(x, y) H_t(y) F(N_t(y)) \mathrm{d}y , \qquad (16.25)$$

where H_t is the temporally varying habitat quality function (compare Sect. 15.1) that determines the fraction of offspring produced at location y that survive to disperse. Function H_t has to be nonnegative and bounded above by unity. When H_t is the characteristic function of some fixed domain Ω, i.e., $H_t(x) = \chi_\Omega$, the model is equivalent to the basic IDE (3.1). When a domain of fixed length moves at constant speed, i.e., $H_t(x) = \chi_{[ct, L_0+ct]}$, we have the model from Sect. 12.3. When the domain length, L_t, varies with time, we may write $H_t(x) = \chi_{[0, L_t]}$. Zhou and Fagan (2017) consider this setting for a population whose habitat is the surface of a wetland. When not only the extent but also the quality vary in space and time, Zhou and Fagan (2017) suggest $H_t(x) = \exp(-x^2/\sigma_t^2)$ for a single wetland, where σ_t^2 is a random variable that indicates the extent in year t. The authors also consider more complex situations with, for example, two adjacent wetlands, modeled by a linear combination of two Gaussian functions, where the extent of each as well as the distance between them can vary over time.

The difficulty in analyzing model (16.25) lies in the variability of the domain with potentially infinite extent. If the support of all functions H_t is uniformly bounded, we can reduce model (16.25) to one on a compact set and obtain the same results as in Hardin et al. (1988a,b). Zhou and Fagan (2017) give conditions on the dispersal kernel and the habitat quality function under which the corresponding results hold even on the entire real line in a T-periodic environment. In particular, they show that, under some conditions, the stability of the zero solution is given by the spectral radius of the period-T-operator, and that the instability of zero implies that the (supremum norm of the) population will eventually be bounded below uniformly by some constant. Zhou and Fagan (2017) manage to calculate persistence conditions explicitly in two special cases. They define the "lower minimal habitat size" as an extension of the critical patch-size (Chap. 3) to periodic environments.

16.4   An Example with the Laplace Kernel

We illustrate some of the theory from the preceding section with a simple example. We assume that in year t, the habitat is an interval of length L_t. Inside the habitat, the growth function is the scaled Beverton–Holt function (2.13) with parameter $R_t > 1$; outside, the growth function is zero. Dispersal follows a Laplace kernel (2.27) with parameter a_t that can be interpreted as the root of the ratio α_t/D_t of the settling rate and the diffusion coefficient in a random walk; see Sect. 7.2. By scaling space in year t with L_t, we can write the equation on the fixed domain $[-1/2, 1/2]$ with kernel parameter $\tilde{a}_t = a_t L_t$ as

$$N_{t+1}(x) = \int_{-1/2}^{1/2} \frac{\tilde{a}_t}{2} e^{-\tilde{a}_t|x-y|} \frac{R_t N_t(y)}{1 + (R_t - 1)N_t(y)} dy .\tag{16.26}$$

In the following, we will drop the tilde to ease notation. We are interested in population persistence. By the theoretical results in the preceding section, we need to study the stability of the zero state. Hence, we linearize the equation at low density. The resulting eigenvalue problem for the integral equation can be turned into an equivalent boundary-value problem for a differential equation (Jacobsen and McAdam 2014; Jacobsen et al. 2015), similar to the procedure in Chap. 3.

The Periodic Case

We consider a two-periodic environment, so that we have four model parameters: $a_{1,2} > 0$ and $R_{1,2} > 1$. The eigenvalue problem for the integral equation is given by

$$\lambda_p \phi(x) = R_1 R_2 \int_{-1/2}^{1/2} \int_{-1/2}^{1/2} \frac{a_1}{2} e^{-a_1|x-y|} \frac{a_2}{2} e^{-a_2|y-z|} \phi(z) dy dz .\tag{16.27}$$

To turn this equation into a boundary-value problem, we follow Jacobsen et al. (2015). We introduce the function

$$\psi(x) = R_1 \int_{-1/2}^{1/2} \frac{a_1}{2} e^{-a_1|x-y|} \phi(y) dy .\tag{16.28}$$

Then ϕ satisfies (16.27) exactly if ϕ and ψ satisfy (16.28) and

$$\phi(x) = \frac{R_2}{\lambda_p} \int_{-1/2}^{1/2} \frac{a_2}{2} e^{-a_2|x-y|} \psi(y) dy .\tag{16.29}$$

Differentiating twice, we find the second-order equations

$$\phi'' = a_2^2 \left(\phi - \frac{R_2}{\lambda_p} \psi \right) \quad \text{and} \quad \psi'' = a_1^2 (\psi - R_1\phi)\tag{16.30}$$

for $x \in (-1/2, 1/2)$; compare (3.10). Differentiating again, these two equations can be turned into a single fourth-order equation for ϕ, namely

$$\phi^{(4)} - (a_1^2 + a_2^2)\phi'' + a_1^2 a_2^2 \left(1 - \frac{R_1 R_2}{\lambda_p} \right) \phi = 0.\tag{16.31}$$

We need to find boundary conditions. Two conditions are obtained exactly as in (3.11) by differentiating (16.28) and (16.29) once. We find

$$\phi'(-1/2) = -a_2\phi(-1/2), \qquad \phi'(1/2) = a_2\phi(1/2), \qquad (16.32)$$

$$\psi'(-1/2) = -a_1\psi(-1/2), \qquad \psi'(1/2) = a_1\psi(1/2). \qquad (16.33)$$

However, we need to find conditions for ϕ, not ψ. Differentiating (16.30) and substituting the above boundary conditions results in

$$\phi'''(-1/2) = a_1\phi''(-1/2) + a_2^2(a_2 - a_1)\phi(-1/2), \qquad (16.34)$$

$$\phi'''(1/2) = -a_1\phi''(1/2) - a_2^2(a_2 - a_1)\phi(1/2). \qquad (16.35)$$

The equation for ϕ has the bi-quadratic characteristic equation $r^4 - ar^2 + d = 0$, where

$$a = a_1^2 + a_2^2 \quad \text{and} \quad d = a_1^2 a_2^2 \left(1 - \frac{R_1 R_2}{\lambda_p}\right). \qquad (16.36)$$

Just as in Chap. 3, we have $\lambda_p < R_1 R_2$, so that $a > 0$ and $d < 0$. We obtain two real and two purely imaginary roots:

$$r_1 = \sqrt{\frac{a + \sqrt{a^2 - 4d}}{2}}, \quad ir_2 = \sqrt{\frac{a - \sqrt{a^2 - 4d}}{2}}, \qquad (16.37)$$

and $r_3 = -r_1$, $r_4 = -r_2$. By symmetry, the eigenfunction can be written as

$$\phi(x) = c_1 \cosh(r_1 x) + c_2 \cos(r_2 x). \qquad (16.38)$$

To satisfy the boundary conditions, coefficients $c_{1,2}$ have to satisfy the equations

$$[a_2 \cosh(r_1/2) + r_1 \sinh(r_1/2)]c_1 + [a_2 \cos(r_2/2) - r_2 \sin(r_2/2)]c_2 = 0,$$

$$\left[(a_1 r_1^2 + a_2^2(a_2 - a_1))\cosh(r_1/2) + r_1^3 \sinh(r_1/2)\right]c_1 \qquad (16.39)$$

$$+ \left[r_2^3 \sin(r_2/2) - (a_1 r_2^2 - a_2^2(a_2 - a_1))\cos(r_2/2)\right]c_2 = 0.$$

For a nonzero solution, the determinant of the coefficient matrix has to vanish. This condition can be evaluated numerically.

We choose the same setup of a two-periodic environment as in the previous section. We write $a_{1,2} = \bar{a} \pm \epsilon_a$ and $R_{1,2} = \bar{R} + \epsilon_R$. Figure 16.3 shows that the dominant eigenvalue λ_p from (16.27) decreases as the variation ϵ_a in the kernel parameter and ϵ_R in the growth rate increases. Instead of the eigenvalue itself, we actually plot the square root of λ_p so that we can compare it with the average per generation rate of increase in the stochastic model below.

Fig. 16.3 Persistence condition for model (16.26) in the two-periodic and random cases. Solid curves show the square root of the eigenvalue, $\sqrt{\lambda_p}$, in (16.27). Stars stand for the numerically obtained value $\Lambda_T \approx \Lambda$ from (16.24). Parameters are $a_{1,2} = 2.7 \pm \epsilon_a$, and $R_{1,2} = \bar{R} \pm \epsilon_R$ with $\epsilon_R = 0$ (solid), $\epsilon_R = 0.5$ (dashed), and $\epsilon_R = 1$ (dash-dot).

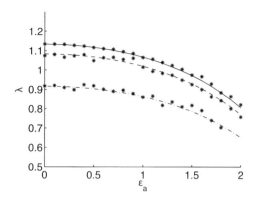

The Stochastic Case

For random variation, we choose binary random variables a_t with values a_1, a_2 and R_t with values R_1, R_2 with equal probability. Then we numerically solve the linear equation

$$n_{t+1}(x) = R_t \int_0^1 \frac{a_t}{2} \exp(-a_t |x - y|) n_t(y) dy \tag{16.40}$$

and approximate Λ from (16.24) by Λ_T for some large value of T. The results for different values of ϵ_a and ϵ_R are plotted as stars in Fig. 16.3. We note that the expected per generation rates of increase (or decrease) in the stochastic and periodic case are very close together. In fact, Jacobsen et al. (2015) find an even better agreement between their numerically calculated value Λ and the analytical expression $\sqrt{\lambda_p}$ in a slightly different setting. We note that persistence is harder to achieve and, in fact, fails, as the variation in each of the two parameters increases.

On a technical note, we found that the FFT algorithm from Sect. 8.2 could not (easily) provide the same accuracy as even the simple direct integration method from Sect. 8.3. Since the equation is linear, solutions grow or decay exponentially. Therefore, as the number of generations in the simulation grows, the values become either very large (if the solution is growing) or very small (if it is decaying), so that accuracy is difficult to maintain for large times. However, since we are interested in the limit of large times, there is some trade-off between accuracy of the computational steps and the number of time steps that one takes. To smooth out the results somewhat, we chose $T = 2000$ and averaged the value of the last 20 time steps.

16.5   Spread Under Temporal Variation

To study the effects of temporal variation on the spread rate of a population, we begin with the work by Neubert et al. (2000) (they provide corrected figures in an erratum) and study the IDE on the real line,

$$
N_{t+1}(x) = Q_t[N_t](x) = \int_{-\infty}^{\infty} K_t(x - y) F_t(N_t(y)) dy \,. \tag{16.41}
$$

Before we discuss results for the stochastic case, we briefly present some explicit results for the periodic case.

We assume that K_t and F_t are T-periodic functions of time. Furthermore, we assume that for each t, the growth function satisfies conditions (F1)–(F4) in Sect. 5.4. We also assume that for each t, the dispersal kernel is continuous, symmetric, and exponentially bounded. Then the period-T operator $\overline{Q}_t = Q_{T-1} \circ \cdots \circ Q_0$ satisfies the conditions of Theorem 5.1. Therefore, there exists a spreading speed, and this speed can be characterized as the slowest traveling-wave speed. Furthermore, the speed is linearly determined.

If we denote this speed by c^*T, where c^* is the average speed per generation, formula (5.17) gives the representation

$$
c^* = \inf_{s>0} \frac{1}{s} \ln \left(\prod_{t=0}^{T-1} R_t M_t(s) \right)^{1/T} = \inf_{s>0} \frac{1}{s} \frac{1}{T} \sum_{t=0}^{T-1} \ln(R_t M_t(s)) \,, \tag{16.42}
$$

where M_t is the moment-generating function of kernel K_t and $R_t = F_t'(0)$. We denote by c_t^* the speed in a constant environment with conditions as in generation t. Since the infimum of the averages is generically greater than the average of the infima (unless the infima all occur at the same location), we have

$$
c^* > \frac{1}{T} \sum_{t=0}^{T-1} \inf_{s>0} \frac{1}{s} \ln(R_t M_t(s)) = \frac{1}{T} \sum_{t=0}^{T-1} c_t^* \,. \tag{16.43}
$$

Hence, the average speed per generation in the periodically varying environment is larger than the average of the speeds, c_t^*, that would occur in each of the corresponding constant environments. This statement can be strengthened as follows. Neubert et al. (2000) define the "instantaneous speed between generations" as

$$
\bar{c}_t = \frac{1}{s^*} \ln(R_t M_t(s^*)) \,, \tag{16.44}
$$

where s^* is the argument that minimizes the expression in (16.42). Then, $\bar{c}_t > c_t^*$, which implies that the instantaneous speed between generations is greater than the asymptotic speed would be in an environment of constant conditions of the most recent generation.

In the particular case that K_t are Gaussian kernels with variance σ_t^2, the procedure in Sect. 5.2 gives the exact expression for the average speed per generation as

$$c^* = \sqrt{2\langle\sigma_t^2\rangle_a \ln(\langle R_t\rangle_g)}, \tag{16.45}$$

where $\langle\cdot\rangle_a$ denotes the arithmetic mean and $\langle\cdot\rangle_g$ the geometric mean.

Stochastic Environments

To consider spread rates for Eq. (16.41) in a stochastically varying environment, we study again the linearized IDE with growth rate $R_t = F_t'(0)$. Neubert et al. (2000) discuss the conditions for which the result is the spread rate in a corresponding nonlinear equation. We can approach the question of spread via the expectation or via direct calculation.

Taking expectations of Eq. (16.41) and assuming that growth and dispersal are uncorrelated with population density, we find

$$E[N_{t+1}](x) = \int_{-\infty}^{\infty} E[R_t K_t(x - y)]E[N_t(y)]dy . \tag{16.46}$$

This is a deterministic equation for the "expectation wave." According to the theory in Chap. 5, there is a spreading speed, c^*. Formula (5.17) applies and results in the expression

$$c^* = \inf_{s>0} \frac{1}{s} \ln(E[R_0 M_0(s)]) , \tag{16.47}$$

where M_0 is the moment-generating function of K_0. When growth and dispersal are uncorrelated, we obtain

$$c^* = \inf_{s>0} \frac{1}{s} \ln(E[R_0]E[M_0(s)]) . \tag{16.48}$$

Positive correlations increase the spreading speed of the expectation wave.

For direct calculations, we choose the initial profile $N_0(x) = \exp(-sx)$ and calculate

$$N_1(x) = \int_{-\infty}^{\infty} R_0 K_0(x - y)e^{-sy}dy = \int_{-\infty}^{\infty} R_0 K_0(z)e^{sz}dy \; e^{-sx} = R_0 M_0(s)e^{-sx} . \tag{16.49}$$

Iteratively, we find

$$N_t(x) = \prod_{j=0}^{t-1} R_j M_j(s) e^{-sx} . \tag{16.50}$$

Just as in Chap. 5, we define the extent of the population as $X_t = \sup_x \{N_t(x) \geq \tilde{N}\}$ for some threshold density, $\tilde{N}$. Unlike in Chap. 5, this quantity is now a random variable. Its explicit expression is

$$X_t = \frac{1}{s} \sum_{j=0}^{t-1} \ln(R_j M_j(s)) - \ln(\tilde{N})/s . \tag{16.51}$$

The mean speed per generation is the random variable

$$C_t = \frac{1}{t}(X_t - X_0) = \frac{1}{t} \sum_{j=0}^{t-1} \frac{1}{s} \ln(R_j M_j(s)) . \tag{16.52}$$

By the central limit theorem, C_t is asymptotically normally distributed with mean and variance given by

$$\mu(s) = \frac{1}{s} E\left(\ln(R_0 M_0(s))\right) \quad \text{and} \quad \sigma^2(s) = \frac{1}{ts^2} \mathrm{Var}(\ln(R_j M_j(s))) . \tag{16.53}$$

As before, the relevant speed is the minimum that occurs with respect to s; we denote the corresponding value by s^*. Hence, the speed that we are interested in is C_t with mean $\mu(s^*)$ and variance $\sigma^2(s^*)$. As $t \to \infty$, we find $\sigma^2(s^*) \to 0$, so that

$$C_t \to \bar{C} = \mu(s^*) = \inf_{s>0} \frac{1}{s} E[\ln(R_0 M_0(s))] . \tag{16.54}$$

Comparing this expression for $\bar{C}$ with c^* from (16.48), we apply Jensen's inequality again and find $\bar{C} \leq c^*$. In other words, the expectation wave is faster than almost every realization of the process. While the speed converges, i.e., $C_t \to \bar{C}$, the spatial extent, X_t, does not converge to $X_0 + \bar{C}t$. The reason for this lack of convergence is that the variance of the expression $C_t t + X_0$ grows linearly in t.

In Fig. 16.4, we plot the front location of two realizations of the process in an environment that switches between two states with probability 1/2. For comparison, we plot the front location in a two-periodic environment, as well as the average speed per generation from (16.45). We observe that one of the stochastic realizations is ahead of the periodic case and the other is behind, but all have the same slope. Hence, while the speed is predictable, the location is not. The plot on the right shows the front location (X_t) and the average speed (C_t) for 15 realizations of the process. Whereas the average speeds converge, the range of front locations spreads over time. Neubert and Parker (2004) review these results on spreading speeds in the context of risk analysis and apply the theory to the invasion of scotch broom.

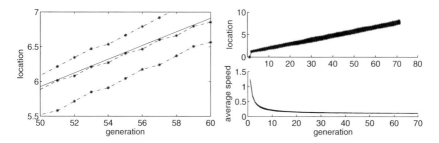

Fig. 16.4 Left: The location of the front as a function of time in a two-periodic environment (dashed) and two realizations in a random environment (dash-dot). The growth functions are Beverton–Holt functions $F(N) = R_j N/(1 + N)$, and the kernels are Gaussian kernels with variance $\sigma_{1,2}^2$. The solid line represents the exact average speed per generation from (16.45). **Right:** The front location, X_t (top), and the average speed, C_t (bottom), for 15 realizations. Parameters are $R_{1,2} = 1.7 \pm 0.5$ and $\sigma_{1,2}^2 = 0.01 \pm 20\%$.

16.6   Further Reading

Lewis and Pacala (2000) and Lewis (2000) formulate a discrete stochastic process for the reproduction and dispersal of individuals and analyze its spreading behavior. They derive a set of moment equations, which turn out to be IDEs. In fact, the equation for the first moment of the linear process is precisely the equation for the expectation wave (16.46). They study permanence of form for the spreading population and use moment closure techniques and comparison methods to bound spread rates. Snyder (2003) continues this theory and shows that stochasticity can slow invasions but concludes that the effect is relatively weak. Kot et al. (2004) link individual-based simulations and IDEs via branching random walks to study the effect of demographic stochasticity on the speed of invasions. They find that stochasticity does not slow the overall asymptotic speed and that accelerating invasions can occur with stochasticity as well.

Several authors consider the spread of structured populations in periodic and stochastic environments. Caswell et al. (2011) extend their previous sensitivity analysis for spread rate in structured population models (see Neubert and Caswell 2000a) to periodic and stochastic environments. Simultaneously but independently, Schreiber and Ryan (2011) derive formulas for invasion speeds for stage-structured IDEs in stochastic environments. They show that invasion speeds are asymptotically normally distributed and that, as is the case for unstructured populations, the variance decreases over time (Fig. 16.4). Increased variation in fecundity decreases invasion speeds, but correlations between fecundity and dispersal determine by how much. Related work for spatial integral projection models can be found in Ellner and Schreiber (2012).

Ding et al. (2013) prove the existence of spreading speeds in time-periodic IDEs in spatially constant or spatially periodic environments. They show that temporal heterogeneity will slow down invasions if space is homogeneous but can speed up invasions if space is also heterogeneous. Jacobsen et al. (2015) determine persistence criteria on a bounded interval with temporally varying unidirectional flow. Bouhours and Lewis (2016) consider a moving-habitat model with stochasticity and determine persistence conditions. Zhou and Fagan (2017) consider temporally varying habitat size and quality and extend the theory by Hardin et al. (1988a) to the case where the habitat may be unbounded. They give several examples of wetland habitats that vary with seasonal rainfall, and they calculate long-term persistence conditions. Reimer et al. (2017) compare and contrast several approaches to determining persistence conditions of populations on a bounded domain under stochasticity. They use individual-based model simulations, Galton–Watson branching processes, and a deterministic IDE. They find that the critical patch-size for the stochastic models is typically larger than that for the deterministic model.

Several authors apply stochastic IDEs to various ecological questions and base their investigation largely on numerical simulations. Mahdjoub and Menu (2008) consider the question of whether and how diapause can affect the population spread of the chestnut weevil (*Curculio elephas*). They consider equations for developing individuals and individuals in diapause. Only the former disperse. They find that prolonged diapause will reduce spread in a constant environment but can increase spread in a temporally varying environment. Gilioli et al. (2013) model the spread of the chestnut gall wasp (*Dryocosmus kuriphilus*) in Europe. They use a deterministic IDE for short-distance dispersal, coupled with a stochastic component for long-distance dispersal, to capture the observed stratified dispersal pattern. Gharouni et al. (2017) formulate a three-stage model for green crab and study the effect of stochastic variation on the spread of the crab against the dominant current. Stochasticity may help the population spread "upstream"; see Sect. 12.2.

Jacobs and Sluckin (2015) study the effect of demographic stochasticity on accelerating invasions. They compare individual-based simulations on a lattice with predictions from a deterministic IDE model. When the dispersal kernel is heavy tailed, the IDE predicts accelerating invasions, but most of the corresponding lattice models appear to have constant-speed invasions.

Chapter 17
Further Topics and Related Models

Abstract Even though IDEs are a relatively recent modeling framework in spatial ecology, their theory and applications contain many more aspects than can fit in a single book. In this final chapter, we mention further topics in the study of IDEs, some related to applications, some to the mathematical theory. We also briefly indicate a number of closely related model formulations and techniques. Some of these models are related in terms of the questions studied, others in terms of the mathematical structure of the equation.

17.1 Further Topics

We begin with a collection of topics that have been explored with IDEs to some extent but that are still far from being completely developed.

Density-Dependent Dispersal

One of the ecologically appealing features of IDEs is that the description of dispersal is not limited to diffusion processes. A histogram of dispersal data can be translated relatively easily into a dispersal kernel; see Lewis et al. (2006) or Lewis et al. (2016) and Sect. 12.7. One limiting assumption is that the dispersal of each individual is independent of all others. While this assumption seems fine in the context of wind-dispersed seeds, it does not necessarily apply to animal dispersal, because animals can interact with conspecifics during dispersal. Few authors propose ways to overcome this difficulty.

The easiest way to include some form of density dependence is to make the *probability of dispersing* a function of local population density. Veit and Lewis (1996) implement this idea in their model for the spread of house finches and explore its effects numerically. When the probability of dispersal increases with density, the instantaneous spread rate is initially slow. It takes more generations than without density-dependent dispersal to achieve a rate close to the asymptotic

F. Lutscher, *Integrodifference Equations in Spatial Ecology*, Interdisciplinary Applied Mathematics 49, https://doi.org/10.1007/978-3-030-29294-2_17

speed. Lutscher (2008) studies the effect of density-dependent dispersal on spread
rates more systematically. He illustrates some mathematical difficulties that can
result from this approach; e.g., the next-generation operator may not preserve
monotonicity (even for monotone growth functions) nor have an obvious invariant
domain (again, for monotone growth functions). In some special cases, Lutscher
(2008) derives explicit expressions for traveling-wave solutions. Dwyer and Morris
(2006) generalize this idea by allowing not only the probability of dispersal but also
the *characteristics of dispersal* (e.g., the variance of the dispersal kernel) to depend
on the population density at the point of origin of the disperser. Their numerical
simulations of spread in a consumer–resource model reveal some intriguing patterns
that still await more systematic analysis. In particular, when the dispersal distance of
the consumer depends on resource density, the instantaneous spread rate fluctuates
widely over time so that it is unclear whether an asymptotic speed exists at all.

Carrillo and Fife (2005) use a completely different strategy. They begin with a
"movement law," a function $g(x, y, [N])$ that describes the number of individuals
that move from $(y, y + \Delta_y)$ to $(x, x + \Delta_x)$ when the population density is given
by $N(x)$. They then write the "balance law" for how the population density changes
due to redistribution (in the absence of birth and death) as

$$N_{t+1}(x) = N_t(x) + \int_{-\infty}^{\infty} g(x, y, [N_t]) \mathrm{d}y - \int_{-\infty}^{\infty} g(y, x, [N_t]) \mathrm{d}y. \qquad (17.1)$$

They derive various forms that function g could have and perform a linear stability
analysis to investigate conditions for pattern formation under congregation behavior.

Nonlocal Interaction

Our formulation of IDE (2.1) assumes that all interactions between individuals are
local. This assumption means that the growth function is evaluated only at the point
where the individuals are located. Individuals of many species interact with one
another over intermediate and even larger spatial scales. For example, trees interact
with others at some distance via an elaborate network of roots and also by shading
from their crowns.

Most reaction–diffusion equations also use local interaction terms, but a substantial theory exists for these equations with nonlocal interaction. The population
density in the per capita growth function at a particular spatial location is replaced
by a suitable average of the density in some neighborhood of that location;
see, e.g., Britton (1989). This nonlocal term makes the analysis of the reaction–
diffusion equation considerably harder since the equation will in general not have a
comparison principle.

The question of nonlocal interaction has received only marginal attention in
IDEs. Merchant (2001) considers the spread of a single species with a nonlocal
growth function. More precisely, for a given population density, $N_t(x)$, Merchant

(2001) defines the effective density via a "competition kernel," $C(z)$, that describes how individuals at distance z affect each other as

$$N_t^{\text{eff}}(x) = \int_{-\infty}^{\infty} C(x - y)N_t(y)\mathrm{d}y. \tag{17.2}$$

The population-level IDE then reads

$$N_{t+1}(x) = \int_{-\infty}^{\infty} K(x, y)\widetilde{F}(N_t^{\text{eff}}(y))N_t(y)\mathrm{d}y, \tag{17.3}$$

where $\widetilde{F}$ is the per capita growth rate. Merchant (2001) simulates and compares deterministic and stochastic versions of this IDE and finds, among other things, that nonlocal interactions can (i) slow the spread of the population and (ii) induce a nonmonotone wave profile even when the function $F(N) = \tilde{F}(N)N$ is monotone.

Aydogmus et al. (2017) also consider nonlocal IDEs. They develop a mathematical framework to determine the bifurcation behavior of such equations on a bounded domain. More specifically, they derive the Stuart–Landau equations, which describe the amplitude of bifurcating solutions. They use these results to study pattern formation in a nonlocal IDE.

The question of when nonlocal models are more appropriate than local models is, of course, one of relative spatial scales. If the spatial interaction scale is much smaller than the dispersal scale, then a local model is appropriate; if the interaction scale is comparable to the dispersal scale, then a nonlocal model is warranted. Merchant (2001) finds that as the interaction scale decreases, the behavior of the nonlocal model approaches that of the local model. The phenomenological model by Deng et al. (2014) for pattern formation in bacterial colonies explicitly considers two different scales: a smaller region of positive interaction and a larger region of negative interaction. The resulting interaction kernel is a "Mexican hat" shape: it has a positive maximum at zero, decreases to negative values for intermediate distances, and approaches zero for large distances. In particular, since the kernel assumes negative values, it cannot be interpreted as a dispersal kernel. Numerical simulations reveal that the model supports the formation of branched patterns observed in experiments.

Optimal Control for IDEs

When we apply population dynamics models to management problems, the questions often involve some aspect of optimality. For example, when harvesting a population, we are not just concerned with setting a maximal harvesting rate to ensure the sustainability of the population, but rather we want to find an optimal harvesting rate that maximizes profit in the long run. While optimal control of finite-dimensional systems is a reasonably well developed field (Lenhart and Workman

2007), optimal control for reaction–diffusion equations is much more elaborate (see, e.g., Kelly et al. 2016), and for IDEs, the field is still in its infancy.

Joshi et al. (2006) consider optimal harvesting strategies for a single, linearly growing population. Their state variable is the population density, satisfying the equation

$$N_{t+1}(x) = (1 - \alpha_t(x)) \int_\Omega K(x, y) R N_t(y) dy . \tag{17.4}$$

The spatially and temporally varying control $\alpha_t(x)$ represents the percentage of the population harvested at location x at the end of generation t. The objective functional

$$J(\alpha) = \sum_{t=0}^{T-1} \int_\Omega \left[A_t e^{-\delta t} \alpha_t(x) \int_\Omega K(x, y) R N_t(y) dy - \frac{B_t}{2} \alpha_t(x)^2 \right] dx \tag{17.5}$$

describes the difference between revenue and cost. The revenue consists of the amount of population harvested, multiplied by the price factor, A_t, and the discount factor, $e^{-\delta t}$. The cost is assumed to be quadratic in the percentage harvested, and B_t is the weight factor. The goal is to find a maximum α^* among all possible controls $\alpha = (\alpha_0(x), \alpha_1(x), \dots, \alpha_{T-1}(x))$ of J, i.e.,

$$J(\alpha^*) = \max_\alpha \{J(\alpha)\} . \tag{17.6}$$

Joshi et al. (2006) prove that an optimal control exists, characterize it in terms of an adjoint problem, and prove its uniqueness for sufficiently large B_t.

Joshi et al. (2007) extend these results to a nonlinearly growing population. Gaff et al. (2007) consider optimal harvesting in the presence of a pathogen. Zhong (2011) compares optimal harvesting of a single species in an IDE with different order of events. Martinez (2015) extends the theory of optimal control to a pest–pathogen system. Kura et al. (2019) formulate an optimal control problem for the release of sterile insects to control an insect pest and prove that an optimal control exists. They study how the optimal control depends on dispersal behavior and other model aspects.

Evolutionary Aspects

For most of this book, we have focused on the ecological dynamics of populations and neglected any possible evolutionary consequences of interactions. Yet, evolution is present in most individual and population processes and should be considered for certain aspects. Since IDEs can represent a wide range of dispersal patterns, a particularly intriguing question is how different dispersal patterns may affect evolutionary processes differently.

In one of the earliest works on the evolution of dispersal in IDEs, Hardin et al. (1990) simulate population dynamics under a few specifically chosen dispersal kernels and measure the success of the dispersal strategy by the persistence and density of the population. They find that a temporally constant environment favors a "stay-in-place" strategy (represented by a delta distribution for the dispersal kernel), whereas a temporarily varying environment favors a "go-everywhere-uniformly" strategy (represented by a spatially uniform dispersal kernel).

Baskett et al. (2007) use a game-theoretic approach to study optimal dispersal behavior in fragmented habitats. They consider an IDE for a marine species with planktonic larvae in a network of reserves, represented by a one-dimensional patchy landscape as in Chap. 15. Among other things, they find that increased fragmentation leads to increased selection pressure for short-distance dispersal.

Lutscher (2008) studies how density-dependent dispersal evolves in a temporally varying habitat. He uses the theory of adaptive dynamics (e.g., Geritz et al. 1998), which is also a game-theoretic framework. Numerical simulations indicate that the rate of density-dependent dispersal converges to an evolutionarily stable state that decreases with the frequency of the temporal variation.

Phillips et al. (2008) study the evolution of the dispersal kernel at the front of the cane toad (*Bufo marinus*) invasion in Australia and relate it to Reid's paradox. Their study relies on individual-based models, and while it derives a dispersal kernel, it does not consider the corresponding IDE.

Williams et al. (2016) consider an IDE for a species spreading in a fragmented habitat, where there is a trade-off between fecundity at low density and competition with others at higher density. They use a game-theoretic approach and approximate analytic solutions to find an evolutionarily stable strategy that depends on the fragmentation level. Typically, larger gaps between patches favor competitively superior individuals, whereas small gaps favor individuals with high fecundity.

Wang et al. (2016) study the evolution of dispersal in a spatially periodic IDE by using tools and approaches from game theory. They consider the probability of dispersal as the trait in question. If that probability is independent of local growth conditions (unconditional dispersal), a population with lower dispersal probability will invade and replace one with higher dispersal probability. Hence, the authors recover the result that the "slower disperser wins," which is well documented and understood in reaction–diffusion equations (Hastings 1983; Hutson et al. 2003). When the dispersal probability depends on local growth conditions (conditional dispersal), the authors find that an "ideal-free strategy" is evolutionarily stable, which is also in analogy with the reaction–diffusion case (Cosner 2014).

Marculis et al. (2017) study the dynamics of so-called neutral genetic patterns for spreading populations. Genetic differences are considered "neutral" if they do not affect the two basic aspects of dispersal and reproduction that are modeled in IDEs. The entire population is divided into neutral fractions, $N_t(x) = \sum_{i=1}^{k} n_t^i(x)$, that satisfy the equation

$$n_{t+1}^i(x) = \int_{-\infty}^{\infty} K(x-y)\widetilde{F}(N_t(y))n_t^i(y)\mathrm{d}y, \qquad (17.7)$$

where $\widetilde{F}$ is the per capita growth function. Due to the neutrality assumption, the total density, N_t, satisfies the basic IDE (2.1) with $F(N) = \widetilde{F}(N)N$. Marculis et al. (2017) show that the genetic pattern of a spreading population with a thin-tailed kernel depends crucially on the growth function. If there is no Allee effect, the different genetic fractions do not mix and the population is dominated by only one such fraction at the leading edge. If there is an Allee effect, the different genetic fractions will mix and will all be present at the leading edge of an invading population. Lewis et al. (2018) extend this analysis to a population on a moving patch. Goodsman et al. (2014) take a simulation approach to study a similar question of neutral genetics during range expansion. They consider a single neutral locus with two alleles and study the dynamics of the three resulting genotypes in a spreading population. They track the degree of heterozygosity in the population. Their results are largely based on stochastic simulations.

Beyond Physical Space

Throughout this book, we have considered a population density. The earliest IDE models were formulated in terms of genetics and tracked the *fraction* of a population with a certain genotype under the assumption that the population density is spatially homogeneous and temporally constant (Slatkin 1973; Weinberger 1982; Lui 1982a,b, 1985, 1986). Their and our independent variable x represents physical space. Many authors consider other "spaces," even in models for ecological processes and definitely in models for evolutionary processes. We only mention a few examples that relate to the material presented in this work; many more might exist.

Cuddington and Hastings (2004) consider a model for the spread of an "ecosystem engineer," i.e., a species that transforms its environment so that its reproductive output is increased. Beavers are a well-known "engineering" species that transform streams of running water into lakes of standing water in which they can build their lodges and be safe from predators. But plants may also engineer their environment. For example, *Spartina alterniflora* is a salt marsh grass from the Atlantic coast that has invaded regions on the Pacific coast, where it increases sedimentation rates, which reduces water flow, so that mudflats are eventually turned into marshes. Cuddington and Hastings (2004) consider habitat quality level as the independent variable, x, in a model for *Spartina*. Their model describes how the distribution of area of a certain habitat quality, $H_t(x)$, and area of a certain quality occupied by the species, $N_t(x)$, changes over time. If habitat quality level follows a shallow gradient, e.g., a gradual transition from mudflat to marshland, then x can be interpreted as physical space. The area occupied by the species grows according to habitat quality. Habitat quality grows with the presence of the species. The authors find that a species that modifies its environment toward higher growth rates for itself may invade more quickly than one that causes no modification.

Lewis et al. (2010) model the outbreak dynamics of mountain pine beetle by considering bark area and tree vigor as independent variables. The density of trees is structured by these two variables. The number of beetles emerging from a tree increases with bark area but decreases with tree vigor. The probability of a tree being killed by attacking beetles increases with the number of beetles attacking and decreases with tree vigor. At low beetle density and high tree vigor, attacking beetles cannot infect any trees; at high beetle density and low tree vigor, an attack may dramatically reduce the number of healthy trees in a stand in a single year.

A number of models consider the independent "space" variable as representing a particular individual trait. For example, quantitative genetics describes the changes of a continuously distributed quantitative phenotypic trait value in a population. A quantitative trait recursion typically tracks the population density, the mean, and the variance of the trait from one generation to the next. Haefner and Dugaw (2000) use a simple model for body size as a trait value. Individuals who are close in body size compete more strongly than individuals whose body sizes differ considerably. Competition can then be expressed by a convolution integral over trait space with a certain weight function ("kernel"), so that the recursion model becomes an IDE. Haefner and Dugaw (2000) emphasize the use of the fast Fourier transform to solve such models numerically.

Trait values are rarely unbounded, as the quantitative genetics approach implies. Hall et al. (2006) use a finite trait space $x \in [-1, 1]$ to indicate the genotype in a population during an invasion. The wild type corresponds to $x = -1$, the alien invader to $x = 1$, and any hybrid to values in between. Each type has a trait-dependent growth rate, and the offspring type is determined by some combination of the parent types and their assumed compatibility. Model outcome varies, depending on parameter choices, from replacement of the native to hybridization to increasing overall diversity.

Britton (2009) studies a host–parasite model in which host and parasite can allocate various amounts of energy into defense and attack, respectively. If these traits are assumed continuous, a system of IDEs results. Britton (2009) analyzes only the difference equations for discrete trait values in detail.

In a completely different context, IDEs arise in the study of directed polymers for a suitable generating function for a logarithmically correlated random energy model (Derrida and Spohn 1988). Webb (2011) studies traveling waves in this equation and proves a number of convergence results.

Statistics and Estimation

Several works use stochastic IDEs in the context of weather phenomena. The authors derive spectral methods for parameter estimation in a hierarchical Bayesian framework. The underlying stochastic IDE is (in our notation) given by

$$N_{t+1}(x) = \int_{\Omega} K(x, y, \theta_K) F(N_t(y), \theta_F) \mathrm{d}y + \eta_{t+1}(x), \qquad (17.8)$$

where η describes the noise and θ_K and θ_F are model parameters. Wikle (2001) and Wikle (2002) both consider cloud intensity as their density to be modeled. They emphasize the importance of spectral methods for dimension reduction. Xu and Wikle (2005) build on this work to obtain "nowcasting" for rainfall activity. Wikle and Holan (2011) use polynomial expansion to estimate parameters for Pacific sea surface temperature. Wikle (2003) uses a reaction–diffusion equation that is discretized in time, so that it looks similar to an IDE. Wikle and Hooten (2006) apply these ideas to invasion biology, namely the invasion of house finches in North America; see Sect. 12.6. Hooten and Wikle (2008) model and estimate parameters for the invasion of the Eurasian Collared Dove (*Streptopelia decaocto*) of North America. Their work is based on reaction–diffusion equations.

Dewar et al. (2009) expand on the idea of data-driven modeling with IDEs by Wikle and coworkers (see above), who assumed that the form of the kernel was known a priori. Dewar et al. (2009) drop this assumption and establish a novel decomposition that allows one to estimate the shape of the kernel at the same time. Scerri et al. (2009, 2011) continue this line of research on estimation by further dimension reduction. Sigrist et al. (2012) apply a linear stochastic IDE to predict rainfall, using a hierarchical Bayesian approach. Zammit-Mangion et al. (2012) use a stochastic point process and a stochastic IDE, together with Bayesian estimation techniques, to analyze the "Afghan war diaries."

Ming and Albrecht (2004) mention IDEs in their work to integrate geographic information system (GIS) into data-driven modeling of biological invasions, but their examples focus on reaction–diffusion equations.

17.2   Related Models

IDEs model the density of a population in discrete time and continuous space. Related models use other combinations of discrete and/or continuous descriptions of space and time. The mathematical challenges and biological applications differ between the various frameworks. We mention continuous-time models for spatial ecology, some mixed continuous-discrete time models, and some discrete-space models. Another set of very closely related models are integral-projection models, which model stage-structured populations with continuous state variable in discrete time. These are in fact IDEs but with a different interpretation and therefore different properties of the "kernel" function. All the models presented here are density based. There is a vast literature on individual-based models applied to spatial ecology. These models are mostly used in simulation studies. Some authors compare the results of individual-based models with corresponding IDEs, e.g., Travis et al. (2011) for invasions.

Models in Continuous Space and Time

Reaction–diffusion equations are the modeling framework most closely related to IDEs in the spirit that we present here. In the simplest scenario, these continuous-time "cousins" of IDEs track the density, $u(\tau, x)$, of a biological population as individuals simultaneously move randomly and interact according to the equation

$$\frac{\partial u}{\partial \tau} = D \frac{\partial^2 u}{\partial x^2} + f(x, u), \tag{17.9}$$

where D is the diffusion coefficient and $f(x, \cdot)$ is the net instantaneous production term. When $f(x, u) = ru$ is linear and spatially homogeneous, the equation can be solved explicitly in terms of the Gaussian kernel. The time-T-map of the solution is an IDE that expresses $u(T + 1, x)$ in terms of $u(T, x)$.

The literature on reaction–diffusion equations is vast. It includes models with nonlinear diffusion, interacting populations, and nonlocal interactions. Cantrell and Cosner (2003) give an excellent treatment of questions in spatial ecology on bounded domains. We are not aware of a single monograph dedicated to spreading phenomena in reaction–diffusion equations, but Lewis et al. (2016) explain the most important phenomena from an ecological perspective, and Zhao (2009) summarizes the mathematics results.

Integrodifferential equations also describe the evolution of a population density in continuous time and space. They are similar to reaction–diffusion equations, except that the (local) diffusion operator is replaced by a nonlocal dispersal operator. When dispersal and population dynamics are independent processes, the basic equation reads

$$\frac{\partial u}{\partial \tau} = D \left(\int K(x, y) u(\tau, y) \mathrm{d}y - u(\tau, x) \right) + f(x, u), \tag{17.10}$$

but other forms are possible when dispersal and reproduction are linked (Medlock and Kot 2003). Here, K is a dispersal kernel as in Sect. 2.3, except that it describes instantaneous dispersal. Integrodifferential equations have not quite received the same level of attention as reaction–diffusion equations, but many aspects have been studied in both. One major difference for analytical purposes is that an integrodifferential equation does not have the same strong compactness properties as a reaction–diffusion equation.

Another closely related modeling approach consists of *continuous-time integral equations* of the form (Diekmann 1978; Thieme 1979; Thieme and Zhao 2003)

$$u(\tau, x) = u_0(\tau, x) + \int_0^\tau \int_\infty^\infty F(u(\tau - s, x - y), s, y) \, \mathrm{d}y \, \mathrm{d}s. \tag{17.11}$$

Here, the integral measures the contribution of an individual at any time prior to τ and any location y on the density at τ and any x. These models are much less prominent in applications to spatial ecology.

Impulsive Reaction–Diffusion Equations

An impulsive differential equation, or hybrid system, combines continuous and discrete dynamics (Bainov and Simeonov 1993; Lakshmikantham and Simeonov 1989). A quantity of interest evolves continuously until a certain condition is reached, at which point there is a discrete impulse to the state variables or an abrupt change in the dynamics. This approach can be used to model seasonal dynamics of biological populations, where the dynamics during the growing season are described by the differential equation and then change from the end of one growing season to the beginning of the next due to the discrete impulse (Geritz and Kisdi 2004; Pachepsky et al. 2008).

Lewis and Li (2012) use this approach to model seasonal reproduction in a biological species in continuous space. They denote the density of individuals in year $t = 0, 1, 2, \ldots$ by $u_t(\tau, x)$, where $0 \leq \tau \leq T$ is the time within a year of length T and x is the spatial variable. Within a given year t, organisms move randomly and die according to the equation

$$\frac{\partial u_t}{\partial \tau} = D \frac{\partial^2 u_t}{\partial x^2} - m_1 u_t - m_2 u_t^2, \qquad 0 \leq \tau \leq T, \tag{17.12}$$

where m_1 and m_2 are the coefficients of the linear and quadratic death rates and D is the diffusion coefficient. From the end of one season to the beginning of the next, the population reproduces according to the map

$$u_{t+1}(x, 0) = g(u_t(x, T)), \tag{17.13}$$

where g is any growth function from Sect. 2.2.

A major difference between this approach and the IDE model framework is that here, individuals can interact during the dispersal phase. The assumption behind a dispersal kernel in an IDE is that individuals do not interact during dispersal. If there is no interaction in the above reaction–diffusion equation (i.e., $m_2 = 0$), the equation can be solved explicitly in terms of the Gaussian kernel; see Chap. 7 and Sect. 12.1. Substituting this explicit expression gives an IDE for $N_t(x) = u_t(x, 0)$.

In fact, Skellam (1951) had already thought of discrete-continuous systems. He gives a diffusion equation for movement and an integral formulation for the generation-to-generation steady state. Borer et al. (2007) use similar hybrid-systems ideas in conjunction with an IDE for the competition between a perennial and an invasive annual grass in California. They use a system of ordinary differential equations to describe the local dynamics during the growth phase (Fig. 2.1) and the usual integral kernel for dispersal between two growth phases.

Vasilyeva et al. (2016) generalize the model by Lewis and Li (2012) by introducing advection into the reaction–diffusion equation and allowing the impulse to be nonlocal. The model by Vasilyeva et al. (2016) is inspired by the drift paradox (Sect. 12.2). The reaction–advection–diffusion equation represents the larval stage of a stream insect, such as a caddisfly or a mayfly, and the nonlocal reproduction map accounts for dispersal of adult flies before ovipositing. When predator and prey interact in a continuous-time equation but predators reproduce only at discrete times, we need to keep track of the location resources during the dispersal process. Wang and Lutscher (2018) propose the first model for this scenario and analyze the conditions for diffusion-driven instabilities. These conditions are surprisingly more complicated than the corresponding results for reaction–diffusion equations or IDEs (Sect. 14.4).

Discrete-Space Models

Models in discrete time and space are known nowadays as *coupled map lattices*. They are used in a wide range of applications, spatial ecology being one of them. One of the first such models, by Roff (1974), studies how dispersal in heterogeneous landscapes may increase population persistence time. Several authors give historical references to coupled map lattices and present them as spatially discrete versions of IDEs, e.g., Brewster and Allen (1997). These authors study a predator–prey insect system with stage structure; see also Legaspi et al. (1998). Gilbert et al. (2004) use a small resolution of 2.5 km for a discrete model of horse chestnut leafminer *Cameraria ohridella* in Germany and compare the model fit with different dispersal kernels. Muthukrishnan et al. (2015) discretize the landscape and dispersal kernel in a study on the spread of *Miscanthus* × *giganteus* along roads. Other authors simply refer to discrete-space models as IDEs, e.g., Best et al. (2007), who simulate a patchy moving-habitat model. Doebeli and Ruxton (1998) study stabilization via pattern formation in discrete space. Abbott and Dwyer (2008) study the spread and synchrony of gypsy moth (*Lymantria dispar*) with stochastic environmental forcing. de Camino-Beck and Lewis (2009) study the spread of an invasive plant (scentless chamomile, *Matricaria perforata*) in a stage-structured model and give references to other recent coupled map lattice models.

Following de Camino-Beck and Lewis (2009), we illustrate this type of model in a one-dimensional domain of discrete locations $x_i = ih$ for some cell size (scale) $h > 0$. Denoting by $N_t(x_i)$ the population density in cell x_i at time t, we obtain the equation

$$N_{t+1}(x_i) = \sum_{x_j \in \Omega} K(x_i, x_j) F_j(x_j), \qquad (17.14)$$

where Ω is the spatial domain of interest. In the classical case, dispersal occurs only between nearest neighbor cells, so that $K = 0$ if $|i - j| \geq 2$, but many other shapes have been considered.

The results by Weinberger (1982) on spreading speeds are formulated on continuous and discrete habitats, so that many of the insights from Chap. 5 carry over to coupled map lattices; see also the review by Zhao (2009).

Similar to the continuous-space hybrid models mentioned above, discrete-space hybrid models have been studied, e.g., by Elderd et al. (2013) for infections.

Integral-Projection Models

Since their first derivation and analysis (Easterling et al. 2000), integral-projection models have become a widely used modeling framework to study structured populations; see Ellner et al. (2016) for a comprehensive introduction and review. The original motivation was the need to model the dynamics of continuously structured populations. It complements the framework of matrix models (Chap. 13), which considers discrete population structure.

We denote by z the continuous variable (e.g., size) that structures the population and by $n(t, z)$ the density of the population with respect to that variable at time (generation) t. We assume that z is in some compact set Ω, e.g., $z \in [z_l, z_u]$, where z_l and z_u represent the lower and upper limits of z. The total number of individuals at time t with size between z_1 and z_2 is given by

$$\int_{z_1}^{z_2} n(t, z)\mathrm{d}z . \tag{17.15}$$

In the most general terms, a linear integral-projection model maps the density from one generation to the next according to the IDE

$$n(t + 1, z) = \int_{\Omega} \widetilde{K}(z', z)n(t, z')\mathrm{d}z' . \tag{17.16}$$

The kernel function, $\widetilde{K}$, is the net result of survival and reproduction and can be written as

$$\widetilde{K}(z', z) = s(z')G(z', z) + F(z', z) , \tag{17.17}$$

where $s(z')$ is the probability that an individual of size z' survives to the next time step, $G(z', z)$ is the size distribution of size-z' individuals that survive to the next time step, and $F(z', z)$ is the production of size-z individuals in generation $t + 1$ from size-z' individuals in generation t.

The model in (17.16) and its nonlinear extension, where $\widetilde{K}$ depends on $n(t, \cdot)$, looks abstractly like IDE (2.1). The two have many similarities, so much so that Ellner et al. (2016) refer to IDEs as "spatial integral-projection models." However, there are also significant differences between the two model types. The similarities are particularly obvious in the mathematical aspects of IDEs on bounded domains. All the considerations on compactness, linearization, spectra, and stability carry over to integral-projection models. Some of the most prominent differences are in the shape of the "kernels" and the mathematical analysis on unbounded domains. A dispersal kernel in an IDE can be symmetric and must satisfy some integration constraints. A kernel function in an integral-projection model typically has no symmetry properties and is not a probability density function. A homogeneous spatial environment in an IDE is worthy of study, yet if a population is "homogeneous" with respect to the structure variable, then it is unstructured and model (17.16) collapses to a simple difference equation. The study of spread in unbounded domains has no meaningful analogue in (17.16).

Since integral-projection models complement matrix models for studying (size-) structured populations, it seems reasonable and natural that they could be generalized to spatial integral-projection models, just as matrix models were generalized to stage-structured IDEs (Chap. 13). The first work in that direction is by Jongejans et al. (2011), who formulate a spatial integral-projection model for an invasive thistle and study the sensitivity of spreading speed to individual and spatial variation.

References

Abbott, K., & Dwyer, G. (2008). Using mechanistic models to understand synchrony in forest insect populations: The North American gypsy moth as a case study. *The American Naturalist, 172*, 613–624.

Adler, F. (1993). Migration alone can produce persistence of host–parasitoid models. *The American Naturalist, 141*, 642–650.

Allee, W. (1949). *Principles of animal ecology*. Christchurch: Saunders.

Allen, L. (2006). *An introduction to mathematical biology*. New York: Pearson.

Allen, L. (2010). *An introduction to stochastic processes with applications to biology*. London: Chapman and Hall/CRC.

Allen, E., Allen, L., & Gilliam, X. (1996). Dispersal and competition models for plants. *Journal of Mathematical Biology, 34*, 455–481.

Allen, J., Brewster, C., & Slone, D. (2001). Spatially explicit ecological models: A spatial convolution approach. *Chaos, Solitons & Fractals, 12*, 333–347.

Alonso, D., Bartumeus, F., & Catalan, J. (2002). Mutual interference between predators can give rise to Turing spatial patterns. *Ecology, 83*(1), 28–34.

Alzoubi, M. (2007). Equilibria in a dispersal model for structured populations. *Turkish Journal of Mathematics, 31*, 421–433.

Alzoubi, M. (2010a). The net reproductive number and bifurcation in an integro-difference system of equations. *Applied Mathematical Sciences, 4*(4), 191–200.

Alzoubi, M. (2010b). Stability and bifurcation in a system of integro-difference equations model. *Applied Mathematical Sciences, 4*(64), 3175–3188.

Amor, D., & Fort, J. (2009). Fronts from two-dimensional dispersal kernels: Beyond the nonoverlapping-generations model. *Physical Review E, 80*, 051918.

Andersen, M. (1991). Properties of some density-dependent integrodifference equation population models. *Mathematical Biosciences, 104*, 135–157.

Aronson, D., & Weinberger, H. F. (1975). Nonlinear diffusion in population genetics, combustion, and nerve pulse propagation. In J. Goldstein (Ed.) *Partial differential equations and related topics. Lecture notes in mathematics* (vol. 446, pp. 5–49). Berlin: Springer.

Assaneo, F., Coutinho, R.M., Lin, Y., Mantilla, C., & Lutscher, F. (2013). Dynamics and coexistence in a system with intraguild mutualism. *Ecological Complexity, 14*, 64–74.

Aydogmus, O., Kang, Y., Kavgaci, M., & Bereketoglu, H. (2017). Dynamical effects of nonlocal interactions in discrete-time growth-dispersal models with logistic-type nonlinearities. *Ecological Complexity, 31*, 88–95.

Baeumer, B., Kovàcs, M., & Meerschaert, M. (2007). Fractional reproduction-dispersal equations and heavy tail dispersal kernels. *Bulletin of Mathematical Biology, 69*, 2281–2297.

F. Lutscher, *Integrodifference Equations in Spatial Ecology*, Interdisciplinary Applied Mathematics 49, https://doi.org/10.1007/978-3-030-29294-2

Bainov, D., & Simeonov, P. (1993). *Impulsive differential equations: Periodic solutions and applications*. Boca Raton: CRC Press.

Barton, N., & Turelli, M. (2011). Spatial waves of advance with bistable dynamics: Cytoplasmic and genetic analogues of Allee effects. *The American Naturalist, 178*, E48–E75.

Baskett, M., Weitz, J., & Levin, S. (2007). The evolution of dispersal in reserve networks. *The American Naturalist, 170*, 59–78.

Bateman, A., Buttenschön, A., Erickson, K., & Marulis, N. (2017). Barnacles vs bullies: Modelling biocontrol of the invasive European green crab using a castrating barnacle parasite. *Theoretical Ecology, 10*, 305–318.

Bateman, A., Neubert, M., Krkošek, M., & Lewis, M. (2015). Generational spreading speed and the dynamics of population range expansion. *The American Naturalist, 186*, 362–375.

Beddington, J. R., Free, C. A., & Lawton, J. H. (1975). Dynamic complexity in predator–prey models framed in difference equations. *Nature, 255*(5503), 58.

Beer, T., & Swaine, M. (1977). On the theory of explosively dispersed seeds. *New Phytologist, 78*, 681–694.

Bellows, T. (1981). The descriptive properties of some models for density dependence. *Journal of Animal Ecology, 50*(1), 139–156.

Berestycki, H., & Hamel, F. (2002). Front propagation in periodic excitable media. *Communications on Pure and Applied Mathematics, 55*(8), 949–1032.

Best, A., Johst, K., Münkemüller, T., & Travis, J. (2007). Which species will successfully track climate change? The influence of intraspecific competition and density dependent dispersal on range shifting dynamics. *Oikos, 116*, 1531–1539.

Beverton, R., & Holt, S. (1957). *On the Dynamics of Exploited Fish Populations. Fisheries Investigation Series* (vol. 2, no. 19). London: Ministry of Agriculture, Fisheries, and Food.

Bianchi, F., Schellhorn, N., & van der Werf, W. (2009). Predicting the time to colonization of the parasitoid *diadegma semiclausum:* The importance of the shape of spatial dispersal kernels for biological control. *Biological Control, 50*, 267–274.

Bocedi, G., Guy Pe'er, G., Heikkinen, R., Matsinos, Y., & Travis, J. (2012). Projecting species' range expansion dynamics: Sources of systematic biases when scaling up patterns and processes. *Methods in Ecology and Evolution, 2*, 1008–1018.

Borer, E., Hosseini, P., Seabloom, E., & Dobson, A. (2007). Pathogen-induced reversal of native dominance in a grassland community. *Proceedings of the National Academy of Sciences, 104*(13), 5473–5478.

Botsford, L. W., Hastings, A., & Gaines, S. D. (2001). Dependence of sustainability on the configuration of marine reserves and larval dispersal distance. *Ecology Letters, 4*, 144–150.

Boucher, D. (1982). The ecology of mutualism. *Annual Review of Ecology and Systematics, 13*, 315–347.

Bouhours, J., & Lewis, M. (2016). Climate change and integrodifference equations in a stochastic environment. *Bulletin of Mathematical Biology, 78*, 1866–1903.

Bourgeois, A. (2016). *Spreading Speeds and Travelling Waves in Integrodifference Equations with Overcompensatory Dynamics*. Master's Thesis, University of Ottawa.

Bourgeois, A., LeBlanc, V., & Lutscher, F. (2018). Spreading phenomena in integrodifference equations with non-monotone growth functions. *SIAM Journal on Applied Mathematics, 78*(6), 2950–2972.

Bourgeois, A., LeBlanc, V., & Lutscher, F. (2019). Dynamical stabilization and traveling waves in integrodifference equations. *Discrete and Continuous Dynamical Systems - Series S*, https://doi.org/10.3934/dcdss.2020117

Bramburger, J., & Lutscher, F. (2019) Analysis of integrodifference equations with a separable dispersal kernel. *Acta Applicandae Mathematicae, 161*(1), 127–151.

Brännström, A., & Sumpter, D. (2005). The role of competition and clustering in population dynamics. *Proceedings of the Royal Society of London B, 272*, 2065–2072.

Brewster, C., & Allen, L. (1997). Spatiotemporal model for studying insect dynamics in large-scale cropping systems. *Environmental Entomology, 26*(3), 473–482.

Brigham, E. (2002). *The fast Fourier transform*. New York: Prentice-Hall.

Britton, N. (1989). Aggregation and the competitive exclusion principle. *Journal of Theoretical Biology, 136*, 57–66.

Britton, N. (2009). Evolution in a host–parasite system. In *Biomat 2008: International Symposium on Mathematical and Computational Biology* (pp. 157–169). Singapore: World Scientific.

Britton-Simmons, K., & Abbott, K. (2008). Short- and long-term effects of disturbance and propagule pressure on a biological invasion. *Journal of Ecology, 96*, 68–77.

Buckley, Y., Brockerhoff, E. G., Langer, L., Ledgard, N. J., North, H. C., & Rees, M. (2005). Slowing down a pine invasion despite uncertainty in demography and dispersal. *Journal of Applied Ecology, 42*, 1020–1030.

Bullock, J., & Clarke, R. (2000). Long distance seed dispersal by wind: Measuring and modelling the tail of the curve. *Oecologia, 124*, 506–521.

Bullock, J., Pywell, R., & Coulson-Phillips, S. (2008). Managing plant population spread: Prediction and analysis using a simple model. *Ecological Applications, 18*(4), 945–953.

Bullock, J., White, S., Prudhomme, C., Tansey, C., Perea, R., & Hooftman, D. (2012). Modelling spread of British wind-dispersed plants under future wind speeds in a changing climate. *Journal of Ecology, 100*, 104–115.

Byers, J., & Pringle, J. (2006). Going against the flow: Retention, range limits and invasions in advective environments. *Marine Ecology Progress Series, 313*, 27–41.

Cantrell, R. S., & Cosner, C. (2003). *Spatial ecology via reaction-diffusion equations. Mathematical and computational biology*. London: Wiley.

Caplat, P., Coutts, S., & Buckley, Y. (2012). Modeling population dynamics, landscape structure, and management decisions for controlling the spread of invasive plants. *Annals of the New York Academy of Sciences, 1249*, 72–83.

Caplat, P., Nathan, R., & Buckley, Y. (2012). Seed terminal velocity, wind turbulence, and demography drive the spread of an invasive tree in an analytical model. *Ecology, 93*(2), 368–377.

Cappaert, D., McCullough, D., Polland, T., & Siegert, N. (2005). Emerald ash borer in North America: A research and regulatory challenge. *American Entomologist, 51*, 152–165.

Carrillo, C., Cherednichenko, K., Britton, N., & Mogie, M. (2009). Dynamic coexistence of sexual and asexual invasion fronts in a system of integro-difference equations. *Bulletin of Mathematical Biology, 71*, 1612–1625.

Carrillo, C., & Fife, P. (2005). Spatial effects in discrete generation population models. *Journal of Mathematical Biology, 50*, 161–188.

Castillo-Chavez, C., Li, B., & Wang, H. (2013). Some recent developments on linear determinacy. *Mathematical Biosciences and Engineering, 10*, 1419–1436.

Caswell, H. (2001). *Matrix population models*. Sunderland: Sinauer Associates.

Caswell, H., Lensink, R., & Neubert, M. (2003). Demography and dispersal: Life table response experiments for invasion speed. *Ecology, 84*(8), 1968–1978.

Caswell, H., Neubert, M., & Hunter, C. (2011). Demography and dispersal: Invasion speeds and sensitivity analysis in periodic and stochastic environments. *Theoretical Ecology, 4*, 407–421.

Chesson, P., & Lee, C. (2005). Families of discrete kernels for modeling dispersal. *Theoretical Population Biology, 67*, 241–256.

Clark, J. (1998). Why trees migrate so fast: Confronting theory with dispersal biology and the paleorecord. *The American Naturalist, 152*(2), 204–224.

Clark, J., Fastie, C., Hurtt, G., Jackson, S., Johnson, C., King, G., et al. (1998a). Reid's paradox of rapid plant migration. *Bioscience, 48*(1), 13–24.

Clark, J., Horváth, L., & Lewis, M. (2001b). On the estimation of spread rate for a biological population. *Statistics & Probability Letters, 51*, 225–234.

Clark, J., Lewis, M., & Horvath, L. (2001a). Invasion by extremes: Population spread with variation in dispersal and reproduction. *The American Naturalist, 157*(5), 537–554.

Clark, J., Lewis, M., McLachlan, J., & HilleRisLambers, J. (2003). Estimating population spread: What can we forecast and how well? *Ecology, 84*(8), 1979–1988.

Clark, J., Macklin, E., & Wood, L. (1998b). Stages and spatial scales of recruitment limitation in southern Appalachian forests. *Ecological Monographs, 68*(2), 213–235.

Clark, J., Silman, M., Kern, R., Macklin, E., & HilleRisLambers, J. (1999). Seed dispersal near and far: Patterns across temperate and tropical forests. *Ecology, 80*(5), 1475–1494.

Cobbold, C., Lewis, M., Lutscher, F., & Roland, J. (2005). How parasitism affects critical patch size in a host–parasitoid system: Application to forest tent caterpillar. *Theoretical Population Biology, 67*(2), 109–125.

Cobbold, C., & Lutscher, F. (2014). Mean occupancy time: Linking mechanistic movement models, population dynamics and landscape ecology to population persistence. *Journal of Mathematical Biology, 68*(3), 549–579.

Collingham, Y., & Huntley, B. (2000). Impacts of habitat fragmentation and patch size upon migration rates. *Ecological Applications, 10*(1), 131–144.

Cosner, C. (2014). Reaction-diffusion-advection models for the effects and evolution of dispersal. *Discrete and Continuous Dynamical Systems - Series B, 35*(5), 1701–1745.

Courchamp, F., Berec, L., & Gascoinge, J. (2008). *Allee effects.* Oxford: Oxford University Press.

Coutinho, R., & Fernandez, B. (2004). Fronts in extended systems of bistable maps coupled via convolutions. *Nonlinearity, 17*, 23–47.

Coutinho, R., Godoy, W., & Kraenkel, R. (2012). Integrodifference model for blowfly invasion. *Theoretical Ecology, 5*, 363–371.

Creegan, P., & Lui, R. (1984). Some remarks about the wave speed and travelling wave solutions of a nonlinear integral operator. *Journal of Mathematical Biology, 20*, 59–68.

Crone, E., Brown, L., Hodgson, J., Lutscher, F., & Schultz, C. (2019). Faster movement in habitat matrix promotes range shifts in heterogeneous landscapes. *Ecology, 100*(7), e02701

Cuddington, K., & Hastings, A. (2004). Invasive engineers. *Ecological Modelling, 178*, 335–347.

Cushing, J. (2014). Backward bifurcations and strong Allee effects in matrix models for the dynamics of structured populations. *Journal of Biological Dynamics, 8*(1), 57–73.

Day, S., & Kalies, W. (2013). Rigorous computation of the global dynamics of integrodifference equations with smooth nonlinearities. *SIAM Journal on Numerical Analysis, 51*(6), 2957–2983.

Day, S., Junge, O., & Mischaikow, K. (2004). A rigorous numerical method for the global analysis of intfinite-dimensional discrete dynamical systems. *SIAM Journal on Applied Dynamical Systems, 3*(2), 117–160.

de Camino-Beck, T., & Lewis, M. (2009). Invasion with stage-structured coupled map lattices: Application to the spread of scentless chamomile. *Ecological Modelling, 220*(23), 3394–3403.

Deng, P., de Vargas Roditi, L., van Ditmarsch, D., & Xavier, J. (2014). The ecological basis of morphogenesis: Branching patterns in swarming colonies of bacteria. *New Journal of Physics, 16*, 015006.

Derrida, B., & Spohn, H. (1988). Polymers on disordered trees, spin glasses, and traveling waves. *Journal of Statistical Physics, 51*(5–6), 817–840.

Dewar, M., Scerri, K., & Kadirkamanathan, V. (2009). Data-driven spatio-temporal modeling using the integro-difference equation. *IEEE Transactions of Signal Processing, 57*(1), 83–91.

Dewhirst, S., & Lutscher, F. (2009). Dispersal in heterogeneous habitats: Thresholds, spatial scales and approximate rates of spread. *Ecology, 90*(5), 1338–1345.

Diekmann, O. (1978). Thresholds and travelling waves for the geographical spread of infection. *Journal of Mathematical Biology, 6*, 109–130.

Ding, W., & Liang, X. (2015). Principal eigenvalues of generalized convolution operators on the circle and spreading speeds of noncompact evolution systems in periodic media. *SIAM Journal on Mathematical Analysis, 47*(1), 855–896.

Ding, W., Liang, X., & Xu, B. (2013). Spreading speeds of n-season spatially periodic integro-difference equations. *Discrete and Continuous Dynamical Systems - Series A, 33*(8), 3443–3472.

Doebeli, M., & Ruxton, G. (1998). Stabilization through spatial pattern formation in metapopulations with long–range dispersal. *Proceedings of the Royal Society of London B, 265*(1403), 1325–1332.

Doedel, E. J. (1981). Auto: A program for the automatic bifurcation analysis of autonomous systems. *Congressus Numerantium, 30*, 265–284.

Drury, K., & Candelaria, J. (2008). Using model identification to analyze spatially explicit data with habitat, and temporal, variability. *Ecological Modelling, 214*, 305–315.

Du, Y. (2006). *Order structure and topological methods in nonlinear partial differential equations. Maximum principles and applications.* Singapore: World Scientific.

Dwyer, G., & Morris, W. (2006). Resource-dependent dispersal and the speed of biological invasions. *The American Naturalist, 167*(2), 165–176.

Easterling, M., Ellner, S., & Dixon, P. (2000). Size-specific sensitivity: Applying a new structured population model. *Ecology, 81*, 694–708.

Edelstein-Keshet, L. (2005). *Mathematical models in biology.* Philadelphia: SIAM.

Einstein, A. (1906). Zur Theorie der Brownschen Bewegung. *Annals of Physics, 19*, 371–381.

Elderd, B., Rehill, B., Haynes, K., & Dwyer, G. (2013). Induced plant defenses, host–pathogen interactions, and forest insect outbreaks. *Ecological Applications, 110*(37), 14978–14983.

Elliott, E., & Cornell, S. (2012). Dispersal polymorphism and the speed of biological invasions. *PLoS ONE, 7*(7), e40496.

Ellner, S. (1984). Asymptotic behavior of some stochastic difference equation population models. *Journal of Mathematical Biology, 19*, 169–200.

Ellner, S., Childs, D., & Rees, M. (2016). *Data-driven modelling of structured populations.* Berlin: Springer.

Ellner, S., & Schreiber, S. (2012). Temporally variable dispersal and demography can accelerate the spread of invading species. *Theoretical Population Biology, 82*(4), 283–298.

Etienne, R., Wertheim, B., Hemerik, L., Schneider, P., & Powell, J. (2002). The interaction between dispersal, the Allee effect and scramble competition affects population dynamics. *Ecological Modelling, 148*, 153–168.

Fagan, W., Lewis, M., Neubert, M., & van den Driessche, P. (2002). Invasion theory and biological control. *Ecology Letters, 5*, 148–157.

Fagan, W., Lewis, M., Neubert, M., Aumann, C., Apple, J., & Bishop, J. (2005). When can herbivores slow or reverse the spread of an invading plant? A test case from mount St. Helens. *The American Naturalist, 166*, 669–685.

Fagan, W., & Lutscher, F. (2006). Average dispersal success: Linking home range, dispersal and metapopulation dynamics to reserve design. *Ecological Applications, 16*(2), 820–828.

Fang, J., & Zhao, X.-Q. (2014). Traveling waves for monotone semiflows with weak compactness. *SIAM Journal on Mathematical Analysis, 46*(6), 3678–3704.

Fang, J., & Zhao, X.-Q. (2015). Bistable traveling waves for monotone semiflows with applications. *Journal of the European Mathematical Society, 17*, 2243–2288.

Fasani, S., & Rinaldi, S. (2011). Factors promoting or inhibiting Turing instability in spatially extended prey–predator systems. *Ecological Modelling, 222*, 3449–3452.

Fedotov, S. (2001). Front propagation into an unstable state of reaction-transport systems. *Physical Review Letters, 86*(5), 926–929.

Fisher, R. (1937). The advance of advantageous genes. *Annals of Eugenics, 7*, 355–369.

Fort, J. (2007). Fronts from complex two-dimensional dispersal kernels: Theory and application to Reid's paradox. *Journal of Applied Physics, 101*, 094701.

Fort, J. (2012). Synthesis between demic and cultural diffusion in the neolithic transition in Europe. *Proceedings of the National Academy of Sciences of the United States of America, 109*(46), 18669–18673.

Fort, J., Pérez-Losada, J., & Isern, N. (2007). Fronts from integrodifference equations and persistence effects on the neolithic transition. *Physical Review E, 76*, 031913.

Fort, J., Pérez-Losada, J., Suñol, J., Escoda, L., & Massaneda, J. (2008). Integro-difference equations for interacting species and the neolithic transition. *New Journal of Physics, 10*, 043045.

Fujiwara, M., Anderson, K., Neubert, M., & Caswell, H. (2006). On the estimation of dispersal kernels from individual mark-recapture data. *Environmental and Ecological Statistics, 13*, 183–197.

Fuller, E., Rush, E., & Pinsky, M. (2015). The persistence of populations facing climate shifts and harvest. *Ecosphere, 6*(9), 153.

Gaff, H., Joshi, H., & Lenhart, S. (2007). Optimal harvesting during an invasion of a sublethal plant pathogen. *Environment and Development Economics, 12*, 673–686.

Gagnon, K., Peacock, S., Yu Jin, Y., & Lewis, M. (2015). Modelling the spread of the invasive alga *codium fragile* driven by long-distance dispersal of buoyant propagules. *Ecological Modelling, 316*, 111–121.

Galliard, J., Marquis, O., & Massot, M. (2010). Cohort variation, climate effects and population dynamics in a short lived lizard. *Journal of Animal Ecology, 79*, 1296–1307.

Garnier, A., & Lecomte, J. (2006). Using a spatial and stage-structured invasion model to assess the spread of feral populations of transgenic oilseed rape. *Ecological Modelling, 194*, 141–149.

Garnier, A., Pivard, S., & Lecomte, J. (2008). Measuring and modelling anthropogenic secondary seed dispersal along roadverges for feral oilseed rape. *Basic and Applied Ecology, 9*, 533–541.

Geritz, S., & Kisdi, É. (2004). On the mechanistic underpinning of discrete-time population models with complex dynamics. *Journal of Theoretical Biology, 228*, 261–269.

Geritz, S., Kisdi, É., Meszéna, G., & Metz, J. (1998). Evolutionarily singular strategies and the adaptive growth and branching of the evolutionary tree. *Evolutionary Ecology, 12*(1), 35–57.

Gharouni, A., Barbeau, M., Chassé, J., Wang, L., & Watmough, J. (2017). Stochastic dispersal increases the rate of upstream spread: A case study with green crabs on the northwest atlantic coast. *PLoS ONE, 12*(9), e0185671.

Gharouni, A., Barbeau, M., Locke, A., Wang, L., & Watmough, J. (2015). Sensitivity of invasion speed to dispersal and demography: An application of spreading speed theory to the green crab invasion on the northwest Atlantic coast. *Marine Ecology Progress Series, 541*, 135–150.

Gilbert, M., Gaffney, E., Bullock, J., & White, S. (2014a). Spreading speeds for plant populations in landscapes with low environmental variation. *Journal of Theoretical Biology, 363*, 436–452.

Gilbert, M., Grégoire, J.-C., Freise, J., & Heitland, W. (2004). Long-distance dispersal and human population density allow the prediction of invasive patterns in the horse chestnut leafminer *cameraria ohridella. Journal of Animal Ecology, 73*, 459–468.

Gilbert, M., White, S., Bullock, J., & Gaffney, E. (2014b). Spreading speeds for stage structured plant populations in fragmented landscapes. *Journal of Theoretical Biology, 349*, 135–149.

Gilbert, M., White, S., Bullock, J., & Gaffney, E. (2017). Speeding up the simulation of population spread models. *Methods in Ecology and Evolution, 8*, 501–510.

Gilioli, G., Pasquali, S., Tramontini, S., & Riolo, F. (2013). Modelling local and long-distance dispersal of invasive chestnut gall wasp in Europe. *Ecological Modelling, 263*, 281–290.

Goodsman, D., Cooke, B., Coltman, D., & Lewis, M. (2014). The genetic signature of rapid range expansions: How dispersal, growth and invasion speed impact heterozygosity and allele surfing. *Theoretical Population Biology, 98*, 1–10.

Goodsman, D., Koch, D., Whitehouse, C., Evenden, M., Cooke, B., & Lewis, M. (2016). Aggregation and a strong Allee effect in a cooperative outbreak insect. *Ecological Applications, 26*, 2623–2636.

Goodsman, D., & Lewis, M. (2016). The minimum founding population in dispersing organisms subject to strong Allee effects. *Methods in Ecology and Evolution, 7*, 1100–1109.

Gouhier, T., Guichard, F., & Menge, B. (2010). Ecological processes can synchronize marine population dynamics over continental scales. *Proceedings of the National Academy of Sciences of the United States of America, 107*(18), 8281–8286.

Gruess, A., Kaplan, D., & Hart, D. (2011). Relative impacts of adult movement, larval dispersal and harvester movement on the effectiveness of reserve networks. *PLoS ONE, 6*(5), e19960.

Haefner, J., & Dugaw, C. (2000). Individual-based models solved using fast Fourier transforms. *Ecological Modelling, 125*, 159–172.

Hall, R., Hastings, A., & Ayres, D. (2006). Explaining the explosion: Modelling hybrid invasions. *Proceedings of the Royal Society B, 273*, 1385–1389.

Hamel, F. (2016). Bistable transition fronts in $\mathbb{R}^n$. *Advances in Mathematics, 289*, 279–344.

Hammerstein, A. (1930). Nichtlineare Integralgleichungen nebst Anwendungen. *Acta Mathematica, 54*, 117–176.

Hardin, D., Takáč, P., & Webb, G. (1988a). Asymptotic properties of a continuous-space discrete-time population model in a random environment. *Journal of Mathematical Biology, 26*, 361–374.

Hardin, D., Takáč, P., & Webb, G. (1988b). A comparison of dispersal strategies for survival of spatially heterogeneous populations. *SIAM Journal on Applied Mathematics, 48*, 1396–1423.

Hardin, D., Takáč, P., & Webb, G. (1990). Dispersion population models discrete in time and continuous in space. *Journal of Mathematical Biology, 28*, 1–20.

Harsch, M., Phillips, A., Zhou, Y., Leung, M.-R., Rinnan, S., & Kot, M. (2017). Moving forward: Insights and applications of moving-habitat models for climate change ecology. *Journal of Ecology, 105*, 1169–1181.

Harsch, M., Zhou, Y., HilleRisLambers, J., & Kot, M. (2014). Keeping pace with climate change: Stage-structured moving-habitat models. *The American Naturalist, 184*(1), 25–37.

Hart, D., & Gardner, R. (1997). A spatial model for the spread of invading organisms subject to competition. *Journal of Mathematical Biology, 35*, 935–948.

Hassell, M. (1975). Density-dependence in single-species populations. *Journal of Animal Ecology, 44*(1), 283–295.

Hassell, M. P. (1978). *The dynamics of arthropod predator–prey systems*. Princeton: Princeton University Press.

Hastings, A. (1983). Can spatial variation alone lead to selection for dispersal? *Theoretical Population Biology, 24*, 244–251.

Heavilin, J., & Powell, J. (2008). A novel method of fitting spatio-temporal models to data, with applications to the dynamics of mountain pine beetles. *Natural Resource Modeling, 21*(4), 489–501.

Holzer, M. (2014). Anomalous spreading in a system of coupled Fisher–KPP equations. *Physica D, 270*, 1–10.

Hooten, M., & Wikle, C. (2008). A hierarchical Bayesian non-linear spatio-temporal model for the spread of invasive species with application to Eurasian Collard-Dove. *Environmental and Ecological Statistics, 15*, 59–70.

Hsu, S.-B., & Zhao, X.-Q. (2008). Spreading speeds and traveling waves for non-monotone integrodifference equations. *SIAM Journal on Mathematical Analysis, 40*(2), 776–789.

Hughes, J., Cobbold, C., Haynes, K., & Dwyer, G. (2015). Effects of forest spatial structure on insect outbreaks: Insights from a host–parasitoid model. *The American Naturalist, 185*(5), E130–E152.

Hurford, A., Hebblewhite, M., & Lewis, M. (2006). A spatially explicit model for an Allee effect: Why wolves recolonize so slowly in Greater Yellowstone. *Theoretical Population Biology, 70*, 244–254.

Hutson, V., Martinez, S., Mischaikow, K., & Vickers, G. (2003). The evolution of dispersal. *Journal of Mathematical Biology, 46*, 483–517.

Iooss, G. (1979). *Bifurcation of maps and applications. Mathematical studies* (vol. 36). Amsterdam: North-Holland.

Isern, N., Fort, J., & Pérez-Losada, J. (2008). Realistic dispersion kernels applied to cohabitation reaction-dispersion equations. *Journal of Statistical Mechanics, 2008*, P10012.

Jacobs, G., & Sluckin, T. (2015). Long-range dispersal, stochasticity and the broken accelerating wave of advance. *Theoretical Population Biology, 100*, 39–55.

Jacobsen, J., & McAdam, T. (2014). A boundary value problem for integrodifference equation models with cyclic kernels. *Discrete & Continuous Dynamical Systems - Series B, 19*(10), 3139–3207.

Jacobsen, J., Jin, Y., & Lewis, M. (2015). Integrodifference models for persistence in temporally varying river environments. *Journal of Mathematical Biology, 70*, 549–590.

Jacquemyn, H., Brys, R., & Neubert, M. (2005). Fire increases invasive spread of *molinia caerulea* mainly through changes in demographic parameters. *Ecological Applications, 55*(6), 449–460.

Jane White, K. A., Lenhart, S., & Martinez, M. V. (2015). Optimal control of integrodifference equations in pest-pathogen systems. *Discrete & Continuous Dynamical Systems - Series B, 20*(6), 1759–1783.

Jin, W., Smith, H., & Thieme, H. (2016). Persistence versus extinction for a class of discrete-time structured population models. *Journal of Mathematical Biology, 72*, 821–850.

Jongejans, E., Shea, K., Skarpaas, O., Kelly, D., & Ellner, S. (2011). Importance of individual and environmental variation for invasive species spread: A spatial integral projection model. *Ecology, 92*(1), 86–97.

Jongejans, E., Shea, K., Skarpaas, O., Kelly, D., Sheppard, A., & Woodburn, T. (2008). Dispersal and demography contributions to population spread of *carduus nutans* in its native and invaded ranges. *Journal of Ecology, 96*, 687–697.

Joshi, H., Lenhart, S., & Gaff, H. (2006). Optimal harvesting in an integro-difference population model. *Optimal Control Applications & Methods, 27*(2), 61–75.

Joshi, H., Lenhart, S., Lou, H., & Gaff, H. (2007). Harvesting control in an integrodifference population model with concave growth term. *Nonlinear Analysis: Hybrid Systems, 1*, 417–429.

Kanary, L., Musgrave, J., Locke, A., Tyson, R., & Lutscher, F. (2014). Modelling the dynamics of invasion and control of competing green crab genotypes. *Theoretical Ecology, 7*(4), 391–404.

Kawasaki, K., & Shigesada, N. (2007). An integrodifference model for biological invasions in a periodically fragmented environment. *Japan Journal of Industrial and Applied Mathematics, 24*, 3–15.

Keener, J. (2000). *Principles of applied mathematics*. Boulder: Westview.

Kelly, Jr., M., Xing, Y., & Lenhart, S. (2016). Optimal fish harvesting for a population modeled by a nonlinear parabolic partial differential equation. *Natural Resource Modeling, 29*(1), 36–70.

Kierstead, H., & Slobodkin, L. B. (1953). The size of water masses containing plankton blooms. *Journal of Marine Research, 12*, 141–147.

Klein, E., Lavigne, C., & Gouyon, P.-H. (2006). Mixing of propagules from discrete sources at long distance: Comparing a dispersal tail to an exponential. *BMC Ecology, 6*, 3.

Kot, M. (1989). Diffusion-driven period-doubling bifurcations. *BioSystems, 22*, 279–287.

Kot, M. (1992). Discrete-time travelling waves: Ecological examples. *Journal of Mathematical Biology , 30*, 413–436.

Kot, M. (2001). *Elements of mathematical ecology*. Cambridge: Cambridge University Press.

Kot, M. (2003). Do invading organisms do the wave? *Canadian Applied Mathematics Quarterly, 10*, 139–170.

Kot, M., Lewis, M., & van den Driessche, P. (1996). Dispersal data and the spread of invading organisms. *Ecology, 77*, 2027–2042.

Kot, M., Medlock, J., Reluga, T., & Walton, D. (2004). Stochasticity, invasions, and branching random walks. *Theoretical Population Biology, 66*, 175–184.

Kot, M., & Neubert, M. (2008). Saddle-point approximations, integrodifference equations, and invasions. *Bulletin of Mathematical Biology, 70*, 1790–1826.

Kot, M., & Phillips, A. (2015). Bounds for the critical speed of climate-driven moving-habitat models. *Mathematical Biosciences, 262*, 65–72.

Kot, M., & Schaffer, W. (1986). Discrete-time growth-dispersal models. *Mathematical Biosciences, 80*, 109–136.

Krasnosel'skii, M. A. (1964). *Positive solutions of operator equations*. Groningen: Noordhoff LTD.

Krasnosel'skii, M. A., & Zabreiko, P. P. (1984). *Geometrical methods of nonlinear analysis*. Berlin: Springer.

Krause, U. (2015). *Positive dynamical systems in discrete time*. Berlin: de Gruyter.

Krkošek, M., Lauzon-Guay, J., & Lewis, M. (2007). Relating dispersal and range expansion of California sea otters. *Theoretical Population Biology, 71*, 401–407.

Krkošek, M., & Lewis, M. (2010). An R_0 theory for source–sink dynamics with application to *dreissena* competition. *Theoretical Ecology, 3*, 25–43.

Kura, K., Khamis, D., El Mouden, C., & Bonsall, M. (2019). Optimal control for disease vector management in SIT models: An integrodifference equation approach. *Journal of Mathematical Biology, 78*, 1821–1839.

Kythe, P., & Puri, P. (2011). *Computational methods for linear integral equations*. New York: Springer Science & Business Media.

Lakshmikantham, V., & Simeonov, P. (1989). *Theory of impulsive differential equations* (vol. 6). Singapore: World scientific.

Lamoureaux, S., Basse, B., Bourdôt, G., & Saville, D. (2015). Comparison of management strategies for controlling *nassella trichotoma* in modified tussock grasslands in New Zealand: A spatial and economic analysis. *Weed Research, 55*, 449–460.

Latore, J., Gould, P., & Mortimer, A. (1998). Spatial dynamics and critical patch size of annual plant populations. *Journal of Theoretical Biology, 190*, 277–285.

Latore, J., Gould, P., & Mortimer, A. (1999). Effects of habitat heterogeneity and dispersal strategies on population persistence in annual plants. *Ecological Modelling, 123*, 127–139.

Le, T., Lutscher, F., & Van Minh, N. (2011). Traveling wave dispersal in partially sedentary age-structured populations. *Acta Mathematica Vietnamica, 2*(36), 319–330.

Le, T., & van Nguyen, M. (2017). Monotone traveling waves in a general discrete model for populations. *Discrete & Continuous Dynamical Systems - Series B, 22*(8), 3221–3234.

Le Corff, J., & Horvitz, C. (2005). Population growth versus population spread of an ant-dispersed neotropical herb with a mixed reproductive strategy. *Ecological Modelling, 188*, 41–51.

Legaspi, Jr., B., Allen, J., Brewster, C., Morales-Ramos, J., & King, E. (1998). Areawide management of the cotton boll weevil: Use of a spatio-temporal model in augmentative biological control. *Ecological Modelling, 110*, 151–164.

Lenhart, S., & Workman, J. (2007). *Optimal control applied to biological models.* London: CRC Press.

Leo, A. (2007). *A Numerical Approach to Calculating Population Spreading Speed.* Master's Thesis, Worcester Polytechnic Institute.

Leung, M.-R., & Kot, M. (2015). Models for the spread of white pine blister rust. *Journal of Theoretical Biology, 382*, 328–336.

Lewis, M. (2000). Spread rate for a nonlinear stochastic invasion. *Journal of Mathematical Biology, 41*, 430–454.

Lewis, M., & Li, B. (2012). Spreading speed, traveling waves and the minimal patch-size in impulsive reaction-diffusion models. *Bulletin of Mathematical Biology, 74*(10), 2383–2402.

Lewis, M., Li, B., & Weinberger, H. (2002). Spreading speed and linear determinacy for two-species competition models. *Journal of Mathematical Biology, 45*, 219–233.

Lewis, M., Marculis, N., & Shen, Z. (2018). Integrodifference equations in the presence of climate change: Persistence criterion, travelling waves and inside dynamics. *Journal of Mathematical Biology, 77*, 1649–1687.

Lewis, M., Nelson, W., & Xu, C. (2010). A structured threshold model for mountain pine beetle outbreak. *Bulletin of Mathematical Biology, 72*, 565–589.

Lewis, M., Neubert, M., Caswell, H., Clark, J., & Shea, K. (2006). A guide to calculating discrete-time invasion rates from data. In M. Cadotte, S. McMahon, & T. Fukami (Eds.), *Conceptual ecology and invasions biology: Reciprocal approaches to nature* (pp. 169–192). Berlin: Springer.

Lewis, M., & Pacala, S. (2000). Modeling and analysis of stochastic invasion processes. *Journal of Mathematical Biology, 41*, 387–429.

Lewis, M., Petrovskii, S., & Potts, J. (2016). *The mathematics behind biological invasions.* Berlin: Springer.

Li, B. (2009). Some remarks on traveling wave solutions in competition models. *Discrete and Continuous Dynamical Systems - Series B, 12*, 389–399.

Li, B. (2012). Traveling wave solutions in a plant population model with a seed bank. *Journal of Mathematical Biology, 65*(5), 855–873.

Li, B., Bewick, S., Barnard, M., & Fagan, W. (2016a). Persistence and spreading speeds of integro-difference equations with an expanding or contracting habitat. *Bulletin of Mathematical Biology, 78*, 1337–1379.

Li, B., Fagan, W., & Meyer, K. (2015). Success, failure, and spreading speeds for invasions on spatial gradients. *Journal of Mathematical Biology, 70*, 265–287.

Li, K., Huang, J., Li, X., & He, Y. (2016b). Asymptotic behavior and uniqueness of traveling wave fronts in a competitive recursion system. *Zeitschrift für Angewandte Mathematik und Physik, 67*(6), 144.

Li, B., Lewis, M., & Weinberger, H. (2009). Existence of traveling waves for integral recursions with nonmonotone growth functions. *Journal of Mathematical Biology, 58*, 323–338.

Li, K., & Li, X. (2012a). Asymptotic behavior and uniqueness of traveling wave solutions in Ricker competition system. *Journal of Mathematical Analysis and Applications, 389*, 486–497.

Li, K., & Li, X. (2012b). Travelling wave solutions in integro-difference competition system. *IMA Journal of Applied Mathematics, 78*(3), 633–650.

Li, B., Weinberger, H., & Lewis, M. (2005). Spreading speeds as slowest wave speeds for cooperative systems. *Mathematical Biosciences, 196*, 82–98.

Liang, X., & Zhao, X.-Q. (2007). Asymptotic speeds of spread and traveling waves for monotone semiflows with applications. *Communications on Pure and Applied Mathematics: A Journal Issued by the Courant Institute of Mathematical Sciences, 60*(1), 1–40.

Liang, X., & Zhao, X.-Q. (2010). Spreading speeds and traveling waves for abstract monostable evolution systems. *Journal of Functional Analysis, 259*, 857–903.

Lin, G. (2015). Traveling wave solutions for integro-difference systems. *Journal of Differential Equations, 258*, 2908–2940.

Lin, H.-T. (1995). On a system of integrodifference equations modelling the propagation of genes. *SIAM Journal on Mathematical Analysis, 26*(1), 35–76.

Lin, G., & Li, W.-T. (2010). Spreading speeds and traveling wavefronts for second order integrodifference equations. *Journal of Mathematical Analysis and Applications, 361*(2), 520–532.

Lin, G., Li, W.-T., & Ruan, S. (2010). Asymptotic stability of monostable wavefronts in discrete-time integral recursions. *Science China: Mathematics, 53*(5), 1185–1194.

Lin, G., Li, W.-T., & Ruan, S. (2011). Spreading speeds and traveling waves in competitive recursion systems. *Journal of Mathematical Biology, 62*(2), 165–201.

Liu, B. R., & Kot, M. (2019). Accelerating invasions and the asymptotics of fat-tailed dispersal. *Journal of Theoretical Biology, 471*, 22–41.

Lockwood, D., Hastings, A., & Botsford, L. (2002). The effects of dispersal patterns on marine reserve: Does the tail wag the dog? *Theoretical Population Biology, 61*, 297–309.

Lotka, A. (1920). Undamped oscillations derived from the law of mass action. *Journal of the American Chemical Society, 42*, 1595–1599.

Lui, R. (1982a). A nonlinear integral operator arising from a model in population genetics, I. Monotone initial data. *SIAM Journal on Mathematical Analysis, 13*(6), 913–937.

Lui, R. (1982b). A nonlinear integral operator arising from a model in population genetics, II. Initial data with compact support. *SIAM Journal on Mathematical Analysis, 13*(6), 938–953.

Lui, R. (1983). Existence and stability of travelling wave solutions of a nonlinear integral operator. *Journal of Mathematical Biology, 16*, 199–220.

Lui, R. (1985). A nonlinear integral operator arising from a model in population genetics, III. Heterozygote inferior case. *SIAM Journal on Mathematical Analysis, 16*(6), 1180–1206.

Lui, R. (1986). A nonlinear integral operator arising from a model in population genetics, IV. clines. *SIAM Journal on Mathematical Analysis, 17*(1), 152–168.

Lui, R. (1989a). Biological growth and spread modeled by systems of recursions. I Mathematical theory. *Mathematical Biosciences, 93*, 269–295.

Lui, R. (1989b). Biological growth and spread modeled by systems of recursions. II Biological theory. *Mathematical Biosciences, 93*, 297–312.

Lutscher, F. (2007). A short note on short dispersal distances. *Bulletin of Mathematical Biology, 69*(5), 1615–1630.

Lutscher, F. (2008). Density-dependent dispersal in integrodifference equations. *Journal of Mathematical Biology, 56*(4), 499–524.

Lutscher, F., & Iljon, T. (2013). Competition, facilitation and the Allee effect. *Oikos, 122*(4), 621–631.

Lutscher, F., & Lewis, M. (2004). Spatially-explicit matrix models. A mathematical analysis of stage-structured integrodifference equations. *Journal of Mathematical Biology, 48*, 293–324.

Lutscher, F., & McCauley, E. (2013). A probabilistic framework for nutrient uptake length. *Theoretical Ecology, 6*(1), 71–86.

Lutscher, F., & Musgrave, J. (2017). Behavioral responses to resource heterogeneity can accelerate biological invasions. *Ecology, 98*(5), 1229–1238.

Lutscher, F., Nisbet, R., & Pachepsky, E. (2010). Population persistence in the face of advection. *Theoretical Ecology, 3*, 271–284.

Lutscher, F., Pachepsky, E., & Lewis, M. (2005). The effect of dispersal patterns on stream populations. *SIAM Review, 47*(4), 749–772.

Lutscher, F., & Petrovskii, S. (2008). The importance of census times in discrete-time growth-dispersal models. *Journal of Biological Dynamics, 2*(1), 55–63.

Lutscher, F., & Van Minh, N. (2013). Spreading speeds and traveling waves in discrete models of biological populations with sessile stages. *Nonlinear Analysis: Real World Applications, 14*(1), 495–506.

Maciel, G., & Lutscher, F. (2013). How individual movement response to habitat edges affects population persistence and spatial spread. *The American Naturalist, 182*(1), 42–52.

Magnus, W., & Winkler, S. (1979). *Hill's equation*. New York: Dover.

Mahdjoub, T., & Menu, F. (2008). Prolonged diapause: A trait increasing invasion speed? *Journal of Theoretical Biology, 251*, 317–330.

Malchow, H., Petrovskii, S., & Venturino, E. (2008). *Spatiotemporal Patterns in Ecology and Epidemiology: Theory, Models, Simulations*. London: Chapman & Hall/CRC Press.

Marchetto, K., Jongejans, E., Shea, K., & Isard, S. (2010). Plant spatial arrangement affects projected invasion speeds of two invasive thistles. *Oikos, 119*, 1426–1468.

Marculis, N., Lewis, M., & Lui, R. (2017). Neutral genetic patterns for expanding populations with nonoverlapping generations. *Bulletin of Mathematical Biology, 79*, 828–852.

Marculis, N., & Lui, R. (2015). Modelling the biological invasion of *Carcinus maenas* (the European green crab). *Journal of Biological Dynamics, 10*(1), 140–163.

Matlaga, D., & Davis, A. (2013). Minimizing invasive potential of *miscanthus × giganteus* grown for bioenergy: Identifying demographic thresholds for population growth and spread. *Journal of Applied Ecology, 50*, 479–487.

May, M. (1973). *Stability and complexity in model ecosystems*. Princeton: Princeton University Press.

May, R. (1975). Biological populations obeying difference equations: Stable points, stable cycles, and chaos. *Journal of Theoretical Biology, 51*, 511–524.

May, R. M., Hassell, M. P., Anderson, R. M., & Tonkyn, D. W. (1981). Density dependence in host–parasitoid models. *Journal of Animal Ecology, 50*(3), 855–865.

Medlock, J., & Kot, M. (2003). Spreading diseases: Integro-differential equations new and old. *Mathematical Biosciences, 184*, 201–222.

Méndez, V., Pujol, T., & Fort, J. (2002). Dispersal probability distributions and the wave-front speed problem. *Physical Review E, 65*, 041109.

Mercader, R., Siegert, N., Liebhold, A., & McCullough, D. (2009). Dispersal of the emerald ash borer, *Agirlus planipennis*, in newly-colonized sites. *Agricultural and Forest Entomology, 11*(4), 421–424.

Merchant, S. (2001). *Analysis of an integrodifference model for biological invasions with a quasi-local interaction*. Master's Thesis, University of British Columbia.

Metz, J., Mollison, D., & van den Bosch, F. (1999). The dynamics of invasion waves. Technical Report, IIASA.

Meyer, K. (2012). *A spatial age-structured model of perennial plants with a seed bank*. Ph.D. Thesis, University of Louisville.

Meyer, K., & Li, B. (2013). A spatial model of plants with an age-structured seed bank and juvenile stage. *SIAM Journal on Applied Mathematics, 73*(4), 1676–1702.

Meyn, S., & Tweedie, R. (2009). *Markov chains and stochastic stability*. Cambridge: Cambridge University Press.

Miller, T., Shaw, K., Inouye, B., & Neubert, M. (2011). Sex-biased dispersal and the speed of two-sex invasions. *The American Naturalist, 177*(5), 549–561.

Miller, T., & Tenhumberg, B. (2010). Contributions of demography and dispersal parameters to the spatial spread of a stage-structured insect invasion. *Ecological Applications, 20*(3), 620–633.

Miller, J., & Zeng, H. (2013). Multidimensional stability of planar traveling waves for an integrodifference model. *Discrete & Continuous Dynamical Systems - B, 18*, 741–751.

Ming, P., & Albrecht, J. (2004). Integrated framework for the simulation of biological invasions in a heterogeneous landscape. *Transactions in GIS, 8*(3), 309–334.

Mistro, D., Rodrigues, L., & Ferreira, W. C. Jr. (2005a). The africanized honey bee dispersal: A mathematical zoom. *Bulletin of Mathematical Biology, 67*, 281–312.

Mistro, D., Rodrigues, L., & Schmid, A. (2005b). A mathematical model for dispersal of an annual plant population with a seed bank. *Ecological Modelling, 188*, 52–61.

Murray, J. D. (2001). *Mathematical biology I: An introduction*. Berlin: Springer.

Murray, J. D. (2002). *Mathematical biology II: Spatial models and biomedical applications*. Berlin: Springer.

Musgrave, J. (2013). *Integrodifference equations in patchy landscapes*. Ph.D. Thesis, University of Ottawa.

Musgrave, J., Girard, A., & Lutscher, F. (2015). Population spread in patchy landscapes under a strong Allee effect. *Theoretical Ecology, 8*(3), 313–326.

Musgrave, J., & Lutscher, F. (2014a). Integrodifference equations in patchy landscapes I: Dispersal kernels. *Journal of Mathematical Biology, 69*(3), 583–615.

Musgrave, J., & Lutscher, F. (2014b). Integrodifference equations in patchy landscapes II: Population level consequences. *Journal of Mathematical Biology, 69*(3), 617–658.

Muthukrishnan, R., West, N. M., Davis, A., Jordan, N., & Forester, J. (2015). Evaluating the role of landscape in the spread of invasive species: The case of the biomass crop *miscanthus × giganteus*. *Ecological Modelling, 317*, 6–15.

Nathan, R., Klein, E., Robledo-Arnuncio, J. J., & Revilla, E. (2012). Dispersal kernels: Review. In J. Clobert, M. Baguette, T. G. Benton, & J. M. Bullock (Eds.), *Dispersal ecology and evolution* (chap. 15.1). Oxford: Oxford University Press.

Nathan, R., Perry, G., Cronin, J., Strand, A., & Cain, M. (2003). Methods for estimating long-distance dispersal. *Oikos, 103*, 261–273.

Neubert, M., & Caswell, H. (2000a). Demography and dispersal: Calculation and sensitivity analysis of invasion speeds for structured populations. *Ecology, 81*(6), 1613–1628.

Neubert, M., & Caswell, H. (2000b). Density-dependent vital rates and their population dynamic consequences. *Journal of Mathematical Biology, 41*, 103–121.

Neubert, M., Caswell, H., & Murray, J. (2002). Transient dynamics and pattern formation: Reactivity is necessary for Turing instabilities. *Mathematical Biosciences, 175*, 1–11.

Neubert, M., & Kot, M. (1992). The subcritical collapse of predator populations in discrete time predator–prey models. *Mathematical Biosciences, 110*, 45–66.

Neubert, M., Kot, M., & Lewis, M. A. (1995). Dispersal and pattern formation in a discrete-time predator–prey model. *Theoretical Population Biology, 48*(1), 7–43.

Neubert, M., Kot, M., & Lewis, M. (2000). Invasion speeds in fluctuating environments. *Proceedings of the Royal Society of London - Series B, 267*, 1603–1610.

Neubert, M., & Parker, I. (2004). Projecting rates of spread for invasive species. *Risk Analysis, 24*(4), 817–831.

Neupane, R., & Powell, J. (2015). Invasion speeds with active dispersers in highly variable landscapes: Multiple scales, homogenization, and the migration of trees. *Journal of Theoretical Biology, 387*, 111–119.

Nicholson, A. (1954). An outline of the dynamics of animal populations. *Australian Journal of Zoology, 2*, 9–65.

Nicholson, A., & Bailey, V. (1935). The balance of animal populations. Part I. *Proceedings of the Zoological Society of London, 105*(3), 551–598.

Okubo, A., & Levin, S. A. (2001). *Diffusion and ecological problems: Modern perspectives*. New York: Springer.

Okubo, A., Maini, P., Williamson, M., & Murray, J. (1989). On the spatial spread of the grey squirrel in Britain. *Proceedings of the Royal Society B, 238*, 113–125.

Othmer, H. (1983). A continuum model for coupled cells. *Journal of Mathematical Biology, 17*, 351–369.

Otto, G. (2017). *Nonspreading solutions in integro-difference models with Allee and overcompensation effects*. Ph.D. Thesis, University of Louisville.

Ovaskainen, O., & Cornell, S. (2003). Biased movement at a boundary and conditional occupancy times for diffusion processes. *Journal of Applied Probability, 40*, 557–580.

Owen, M., & Lewis, M. (2001). How predation can slow, stop, or reverse a prey invasion. *Bulletin of Mathematical Biology, 63*, 655–684.

Pachepsky, E., & Levine, J. (2011). Density dependence slows invader spread in fragmented landscapes. *The American Naturalist, 177*(1), 18–28.

Pachepsky, E., Nisbet, R. M., & Murdoch, W. W. (2008). Between discrete and continuous: Consumer-resource dynamics with synchronized reproduction. *Ecology, 89*, 280–288.

Pan, S., & Lin, G. (2011). Propagation of second order integrodifference equations with local monotonicity. *Nonlinear Analysis: Real World Applications, 12*, 535–544.

Pan, S., & Lin, G. (2014). Coinvasion-coexistence travelling wave solutions of an integro-difference competition system. *Journal of Difference Equations and Applications, 20*(4), 511–525.

Pan, S., & Yang, P. (2014). Traveling wave solutions in a Lotka-Volterra type competition recursion. *Advances in Difference Equations, 2014*(1), 173.

Pan, S., & Zhang, P.-A. (2011). Bistable wave fronts in integrodifference equations. *Abstract and Applied Analysis, 2011*, 230851.

Pavliotis, G., & Stuart, A. (2008). *Multiscale methods: Averaging and homogenization*. New York: Springer.

Petrov, V. (1975). *Sums of independent random variables*. Berlin: Springer.

Petrovskii, S., & Morozov, A. (2009). Dispersal in a statistically structured population: Fat tails revisited. *The American Naturalist, 173*(2), 278–289.

Phillips, B., Brown, G., Travis, J., & Shine, R. (2008). Reid's paradox revisited: The evolution of dispersal kernels during range expansion. *The American Naturalist, 172*, S34–S48.

Phillips, A., & Kot, M. (2015). Persistence in a two-dimensional moving-habitat model. *Bulletin of Mathematical Biology, 77*(11), 2125–2159.

Pielaat, A., Lewis, M., Lele, S., & de Camino-Beck, T. (2006). Sequential sampling design for catching the tail of dispersal kernels. *Ecological Modelling, 190*, 205–220.

Pittman, S., Muthukrishnan, R., West, N., Davis, A., Jordan, N., & Forester, J. (2015). Mitigating the potential for invasive spread of the exotic biofuel crop, *miscanthus × giganteus*. *Biological Invasions, 17*, 3247–3261.

Powell, J. (2009). Spatiotemporal models in ecology: An introduction to integro-difference equations. Technical Report, Utah State University.

Powell, J., & Bentz, B. (2014). Phenology and density-dependent dispersal predict patterns of mountain pine beetle (*dendroctonus ponderosae*) impact. *Ecological Modelling, 273*, 173–185.

Powell, J., Slapničar, I., & van der Werf, W. (2005). Epidemic spread of a lesion-forming plant pathogen – analysis of a mechanistic model with infinite age structure. *Linear Algebra and its Applications, 398*, 117–140.

Powell, J., & Zimmermann, N. (2004). Multiscale analysis of active seed dispersal contributed to resolving Reid's paradox. *Ecology, 85*(2), 490–506.

Pringle, J., Lutscher, F., & Glick, E. (2009). Going against the flow: The effect of non-Gaussian dispersal kernels and reproduction over multiple generations. *Marine Ecology Progress Series, 337*, 13–17.

Radcliffe, J., & Rass, L. (1997). Discrete time spatial models arising in genetics, evolutionary game theory, and branching processes. *Mathematical Biosciences, 140*, 101–129.

Ramanantoanina, A., & Hui, C. (2016). Formulating spread of species with habitat dependent growth and dispersal in heterogeneous landscapes. *Mathematical Biosciences, 275*, 51–56.

Ramanantoanina, A., Ouhinou, A., & Hui, C. (2014). Spatial assortment of mixed propagules explains the acceleration of range expansion. *PLoS ONE, 9*(8), e103409.

Ramanantoanina, A., Ouhinou, A., & Hui, C. (2015). Correction: Spatial assortment of mixed propagules explains the acceleration of range expansion. *PLoS ONE, 10*(8), e0136479.

Reimer, J., Bonsall, M., & Maini, P. (2016). Approximating the critical patch-size of integrodifference equations. *Bulletin of Mathematical Biology, 78*, 72–109.

Reimer, J., Bonsall, M., & Maini, P. (2017). The critical patch-size of stochastic population models. *Journal of Mathematical Biology, 74*, 755–782.

Ricker, W. (1954). Stock and recruitment. *Journal of Fisheries Research Board of Canada, 11*, 559–632.

Rietkerk, M., & van de Koppel, J. (2008). Regular pattern formation in real ecosystems. *Trends in Ecology & Evolution, 23*(3), 169–175.

Rinnan, D. S. (2017). The dispersal success and persistence of populations with asymmetric dispersal. *Theoretical Ecology, 11*(1), 55–69.

Robbins, T. (2004). *Seed dispersal and biological invasion: A mathematical analysis.* Ph.D. Thesis, University of Utah.

Robertson, S. (2009). *Spatial patterns in stage-structured populations with density dependent dispersal.* Ph.D. Thesis, The University of Arizona.

Robertson, S., & Cushing, J. (2011). Spatial segregation in stage-structured populations with an application to *Tribolium. Journal of Biological Dynamics, 5*(5), 398–409.

Robertson, S., & Cushing, J. (2012). A bifurcation analysis of stage-structured density dependent integrodifference equations. *Journal of Mathematical Analysis and Applications, 288*, 490–499.

Robertson, S., Cushing, J., & Costantino, R. (2012). Life stages: Interactions and spatial patterns. *Bulletin of Mathematical Biology, 74*, 491–508.

Robinet, C., Kehlenbeck, H., Kriticos, D. J., Baker, R. H. A., Battisti, A., Brunel, S., et al. (2012). A suite of models to support the quantitative assessment of spread in pest risk analysis. *PLoS ONE, 7*(10), e43366.

Rodriguez, M. (2010). A modeling framework for assessing long-distance dispersal and loss of connectivity in stream fish. *American Fisheries Society Symposium, 73*, 263–279.

Roff, D. (1974). Spatial heterogeneity and the persistence of populations. *Oecologia, 15*, 245–258.

Samia, Y., & Lutscher, F. (2012). Persistence probabilities for stream populations. *Bulletin of Mathematical Biology, 74*(7), 1629–1650.

Sandefur, J. (2018). A unifying approach to discrete single-species populations models. *Discrete & Continuous Dynamical Systems - Series B, 23*, 493–508.

Santini, L., Cornulier, T., Bullock, J. M., Palmer, S., White, S., Hodgson, J. A., et al. (2016). A trait-based approach for predicting species responses to environmental change from sparse data: How well might terrestrial mammals track climate change? *Global Change Biology, 22*, 2415–2424.

Scerri, K., Dewar, M., Aram, P., Freestone, D., Kadirkamanathan, V., & Grayden, D. (2011). Balanced reduction of an IDE-based spatio-temporal model. In *Computational Tools 2011.*

Scerri, K., Dewar, M., & Kadirkamanathan, V. (2009). Estimation and model selection for an IDE-based spatio-temporal model. *IEEE Transactions on Signal Processing, 57*(2), 482–492.

Scheltema, R. (1986). On dispersal and planktonic larvae of benthic invertebrates: An eclectic overview and summary of problems. *Bulletin of Marine Science, 39*, 290–322.

Schofield, P. (2002). Spatially explicit models of Turelli–Hoffmann wolbachia invasive wave fronts. *Journal of Theoretical Biology, 215*, 121–131.

Schreiber, S. (2003). Allee effects, extinctions, and chaotic transients in simple population models. *Theoretical Population Biology, 64*, 201–209.

Schreiber, S., & Ryan, M. (2011). Invasion speeds for structured populations in fluctuating environments. *Theoretical Ecology, 4*(4), 423–434.

Shea, K., Jongejans, E., Skarpaas, O., Kelly, D., & Sheppard, A. (2010). Optimal management strategies to control local population growth or population spread may not be the same. *Ecological Applications, 20*(4), 1148–1161.

Sherratt, J., Eagan, B., & Lewis, M. (1997). Oscillations and chaos behind predator–prey invasion: Mathematical artifact or ecological reality? *Philosophical Transactions of the Royal Society of London B, 352*, 21–38.

Shigesada, N., Kawasaki, K., & Teramoto, E. (1986). Traveling periodic waves in heterogeneous environments. *Theoretical Population Biology, 30*, 143–160.

Sigrist, F., Künsch, H., & Stahel, W. (2012). A dynamic nonstationary spatio-temporal model for short term prediction of precipitation. *The Annals of Applied Statistics, 6*(4), 1452–1477.

Skalski, G., & Gilliam, J. (2003). A diffusion-based theory of organism dispersal in heterogeneous populations. *The American Naturalist, 161*(3), 441–458.

Skarpaas, O., & Shea, K. (2007). Dispersal patterns, dispersal mechanisms, and invasion wave speeds for invasive thistles. *The American Naturalist, 170*(3), 421–430.

Skellam, J. G. (1951). Random dispersal in theoretical populations. *Biometrika, 38*, 196–218.

Skelsey, P., Kessel, G., Rossing, W., & van der Werf, W. (2009a). Parameterization and evaluation of a spatiotemporal model of the potato late blight pathosystem. *Phytopathology, 99*, 290–300.

Skelsey, P., Rossing, W., Kessel, G., Powell, J., & van der Werf, W. (2005). Influence of host diversity on development of epidemics: An evaluation and elaboration of mixture theory. *Phytopatology, 95*(4), 328–338.

Skelsey, P., Rossing, W., Kessel, G., & van der Werf, W. (2009b). Scenario approach for assessing the utility of dispersal information in decision support for aerially spread plant pathogens, applied to *phytophthora infestans. Phytopathology, 99*, 887–895.

Skelsey, P., Rossing, W., Kessel, G., & van der Werf, W. (2010). Invasion of *phytophthora infestans* at the landscape level: How do spatial scale and weather modulate the consequences of spatial heterogeneity in host resistance? *Phytopathology, 100*, 1146–1161.

Slatkin, M. (1973). Gene flow and selection in a cline. *Genetics, 75*, 733–756.

Slone, D. (2011). Increasing accuracy of dispersal kernels in grid-based population models. *Ecological Modelling, 222*, 573–579.

Smith, C., Giladi, I., & Lee, Y.-S. (2009). A reanalysis of competing hypotheses for the spread of the California sea otter. *Ecology, 90*(9), 2503–2512.

Snäll, T., O'Hara, R., & Arjas, E. (2007). A mathematical and statistical framework for modelling dispersal. *Oikos, 116*, 1037–1050.

Snyder, R. (2003). How denographic stochasticity can slow biological invasions. *Ecology, 84*(5), 1333–1339.

Soons, M., & Bullock, J. (2008). Non-random seed abscission, long-distance wind dispersal and plant migration rates. *Journal of Ecology, 96*, 581–590.

Stamp, N., & Lucas, J. (1983). Ecological correlates of explosive seed dispersal. *Oecologia, 59*, 272–278.

Stover, J. P., Kendall, B. E., & Nisbet, R. M. (2014). Consequences of dispersal heterogeneity for population spread and persistence. *Bulletin of Mathematical Biology, 76*, 2681–2710.

Sullivan, L., Li, B., Miller, T., Neubert, M., & Shaw, A. (2017). Density dependence in demography and dispersal generates fluctuating invasion speeds. *Proceedings of the National Academy of Sciences, 114*(19), 5053–5058.

Taylor, C., & Hastings, A. (2005). Allee effects in biological invasions. *Ecology Letters, 8*, 895–908.

Thieme, H. (1979). Density-dependent regulation of spatially distributed populations and their asymptotic speed of spread. *Journal of Mathematical Biology, 8*, 173–187.

Thieme, H., & Zhao, X.-Q. (2003). Asymptotic spreads of speed and traveling waves for integral equations and delayed reaction-diffusion models. *Journal of Differential Equations, 195*, 430–470.

Tilman, D. (1982). *Resource competition and community structure*. Princeton: Princeton University Press.

Travis, J., Harris, C., Park, K., & Bullock, J. (2011). Improving prediction and management of range expansions by combining analytical and individual-based modelling approaches. *Methods in Ecology and Evolution, 2*, 477–488.

Tufto, J., Ringsby, T.-H., Dhondt, A., Adriaensen, F., & Matthysen, E. (2005). A parametric model for estimation of dispersal patterns applied to five passerine spatially structured populations. *The American Naturalist, 165*, E13–E26.

Turchin, P. (1998). *Quantitative analysis of movement: Measuring and modeling population redistribution of plants and animals.* Sunderland, MA: Sinauer Associates.

Turing, A. (1952). The chemical basis of morphogenesis. *Philosophical Transactions of the Royal Society B, 237*, 5–72.

Tyson, R., Wilson, J., & Lane, W. (2011). Beyond diffusion: Modelling local and long-distance dispersal for organisms exhibiting intensive and extensive search modes. *Theoretical Population Biology, 79*, 70–81.

van den Bosch, F., Metz, J., & Diekmann, O. (1990). The velocity of spatial population expansion. *Journal of Mathematical Biology, 28*, 529–565.

Van Kirk, R. (1995). *Integrodifference models for biological growth and dispersal.* Ph.D. Thesis, University of Utah.

Van Kirk, R., & Lewis, M. (1997). Integrodifference models for persistence in fragmented habitats. *Bulletin of Mathematical Biology, 59*(1), 107–137.

Van Kirk, R., & Lewis, M. (1999). Edge permeability and population persistence in isolated habitat patches. *Natural Resource Modeling, 12*, 37–64.

Vasilyeva, O., Lutscher, F., & Lewis, M. (2016). Analysis of spread and persistence for stream insects with winged adult stages. *Journal of Mathematical Biology, 72*(4), 851–875.

Veit, R. R., & Lewis, M. A. (1996). Dispersal, population growth, and the Allee effect: Dynamics of the house finch invasion in eastern North America. *The American Naturalist, 148*(2), 255–274.

Vellend, M., Knight, T., & Drake, J. (2006). Antagonistic effects of seed dispersal and herbivory on plant migration. *Ecology Letters, 9*, 319–326.

Verhulst, P.-F. (1838). Notice sur la loi que la population poursuit dans son accroissement. *Correspondance mathamatique et physique, 10*, 113–121.

Volkov, D., & Lui, R. (2007). Spreading speed and traveling wave solutions of a partially sedentary population. *IMA Journal of Applied Mathematics, 72*(6), 801–816.

Volterra, V. (1926). Variazioni e fluttuazioni del numero d'individui in specie animali conviventi. *Mémoires de l'Académie Lincei Roma, 2*(31–113).

Wang, H., & Castillo-Chavez, C. (2012). Spreading speeds and traveling waves for non-cooperative integro-difference systems. *Discrete and Continuous Dynamical Systems - Series B, 17*(6), 2243–2266.

Wang, J., Cui-Ping Cheng, C.-P., & Shuibo Huang, S. (2016). Evolution of dispersal in a spatially periodic integrodifference model. *Nonlinear Analysis: Real World Applications, 32*, 10–34.

Wang, M.-H., Kot, M., & Neubert, M. (2002). Integrodifference equations, Allee effects, and invasions. *Journal of Mathematical Biology, 44*, 150–168.

Wang, M.-H., Kot, M., & Neubert, M. (2013). Erratum to: Integrodifference equations, Allee effects, and invasions. *Journal of Mathematical Biology, 66*, 1339–1340.

Wang, X., & Lutscher, F. (2019). Turing patterns in a predator–prey model with seasonality. *Journal of Mathematical Biology, 78*(3), 711–737.

Warner, D., & Shine, R. (2008). Determinants of dispersal distance in free-ranging juvenile lizards. *Ethology, 114*(4), 361–368.

Watkinson, A., Freckleton, R., & Forrester, L. (2000). Population dynamics of *vulpia ciliata*: Regional, patch and local dynamics. *Journal of Ecology, 88*, 1012–1029.

Webb, C. (2011). Exact asymptotics of the freezing transition of a logarithmically correlated energy model. *Journal of Statistical Physics, 145*(6) 1595–1619.

Wei, H., & Lutscher, F. (2013). From individual movement rules to population level patterns: The case of central-place foragers. In M. Lewis, P. Maini, & S. Petrovskii (Eds.), *Dispersal, individual movement and spatial ecology. Lecture notes in mathematics* (vol. 2071). Berlin: Springer.

Weinberger, H. (1978). Asymptotic behavior of a model in population genetics. In J. Chadam (Ed.), *Nonlinear partial differential equations and applications* (vol. 648). Berlin: Springer.

Weinberger, H. (1982). Long-time behavior of a class of biological models. *SIAM Journal on Mathematical Analysis, 13*, 353–396.

Weinberger, H. (2002). On spreading speeds and traveling waves for growth and migration models in a periodic habitat. *Journal of Mathematical Biology, 45*, 511–548.

Weinberger, H., Kawasaki, K., & Shigesada, N. (2008). Spreading speeds of spatially-periodic integro-difference models for populations with non-monotone recruitment functions. *Journal of Mathematical Biology, 57*, 387–411.

Weinberger, H., Lewis, M., & Li, B. (2002). Analysis of linear determinacy for spread in cooperative models. *Journal of Mathematical Biology, 45*, 183–218.

Weinberger, H., Lewis, M., & Li, B. (2007). Anomalous spreading speeds of cooperative recursion systems. *Journal of Mathematical Biology, 55*(2), 207–222.

Weinberger, H., & Zhao, X.-Q. (2010). An extension of the formula for spreading speeds. *Mathematical Biosciences and Engineering, 7*(1), 187–194.

Westerberg, L., & Wennergren, U. (2003). Predicting the spatial distribution of a population in a heterogeneous landscape. *Ecological Modelling, 166*, 53–65.

White, S., & White, K. (2005). Relating coupled map lattices to integro-difference equations: Dispersal-driven instabilities in coupled map lattices. *Journal of Theoretical Biology, 235*, 463–475.

Wikle, C. (2001). A kernel-based spectral approach for spatio-temporal dynamic models. In *Proceedings of the 1st Spanish Workshop on Spatio-Temporal Modelling of Environmental Processes (METMA)* (pp. 167–180)

Wikle, C. (2002). A kernel-based spectral model for non-Gaussian spatio-temporal processes. *Statistical Modelling, 2*(4), 299–314.

Wikle, C. (2003). Hierarchical Bayesian models for predicting the spread of ecological processes. *Ecology, 84*(6), 1382–1394.

Wikle, C., & Holan, S. (2011). Polynomial nonlinear spatio-temporal integro-difference equation models. *Journal of Time Series Analysis, 32*(4), 339–350.

Wikle, C., & Hooten, M. (2006). Hierarchical Bayesian spatio-temporal models for population spread. In J. Clark, & A. Gelfand (Eds.), *Hierarchical modelling for the environmental sciences*. Oxford: Oxford University Press.

Williams, J., Snyder, R., & Levine, J. (2016). The influence of evolution on population spread through patchy landscapes. *The American Naturalist, 188*(1), 15–26.

Wright, R., & Hastings, A. (2007). Spontaneous patchiness in a host–parasitoid integrodifference model. *Bulletin of Mathematical Biology, 69*, 2693–2709.

Wu, R., & Zhao, X.-Q. (2018). Propagation dynamics for a spatially periodic integrodifference competition model. *Journal of Differential Equations, 264*, 6507–6534.

Xu, K., Wikle, C., & Fox, N. I. (2005). A kernel-based spectral approach for spatio-temporal dynamical model for nowcasting weather radar reflectivities. *Journal of the American Statistical Association, 100*, 1133–1144.

Yamamura, K. (2002). Dispersal distance of heterogeneous populations. *Population Ecology, 44*, 93–101.

Yi, T., & Zou, X. (2015). Asymptotic behavior, spreading speeds and taveling waves of nonmonotone dynamical systems. *SIAM Journal on Mathematical Analysis, 47*(4), 3005–3034.

Yu, Z.-X., & Yuan, R. (2012). Properties of traveling waves for integrodifference equations with nonmonotone growth functions. *Zeitschrift für Angewandte Mathematik und Physik, 63*, 249–259.

Zammit-Mangion, A., Dewar, M., Kadirkamanathan, V., & Sanguinetti, G. (2012). Point process modelling of the Afghan war diary. *Proceedings of the National Academy of Sciences, 109*, 12414–12419.

Zhang, R., Jongejans, E., & Shea, K. (2011). Warming increases the spread of an invasive thistle. *PLoS ONE, 6*(6), e21725.

Zhang, Y., & Zhao, X.-Q. (2012). Bistable travelling waves in competitive recursion systems. *Journal of Differential Equations, 252*, 2630–2647.

Zhao, X.-Q. (1996). Global attractivity and stability in some monotone discrete dynamical systems. *Bulletin of the Australian Mathematical Society, 53*, 305–324.

Zhao, X.-Q. (2003). *Dynamical systems in population biology*. *CMS books in mathematics*. Berlin: Springer.

Zhao, X.-Q. (2009). Spatial dynamics of some evolution systems in biology. In Y. Du, H. Ishii, & W.-Y. Lin (Eds.), *Recent progress on reaction-diffusion systems and viscosity solutions* (pp. 332–363). Singapore: World Scientific.

Zhong, P. (2011). *Optimal Theory Applied in Integrodifference Equation Models and in a Cholera Differential Equation Model*. Ph.D. Thesis, University of Tennessee.

Zhou, Y., & Fagan, W. (2017). A discrete-time model for population persistence in habitats with time-varying sizes. *Journal of Mathematical Biology, 75*(3), 649–704.

Zhou, Y., & Kot, M. (2011). Discrete-time growth-dispersal models with shifting species ranges. *Theoretical Ecology, 4*, 13–25.

Zhou, Y., & Kot, M. (2013). Life on the move: Modeling the effects of climate-driven range shifts with integrodifference equations. In *Dispersal, individual movement and spatial ecology* (pp. 263–292). Berlin: Springer.

Index

© Springer Nature Switzerland AG 2019
F. Lutscher, *Integrodifference Equations in Spatial Ecology*, Interdisciplinary
Applied Mathematics 49, https://doi.org/10.1007/978-3-030-29294-2

Printed in the United States
By Bookmasters